Wildlife Trade in India

Asha Poonia • Sandeep Kumar Gupta
Vikram Delu
Editors

Wildlife Trade in India

Conservation and Commerce

Springer

Editors
Asha Poonia
Faculty of Life Sciences
Chaudhary Bansi Lal University
Bhiwani, Haryana, India

Sandeep Kumar Gupta
Scientist -F
Animal Ecology & Conservation Biology Lab
Wildlife Institute of India
Dehradun, Uttarakhand, India

Vikram Delu
Scientist -B
Zoological Survey of India
Arunachal Pradesh Regional Centre
Itanagar, Arunachal Pradesh, India

ISBN 978-981-95-3693-1 ISBN 978-981-95-3694-8 (eBook)
https://doi.org/10.1007/978-981-95-3694-8

This Springer imprint is published by the registered company Springer Nature Singapore Pte Ltd.
The registered company address is: 152 Beach Road, #21-01/04 Gateway East, Singapore 189721, Singapore

This book is dedicated to all those who tirelessly work toward the protection and conservation of India's wildlife, ensuring a sustainable future for generations to come.

Foreword

India has always been a land where humans and animals have lived together—sometimes in reverence, sometimes in fear, but always in close proximity. From the deer that roamed the forests of our ancient ashrams to the peacocks dancing in monsoon fields, our stories, temples, and traditions are filled with wildlife. Even gods and goddesses in Hindu mythology ride on animals—the tiger, the mouse, the eagle, the swan. This deep connection between nature and our way of life is not just part of our past; it is meant to guide our future.

Yet, somewhere along the journey of development, urbanization, and growing markets, this sacred balance has been disturbed. The forests have shrunk. Rivers have been polluted. And many animals are being stolen from their homes—hunted, smuggled, and sold like goods. Some are taken for their skin or bones, some for medicine, some just to be locked up in cages as exotic pets. This is what we call wildlife trade—and when it happens illegally, it becomes a threat not just to the animals but to all of us.

This book is a heartfelt attempt to bring together the knowledge, the facts, the science, and the voices of people from across the country who are working to understand and stop this harmful trade. It is not just a book for experts or forest officers, it is for every Indian who has ever stopped to watch a singing bird, who has seen a monkey leap across a rooftop, who has read about a tiger in the newspaper, or who has prayed to Lord Ganesha with love. Because saving wildlife is not the job of a few—it is the responsibility of us all.

The chapters in this book take us on a journey—from the sacred groves of the past to the bustling online markets of today, from the borders where animals are smuggled, to the courtrooms where justice is delayed, from the villages where poachers live in poverty, to the laboratories using DNA to catch criminals. It is about elephants and pangolins, parrots and turtles, and even the invisible viruses that can jump from wild animals to humans—like COVID-19. Importantly, this book also shows hope—stories of species that have come back from the brink, of communities who have chosen conservation over consumption, and of technologies that can help us protect forests better. These stories are Indian in spirit and global in importance.

Whether a school student or a farmer, a bureaucrat or a homemaker, a forest guard or a policymaker—this book offers the reader something: knowledge, awareness, and most importantly, the inspiration to act. The fate of our wildlife depends not only on laws and fences but on love and understanding.

Let this book be a lamp—to light the path toward a future where India remains not just a land of tigers and peacocks but a country where wildlife is safe, respected, and celebrated.

Gobind Sagar Bhardwaj
Director Wildlife Institute of India
Dehradun, Uttarakhand, India

Foreword

India with its rich biodiversity holds within its forests, grasslands, wetlands, and oceans an extraordinary array of life forms found nowhere else on the Earth. From the majestic Bengal tiger to the shy pangolin, from vibrantly colored parrots to delicate seahorses, our country is a sanctuary for thousands of species, many of which are vital to our ecological and cultural heritage. However, this treasure trove of biodiversity is under threat. The rising demand for wild animals and their parts, whether for luxury goods, traditional medicines, exotic pets, or international markets, has placed immense pressure on wildlife populations. The trade in wildlife, both legal and illegal, has grown more sophisticated, more global, and more destructive than ever before. It has far-reaching consequences, not only for the species being traded but also for the ecosystems they inhabit and the communities that cohabit.

This book, *Wildlife Trade in India: Conservation and Commerce*, comes at a critical juncture. It offers comprehensive overviews in the Indian context. By bringing together diverse fields, this volume paints a multidimensional picture of a crisis that needs to be urgently attended. The authors explore everything from traditional cultural practices to modern e-commerce networks, from the role of frontline enforcement to the ethics of pet ownership, and from zoonotic disease emergence to solutions rooted in sustainable livelihoods. The book also explores possibilities of species recovery, community conservation, etc. Besides, this book is meaningful for every citizen of India. Understanding wildlife trade is not a niche concern, it is central to our national interest, public health, and the moral legacy. I hope this book serves as both a mirror and a roadmap for a sustainable future.

Deepti Dharmani
Vice Chancellor, Chaudhary Bansi Lal
University, Bhiwani, Haryana, India

Preface

India's relationship with wildlife has always been one of paradoxes. On the one hand, our myths, folklore, and rituals are overflowing with animals—Lord Ganesha with his elephant head, Goddess Durga astride a tiger, the mouse as a divine vehicle, and the cobra celebrated during Nag Panchami. On the other hand, the same creatures too often find themselves in poachers' traps, illegal cages, or smuggled across borders. This contradiction—of reverence and exploitation—frames the contemporary story of wildlife trade in India.

The present volume, *Wildlife Trade in India: Conservation and Commerce*, attempts to unpack these contradictions with academic precision and policy relevance. It is not a casual walk through the literature, but rather a structured journey through the multiple dimensions of trade—cultural, ecological, economic, and legal. In recent years, India has not only amended its Wildlife (Protection) Act, 1972 to align with evolving CITES commitments but has also witnessed unprecedented seizures of species ranging from star tortoises and rose-ringed parakeets to pangolin scales. Such events remind us that laws, however progressive, cannot alone stem the tide unless enforcement and public support keep pace.

The book is organized into thematic sections, each illuminating a critical dimension of wildlife trade. We begin with a historical and cultural framing—because today's online pet markets cannot be understood without yesterday's traditions of falconry or temple rituals. The legal and enforcement section provides an unflinching review of India's conservation architecture: progressive in vision, but often porous in practice. Our contributors draw attention to practical hurdles—whether it is underfunded forest departments, delays in wildlife crime trials, or the shadow of corruption that allows some offenders to slip through nets designed to protect endangered species.

The ecological section underscores the cost of this trade—species driven toward extinction, ecosystems destabilized, and zoonotic diseases knocking at our doors. COVID-19 was a global wake-up call, but India has long known how porous the barrier between humans and wildlife can be—Nipah outbreaks in Kerala, for instance, are stark reminders. When pangolins, bats, and primates are commodified, the risks transcend conservation and enter the realm of public health security.

The volume then turns to socioeconomic drivers. After all, wildlife trade is not only sustained by wealthy consumers demanding exotic pets or trophies but also by local communities who often see little choice beyond poaching. Understanding these dynamics is essential for crafting alternatives. Encouragingly, several chapters highlight community-led conservation and livelihood models—reminding us that enforcement cannot succeed without empathy and economic options.

Equally forward-looking are the chapters on technology: DNA barcoding to trace origins of seized body parts, drones patrolling vulnerable landscapes, and artificial intelligence scanning online platforms for illicit trade. India's forensic laboratories, once peripheral to conservation, are now central to unraveling trafficking routes. It is science in service of law, and a powerful example of interdisciplinary problem-solving.

The book concludes with stories of hope: gharial reintroduction successes, vulture population recoveries after bans on harmful veterinary drugs, and community-managed forests where biodiversity thrives. These accounts demonstrate that India's conservation story is not just one of crisis but also of resilience and innovation.

As editors, we are proud of the diversity of voices represented here—senior scholars with decades of expertise, early-career researchers with fresh insights, and practitioners at the enforcement and community levels. Their combined scholarship affirms that wildlife trade is not a single-discipline issue but one that demands legal clarity, ecological science, social empathy, and political will.

We extend our deepest gratitude to our contributors, our institutions, and Springer Publications for supporting this collective vision. Above all, we thank you, the reader. Because the fate of India's wildlife will not be decided in courtrooms alone, nor in the forests, but in the choices that ordinary citizens make every day.

If this book makes you pause before admiring a caged parakeet in a marketplace, or question why pangolin seizures keep making the news despite their strict protection under CITES Appendix I, or inspires a student to take up wildlife forensics as a career, then our effort has been worthwhile. Wildlife trade may be commerce, but conservation, ultimately, is about compassion and justice. And perhaps India—with all its contradictions, its biodiversity, and its cultural depth—is uniquely placed to show the world how reverence for animals can be translated into policy, and how policy can, in turn, protect reverence.

Bhiwani, Haryana, India — Asha Poonia
Dehradun, Uttarakhand, India — Sandeep Kumar Gupta
Itanagar, Arunachal Pradesh, India — Vikram Delu

Acknowledgments

The journey of preparing this volume, *Wildlife Trade in India: Conservation and Commerce*, has been one of deep learning, collaboration, and reflection. We, the editors, wish to record our sincere gratitude to all those who made this work possible.

Our foremost thanks go to the contributing authors, whose scholarship, dedication, and patience have shaped this book into a collective intellectual endeavor. Their expertise spans law, ecology, forensic science, economics, and community engagement, and it is this diversity that gives the volume its strength. We also extend our appreciation to Springer Publications for their professional guidance, which has steered this manuscript from conception to completion.

We gratefully acknowledge our parent institutions—*Chaudhary Bansi Lal University, Bhiwani; the Wildlife Institute of India, Dehradun; and the Zoological Survey of India*—for providing not only academic support but also the intellectual environment in which our ideas could take root and grow. Without the encouragement and infrastructure of these institutions, this book would not have been possible.

Our gratitude is not limited to people and organizations alone. We owe a profound debt to *nature itself—the forests, rivers, mountains, and oceans of India, and to the countless wild beings, from the smallest microorganisms to the largest mammals—whose existence inspires our science and whose silent resilience has taught us more about selfless love and survival than any textbook could.* In acknowledging them, we also remind ourselves of the very purpose of this work: to ensure their protection for future generations.

On a personal note, each of us owe much to our families, colleagues, and mentors who offered encouragement and understanding during the many hours devoted to this project. Their unseen contributions are etched into these pages.

Finally, we wish to dedicate this effort to all those who believe that conservation is not simply a profession but a shared responsibility, and that science, when coupled with compassion, can illuminate the path to a sustainable and just future.

Contents

Contributors

Daamini Arora PG Department of Zoology, Dev Samaj College for Women, Ferozepur, Punjab, India

Deepak Rai Babbar Department of Zoology, Kurukshetra University, Kurukshetra, Haryana, India

Priya Battu Regeneration Biology Lab, IISER Mohali, Mohali, Punjab, India

Kavita Bisht Laboratory for the Conservation of Endangered Species (LaCONES), CSIR-Centre for Cellular and Molecular Biology, Hyderabad, Telangana, India

Priyanshi Chaudhary School of Forensic Science, Uttar Pradesh State Institute of Forensic Science, Aurawan, Lucknow, Uttar Pradesh, India

Bharti Chauhan Faculty of Life Science, Chaudhary Bansi Lal University, Bhiwani, Haryana, India

Shrey Dandriyal Wildlife Institute of India, Chandrabani, Dehradun, Uttarakhand, India

Vikram Delu Scientist -B, Zoological Survey of India, Arunachal Pradesh Regional Centre, Itanagar, Arunachal Pradesh, India

Preeti Devi Faculty of Commerce and Management, Chaudhary Bansi Lal University, Bhiwani, Haryana, India

Kumudani Bala Gautam Department of Biosciences, Graphic Era (Deemed to be University), Dehradun, Uttarakhand, India

Anil Ghanghas Department of Laws, Chaudhary Devi Lal University, Sirsa, Haryana, India

Harsh Gulati Department of Zoology, School of Bioengineering and Biosciences, Lovely Professional University, Phagwara, Punjab, India

Sandeep Kumar Gupta Scientist -F, Animal Ecology & Conservation Biology Lab, Wildlife Institute of India, Dehradun, Uttarakhand, India

Shantabala Devi Gurumayum Zoological Survey of India, Government of India, Arunachal Pradesh Regional Centre, Itanagar, India

Monika Jangra Department of Zoology, Chaudhary Bansi Lal University, Bhiwani, Haryana, India

Sharda Kalra Department of Zoology, Chaudhary Bansi Lal University, Bhiwani, Haryana, India

Rajinder Kaur Department of Zoology, Government Mohindra College, Patiala, Punjab, India

Ramneek Kaur PG Department of Zoology, Dev Samaj College for Women, Ferozepur, Punjab, India

Akhilesh Kukreti Government PG College Karanprayag, Chamoli, Uttarakhand, India

Ajit Kumar School of Forensic Science, Uttar Pradesh State Institute of Forensic Science, Aurawan, Lucknow, Uttar Pradesh, India

Amit Kumar Department of Zoology, Chaudhary Bansi Lal University, Bhiwani, Haryana, India

Diksha Kumari Defence Institute of Physiology & Allied Sciences (DIPAS), New Delhi, India

Angel Maria Lorance Department of Zoology, School of Bioengineering and Biosciences, Lovely Professional University, Phagwara, Punjab, India

Yomto Mayi Zoological Survey of India, Government of India, Arunachal Pradesh Regional Centre, Itanagar, India

Surender Mehra NITI Aayog, Government of India, New Delhi, India

Aarush S. Nair Department of Zoology, School of Bioengineering and Biosciences, Lovely Professional University, Phagwara, Punjab, India

Neelam Pawar Department of Pharmaceutical Sciences, Chaudhary Bansi Lal University, Bhiwani, Haryana, India

Asha Poonia Faculty of Life Sciences, Chaudhary Bansi Lal University, Bhiwani, Haryana, India

Awasthi Rajan Scientific Advisor, Mangrove Foundation of India, Tiruvallur, Tamil Nadu, India

Sarita Rana Department of Zoology, Kurukshetra University, Kurukshetra, Haryana, India

Preeti Rani Department of Zoology, School of Bioengineering and Biosciences, Lovely Professional University, Phagwara, Punjab, India

Kapish Samota Department of Zoology, Chaudhary Bansi Lal University, Bhiwani, Haryana, India

Komalpreet Kaur Sandhu Department of Zoology, Akal University, Talwandi Sabo, Bathinda, Punjab, India

Abhinav Saxena Department of Zoology and Environmental Sciences, Punjabi University, Patiala, Punjab, India

Parvati Sharma Department of Zoology, Chaudhary Bansi Lal University, Bhiwani, Haryana, India

Pranjal Sharma School of Forensic Science, Uttar Pradesh State Institute of Forensic Science, Aurawan, Lucknow, Uttar Pradesh, India

Radhika Sharma Department of Biology and Environmental Sciences, College of Basic Sciences, Chaudhary Sarwan Kumar Himachal Pradesh Krishi Vishvavidyalaya, Palampur, Himachal Pradesh, India

Devinder Singh Department of Laws, Panjab University, Chandigarh, India

Ekashvi Thakur Formerly at Department of Zoology, Panjab University, Chandigarh, India

R. Urmila Department of Zoology, Chaudhary Bansi Lal University, Bhiwani, Haryana, India

Ravikant Verma Department of Zoology, CCS Haryana Agricultural University, Hisar, Haryana, India

Abbreviations

AI	Artificial Intelligence
AML	Anti-Money Laundering
ASEAN-WEN	Association of Southeast Asian Nations Wildlife Enforcement Network
AWBI	Animal Welfare Board of India
BDA	Biological Diversity Act, 2002 (India)
CAP	Conservation Action Plan
CBD	Convention on Biological Diversity
CBIC	Central Board of Indirect Taxes and Customs
CBO	Community-Based Organization
CFSL	Central Forensic Science Laboratory (India)
CITES	Convention on International Trade in Endangered Species of Wild Fauna and Flora
CITES CoP	Conference of the Parties to CITES
CMFRI	Central Marine Fisheries Research Institute
CMS	Convention on Migratory Species (Bonn Convention)
COI	Cytochrome c Oxidase Subunit I
CPB	Cartagena Protocol on Biosafety
CrPC	Code of Criminal Procedure (India)
CRZ	Coastal Regulation Zone
CSR	Corporate Social Responsibility
cyt b	Cytochrome b gene
DGFT	Directorate General of Foreign Trade
DL	Deep Learning
DNA	Deoxyribonucleic Acid
DRI	Directorate of Revenue Intelligence
EIA	Environmental Impact Assessment
EoI	Expression of Interest
ES	Ecosystem Services
ESA	Ecologically Sensitive Area

FCA	Forest (Conservation) Act, 1980 (India)
FGD	Focus Group Discussion
FIR	First Information Report
FSL	Forensic Science Laboratory
GoI	Government of India
IBA	Important Bird and Biodiversity Area
ICAR	Indian Council of Agricultural Research
ICMR	Indian Council of Medical Research
IFA	Indian Forest Act, 1927 (India)
INTERPOL	International Criminal Police Organization
IPC	Indian Penal Code (India)
IPR	Intellectual Property Rights
IT Act	Information Technology Act, 2000 (India)
IUCN	International Union for Conservation of Nature
IUCN SSC	Species Survival Commission (IUCN)
KBAs	Key Biodiversity Areas
KII	Key Informant Interview
ML	Machine Learning
MoEFCC	Ministry of Environment, Forest and Climate Change (Government of India)
MPA	Marine Protected Area
MPEDA	Marine Products Export Development Authority
MSME	Micro, Small and Medium Enterprises
mtDNA	Mitochondrial DNA
NBSAP	National Biodiversity Strategy and Action Plan
NBWL	National Board for Wildlife
NCRP	National Cyber Crime Reporting Portal (India)
NGO	Nongovernmental Organisation
NGS	Next-Generation Sequencing
NGT	National Green Tribunal
NIA	National Investigation Agency
NP	Nagoya Protocol on Access and Benefit-sharing
NPP	Net Primary Productivity
NTCA	National Tiger Conservation Authority
PA	Protected Area
PCR	Polymerase Chain Reaction
PPP	Public–Private Partnership
PRA	Participatory Rural Appraisal
qPCR	Quantitative PCR
Red List	IUCN Red List of Threatened Species
RS	Remote Sensing
RT-PCR	Reverse Transcription PCR
SAP	Species Action Plan
SAWEN	South Asia Wildlife Enforcement Network

SDG 14	Sustainable Development Goal 14: Life Below Water
SDG 15	Sustainable Development Goal 15: Life on Land
SDGs	Sustainable Development Goals
SDM	Species Distribution Modelling
SHG	Self-Help Group
SIT	Special Investigation Team
SNP	Single Nucleotide Polymorphism
SOP	Standard Operating Procedure
SPS	Sanitary and Phytosanitary (WTO Agreement)
SSC	Species Survival Commission (IUCN)
STR	Short Tandem Repeat
TBT	Technical Barriers to Trade (WTO Agreement)
TRAFFIC	Trade Records Analysis of Flora and Fauna in Commerce
UAS	Unmanned Aircraft System
UAV	Unmanned Aerial Vehicle (drone)
UNCTAD	United Nations Conference on Trade and Development
UNEP	United Nations Environment Programme
UNODC	United Nations Office on Drugs and Crime
USGS	United States Geological Survey
WCCB	Wildlife Crime Control Bureau (India)
WHO	World Health Organization
WII	Wildlife Institute of India, Dehradun
WPA	Wildlife (Protection) Act, 1972 (India)
WTO	World Trade Organization
WWF	World Wide Fund for Nature

Part I
Introduction to Wildlife Trade in India

Chapter 1
Historical and Cultural Dimensions of Wildlife Trade in India

Komalpreet Kaur Sandhu, Ramneek Kaur, and Daamini Arora

Contents

Abstract The wildlife trade in India has a deep historical and cultural significance, shaped by centuries of indigenous practices, colonial influences, and evolving global demands. This chapter explores the historical and cultural dimensions of wildlife trade in India, focusing on the role of traditional hunting, religious beliefs, and socio-economic factors that have contributed to the exploitation of wildlife. From ancient

K. K. Sandhu (✉)
Department of Zoology, Akal University, Talwandi Sabo, Bathinda, Punjab, India

R. Kaur · D. Arora
PG Department of Zoology, Dev Samaj College for Women, Ferozepur, Punjab, India

A. Poonia et al. (eds.), *Wildlife Trade in India*,
https://doi.org/10.1007/978-981-95-3694-8_1

times, wildlife products such as ivory, animal skins, and medicinal herbs were valued for religious rituals, royal patronage, and trade. The colonial era introduced new forms of exploitation, with hunting as both sport and economic activity. In contemporary India, the illegal wildlife trade remains a major threat to biodiversity, fueled by both local traditions and international demand for exotic animal products. The chapter also examines the intersection of wildlife conservation and cultural practices, highlighting efforts to reconcile traditional beliefs with modern conservation policies. Through this exploration, the chapter sheds light on the complex relationships between culture, history, and wildlife trade, offering insights into how these factors continue to influence wildlife protection efforts nowadays.

Keywords Wildlife trade · India · Cultural heritage · Colonial history · Biodiversity Conservation · Illegal trade

1.1 Introduction

Wildlife and human culture has always been an inseparable. In childhood, children are introduced to animals via poems, stories, toys and educational materials. Now-a-days, youngsters are not much interacted with the natural world. Even now children are playing with plastic toys instead of wooden or earthen toys, they only mesmerize wildlife in T.V., mobile phones or in zoos. As reported by Rajak and his co workers (Rajak & Chatruvedi, 2023a). They never experience a real wildlife. Also ruthless human activities such as deforestation, pollution and habitat destruction led to the decline in wildlife population. Many species become extinct and many at the risk of extinction. These threats to wildlife is all over the world including India (Klappenbach, 2014). A rich and beautiful wildlife of India also facing the similar threats of destruction and extinction. For the sake of development, humans continuously destroying forests, wetlands, grasslands and other natural habitats. Since industrial evolution, these selfish human activities increases much and endangering environment, wildlife and human well-being. Now, most of Earth's water bodies are polluted, many wetlands are lost and more than 50% of ice-free land, e.g., Antarctica and Greenland is changed significantly. This environmental destruction has left 5 lakh species of plants and animals, and 5 lakh species of insects at the of extinction in coming ages. If today, everyone focusses on conservation of nature then many of species can be protected from extinction (Rajak & Chatruvedi, 2023a). The importance of wildlife for humans on Earth is significant and large number of evidences prove that human activities causing unparallel harm to the wildlife at a large extent. India is home for various remarkable wildlife species include Bengal tiger, Asiatic lion, one-horned rhino, snow leopard, Indian pangolin, Indian peafowl, Kashmir stag, red panda India bison, Asian elephant, sloth bear, South Asian river dolphin, spotted deer, nilgai, swan, parrot, rock pigeon and many more. India has about 45,500 species of plants, 423 species of mammals, 1233 species of Aves, 526 species of reptiles, 342 species of amphibians, 3022 species of Pisces and 83,436 species of invertebrates including 63,423 species of insects. If humans are not interfering then these animals live with peace and harmony according to the natural laws but for the sake of development, human activities disturb the whole ecosystem which results in extinction of various species and many species become endangered. According to the International Union for Conservation of Nature (IUCN), about 900 animal species including leopard, Kashmir stag etc. and 600 plant species including ebony tree, bird's foot etc. are endangered in India. Human activities such as destruction of wildlife habitat and poaching of animals continuously increasing the risk of extinction of various species (Aggarwal, 2006).

It is always been a complicated relationship between efforts for wildlife conservation and local communities, especially in Asia where forests have traditional cultural values and life of these communities depend on forests in many ways. The local communities depend on forest resources which are essential for their livelihood but when the conservation policies, includes the creation of protected areas in the form of national parks and wildlife sanctuaries for the protection of wildlife made by state or centre government deprive these local communities to utilize forest resources. These people show resentment towards these conservation policies and this leads to the conflicts between local communities and officers which deals with these policies, and this results in human injuries, loss of crops and livestock. This discussion also highlights the counter debate between land-sharing i.e. the conservation of biodiversity within agricultural land areas and land-sparing i.e. agricultural practices must not include in protected areas for conservation of wildlife. These debates or discussions give more importance to agricultural practices for food production rather than considering the needs of people who depend on the forests for their essential needs. There must be a peaceful solution for conservation of wildlife, agricultural practices and needs of local communities because each of them is important. From the colonial era to post independence era, there always a incidences of wildlife exploitation and also policies made to protect wildlife. In India, the wildlife always encircles in various traditions, politics but after independence the nation is striving for developing a balance between development and biodiversity conservation (Rajak & Chatruvedi, 2023a).

1.2 Definition of Wildlife

The 'wildlife' means naturally occurring or undomesticated animals (fauna) including Pisces, mammals, Aves etc. and uncultivated plants (flora) that exists in their natural habitat without any kind of human interference. The wild species are not dependent on humans for their protection, food and shelter (Tian et al., 2023).

(i) **Wildlife Protection Act, 1972:**

Wildlife means any animal species or plant species either terrestrial or aquatic which forms a part of any habitat (Wildlife Protection Act, 1972. Sec. 2 (37)).

(ii) **Usher's View, 1986:**

Wildlife is defined as non-introduced and undomesticated (Usher, 1986).

(iii) **Yarrow's View, 2009:**

Wildlife refers to all the living organisms that are not controlled by the humans (Yarrow, 2009).

(iv) **Sharma et al.'s View, 2011:**

Wildlife considers as uncultivated plant species and undomesticated animal species that are living in their natural habitat (Sharma et al., 2011).

1.3 Values of Wildlife

The wildlife is an integral part of man's life with significant values. The wildlife is important for humans as well as for environment. The huge efforts are required for the conservation of wildlife (Alok, 2018) as wildlife has certain ethical, ecological, cultural, scientific and medicinal values as shown in Fig. 1.1.

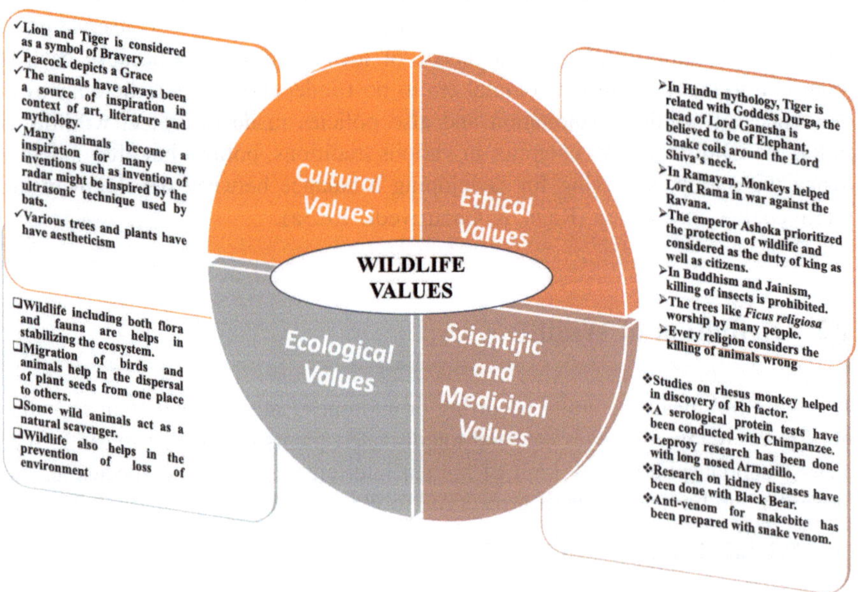

Fig. 1.1 Figure showing various kinds of values of wildlife

1.4 Historical Roots of Wildlife Exploitation from Colonial to Post-Independence Eras

1.4.1 Exploitation of Wildlife in Colonial Era (1858–1947, British Rule)

India had rich and diverse wildlife before the arrival of Britishers but their policies and rules resulted in declination of huge number of wild animals due to increasing demand for land and requirement new and powerful weapons. This triggered in the hunting of large number of wild animals including elephants, lions, tigers, leopards and many other animals without any kind of monitoring. In early nineteenth century, the East India Company extended the colonial control over forests which led to the provision of forest resources and wildlife. The British officials along with the East India Company promoted the hunting of various carnivores' animals including tigers by announcing rewards for both Indian and European hunters. The Britisher did various things for the sake of showing colonial power, giving privileged to many Europeans and set standards for status symbol like shooting of wild animals which accelerated the poaching of wild animals. The poaching of tiger also increased for their skin; people started selling tiger's skin for their own financial benefit. They also extended land areas to forests for development which abolished the various wildlife species including both flora and fauna (Rajak & Chatruvedi, 2023b). Throughout nineteenth century, Britishers rewarded anyone who killed any roaming creature. In 1822, the administration of East India Company offered rewards of approximately 38,000/– for the sole extinction of about 5500 tigers in the Bengal Presidency. At the same time, the government of Bombay started offering rewards for killing of tigers and other undomesticated creatures. In early twentieth century, various tour companies and local guides started earning money by assisting hunters for kills. In 1904, the British Indian Newspaper promoted Cordite Rifles as a suitable weapon for killing powerful wild animals like tigers. In 1912, the business of about 43,000 tiger skins was done to earn a alluring profit. At early twentieth century, about 40,000 were present which declined drastically to 4000 in 1964. Similar impact also seen on other wild animals (Sharma & Tripathi, 2024).

1.4.2 Conservation of Wildlife in Colonial Era

In nineteenth century, wildlife conservationists initiated the movement for the protection of wildlife in India, especially the preservation of endangered species and establishment of wildlife sanctuaries and national parks. Although protection of wildlife was not a matter of concern during the British period, but these movements forced them to take step towards wildlife conservation (Guha & Gadgil, 1989). Then British government also started realising the importance of wildlife in balancing of ecosystem and also for tourism industry. The government made a various policies including Indian Forest Act, 1878 to regulate use of forest resources and wildlife. By the implementation of this act, the forests became under the direct control of government and this affects the farmers, local communities and tribal people who depend on the forests resources and use forest land for semi-agriculture practices. The colonial government's approach towards wildlife conservation didn't consider the cultural and traditional knowledge of local communities. The government's policies favoured wealthy and elite class people which led to the unequal distribution of resources. This resulted in social and political disturbance in colonial country. In 1887, the colonial British government enforced a law of the Wild Birds Protection Act. It is a first legal law passed by the British government for wildlife conservation in India. According to this act, no one could capture or killed the birds during the breeding season. In 1912, the British government enforces the Wild Bird and Animal Protection Act. This act prohibited the hunting or poaching of wild birds and wild animals. By understanding the importance of wildlife conservation, this law also amended in 1935 along with various other policies. In 1873, the Madras Elephant Preservation Act was a remarkable act in history of conservation of wildlife. It was first act that was made to protect a particular species from destruction. This act banned the hunting of elephants and there was also a penalties and punishments for violators. To protect the elephants from killing or capturing across the country, the Central Government's Elephant Preservation Act in 1879 was passed. In 1972, two important acts were passed to conserve forest resources and wildlife. First one was the Indian Forest Act, 1927 to regulate the transport of forest produce and also put duties on various forest products. Although, it's main objective to collect revenue from forest resources but either directly or indirectly this helped in conserving forests and wildlife, and protecting environment. The second act was Wildlife Protection Act, 1972 which included protected areas like Kaziranga Sanctuary, Corbet National Park and others. The wildlife policies made during colonial period was driven by the unchecked hunting of wildlife. The hunting of wild animals during nineteenth century results in decline of large number of animals. However, the conservation policies during colonial period helped in formation of protected areas for wildlife conservation (Rajak & Chatruvedi, 2023b).

1.4.3 Exploitation of Wildlife in Post-Independence Era (1947–Present)

Even after the independence, the attitude of public towards wildlife conservation didn't change much. Many people belonging to different classes still focused on hunting. Wildlife trade was still of a great profit. Between 1950 to 1960, huge number of monkeys exported from India to United States. Also skins of snakes, leopards and tigers were exported in exchange of foreign currency. Rhino's horn and Elephants' ivory also traded a lot. The arrival of modern technology like long range rifles, automobiles fasten the decline of wildlife in India. When government provided farmers with guns to protect their crops but this only resulted in large scale demolition of wildlife. In 1951, Cheetahs became extinct who once roamed freely in India. One more reason for wildlife exploitation in India after independence was continuous increase population, and urban and agricultural expansion which led to the habitat loss for numerous wildlife species. Fast growing industrialisation also altering the environment at great extent. After independence, it became a difficult for government to balance economy development and protection of biodiversity (Singh & Singh, 2020).

1.4.4 Conservation of Wildlife in Post-Independence Era

Just after independence, there was no rules and policies were made for wildlife conservation. Overtime, the importance of conservation of wildlife became a concern for government. During last 30 years of twentieth century, the government of India framed a various policy in context of conservation of wildlife. India took many steps towards the conservation of wildlife and protect of natural resources and environment. In 1950, the Elephant Preservation Act and the Rhinoceros Preservation Act were passed to protect these threatened species from exploitation. In 1951, the Advisory Committee consisted of renowned wildlife conservationists and other members suggested many ways to establish national parks and wildlife sanctuaries. The Bharatpur Bird Sanctuary and Sunderbans National Park was marked as a great decision for the protection of numerous bird species and Bengal tiger. In December, 1952 the inauguration of Indian Board for Wildlife was held at Mysore (Rajak & Chatruvedi, 2023a). The establishment of Wildlife Protection Act, 1972 and Project Tiger, 1973 was a significant effort in the history of conservation of wildlife. The Project Tiger mainly aimed to protect endangered tigers from extinction (Project Tiger, Government of India). In 1983, the National Wildlife Action Plan was approved to track the declining wildlife population. In 1991–1992, Project Elephant was launched by the Government of India to protect elephants and their habitats by providing funds and technical support (Project Elephant, Government of India). In 2002, the National Biodiversity Act was introduced by the ministry of Environment and Forest to regulate the usage of biological resources and its conservation. In 2006, the Integrated Development of Wildlife Habitats Program was launched by the government of India to provide funds management and protection of biosphere reserves, national parks and wildlife sanctuaries (National Biodiversity Act, 2002, Government of India). In 2010, the National Wildlife Plan for 2017–2031 was launched by the ministry of Environment and Forests. To re-establish the Cheetahs in India, 8 Cheetahs from Namibia released to Kuno National Park on 17 September, 2022. The government of India also doing a efforts to develop a sustainable tourism practices to raise awareness about the protection of wildlife and this also provide the financial help for conservation of wildlife. The government, wildlife conservationists, various other institutes doing numerous efforts to protect wildlife at own's level and also in collaboration with various bodies at global level (Mishra, 2021).

1.5 The Wildlife Protection Act, 1972

The wildlife protection act, 1972 is the only conservation policy exist in independent India after the implementation of constitution. The strong policies were needed to protect the rapid decrease of wildlife and natural resources. The wildlife protection act, 1972 was important initial step which include the establishment and protection of national parks and wildlife sanctuaries. The wildlife protection act, 1972 and its amendments till 2006 provide a legal structure of the wildlife conservation in India.

1.5.1 Need, Scopes and Objectives of the Wildlife Protection Act, 1972

India is a country with rich and diverse wildlife. Overtime, there was a drastic declination of both flora and fauna which disturbs the ecosystem balance. There was a need of strong legislation to protect wildlife and conserve natural resources. Only five national parks were present before the implementation of the wildlife protection act, 1972. As shown in Fig. 1.2 scopes of this act include Protection of endangered or critically endangered species including all the species in the specified protected areas. The objectives of Wildlife Protection Act, 1972 demonstrated a uniform Laws, policies in our country, controlling illegal trade and providing habitat in form of protected areas for the better conservation (Fig. 1.2). There are various significant gaps exists in conservation of wildlife in India and each gap required a

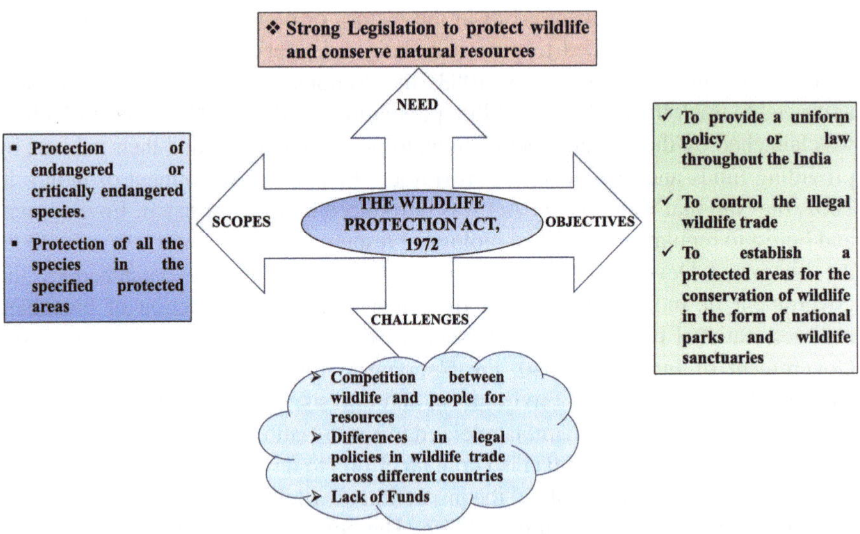

Fig. 1.2 Figure showing need, scope, objectives and challenges in The Wildlife Protection Act, 1972

attention to improve a conservation approach. Hence, There are also so many challenges for fulfilling the objectives of Wildlife given in Fig. 1.2 above. For example, if we discuss about the challenge of differences in legal policies in wildlife trade across different countries, there is an example Star tortoise included in protected species under Wildlife Protection Act, 1972 in India but in other countries it's trade is very common. This act does not include the trade of foreign species. In India, there are various unlicensed markets where plants, animals and birds traded openly though illegally. The sellers also trade endangered species to make huge profits (Choudhary, 2024).

1.5.2 Amendments in the Wildlife Protection Act, 1972

(i) **The Wildlife Protection (Amendment) Act, 1982**

- This allowed states to regulate the hunting of wild animals that are believed to be harmful to crops.
- In this amendment, penalties are also increased for violators.

(ii) **The Wildlife Protection (Amendment) Act, 1991**

- In this amendment, the conservation measures for illegal wildlife trade was became more strict.
- This also included the regulation of zoos according to conservation norms.

(iii) **The Wildlife Protection (Amendment) Act, 2002**

- In this amendment, local communities were also included to aid the conservation of wildlife.
- New protected areas were also introduced in this amendment.
- Conservation measure to increase the protection of tigers was also strengthened.

(iv) **The Wildlife Protection (Amendment) Act, 2006**

- This amendment mainly focused on the conservation of natural habitats of wildlife.
- Establishment of Tiger Conservation Authority was also done to conserve tiger population.

(v) **The Wildlife Protection (Amendment) Act, 2022**

- In this recent amendment, the wildlife protection act co-ordinated with the provisions of the Convention on International Trade in Endangered Species of Wild Fauna and Flora (CITES).
- This also included the regulation of invading foreign species.
- This amendment also gave importance to the sustainable usage of resources (The Wildlife Protection Act, 1972, Government of India).

1.5.3 Suggestions to Improve Wildlife Protection Act, 1972

- Increase penalties and punishments for violators.
- Raise public awareness about wildlife conservation.
- More collaborations between wildlife authorities and law enforcement.
- Improvement of governance.
- More technological surveillance which includes installation of voice aided CCTV cameras in protected areas, drones and satellite imaging.
- Collabs with other countries to make laws for protecting common wildlife species.
- Establishment of wildlife laboratories nearby protected areas and forests for effective investigation of wildlife crime.
- Expanding veterinary hospitals to treat wildlife diseases more efficiently.
- More strict laws are required to prohibit illegal wildlife trade.
- Encouragement of sustainable tourism practices (Choudhary, 2024).

The Wildlife Protection Act, 1972 is an astonishing approach of India's conservation policy. It provisioned the establishment of protected areas including national parks and wildlife sanctuaries for the protection of wildlife. This act and it's amendments provide a legal framework for wildlife conservation in India.

1.6 Influence of Religious and Cultural Practices on Wildlife Use

Religious and cultural practices have a significant role in influencing the human's interactions with the wildlife (Mehta & Shah, 2019). These practices are deeply rooted in beliefs, traditions, and values, decides how societies view and use animals for food, clothing, rituals, and medicinal purposes. Such practices often intertwined with the local ecosystem, creating a complex dynamic between conservation efforts, biodiversity, and the sustainability of wildlife resources. While Discussing Religious beliefs impacting positively wildlife, Hinduism and Buddhism practicing a concept of non-violence to animals. Various animals are worshiped and considered sacred (Thompson, 2004). On the other hand, religious rituals have not been considered complete without the sacrifice of animals (Miller, 2010). Some tribes are practicing hunting (Adams & Mulligan, 2019), Captivity of animals in religious ceremonies has profound negative impact on wildlife leading to fragile ecosystems (Rohner, 2016) as shown in Fig. 1.3.

Cultural Practices also had a significant impact on wildlife use in all over the world, often influencing hunting, harvesting, and consumption practices. These cultural practices can either promote sustainability or contribute to the overexploitation of wildlife. Lema and his co-workers (Lema et al., 2025) reported various ethnozoological relationships between human and animals. They reported

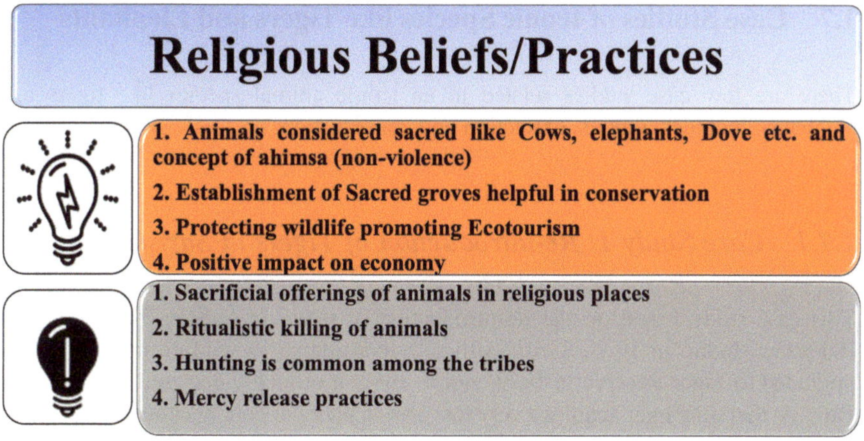

Fig. 1.3 Figure explores the role of religious practices highlighting both the positive and negative consequences for wildlife

Fig. 1.4 (**a**) Hunter with killed warthogs (**b**) Tusks of Elephant. (Source: Lema et al. (2025))

that indigenous people in Ethiopia used various parts of animals for preparing conventional medicines to cure various human ailments, rabbits killed for meat (Fig. 1.4).

In another latest study by Mengistu and his co-workers (Mengistu et al., 2024) in Southern Ethiopia showed that animal's body parts like horns used for making musical instruments, bone spoons, Animal Hide's Drum, Animal skin for making cover of Chair, Animal Fur for making traditional cape and other various decorative material during celebrations. Various studies emphasized the various ethnic communities' cultural relevance and knowledge systems related to medicinal animals.

1.7 Case Studies of Iconic Species like Tigers and Elephants

There are few case studies related to an iconic animal species like tigers and elephants, includes, threats, extinction and conservation.

1.7.1 Case Study 1: Re-introduction of Tigers in Sariska

This case study based on the re-introduction of tigers in Sariska Tiger Reserve, Rajasthan, India. In 1995, Sariska officially considered as wildlife reserve. It was upgraded to Tiger Reserve in 1979, and in 1982, it officially declared as a National Park. At that time tiger acquire a very less area in national park because of the human movement for various activities such as grazing, selling milk and others. The loss of tigers in Sariska was mainly occurred after independence. Sariska was a able to attracted a tourists especially due to the presence tigers. With time poaching of wild animals increased in protected areas which also impacted a tigers in Sariska Wildlife Park (Shahabuddin, 2010). In 2005, it was officially declared that tigers were locally extinct from Sariska (Sebastian, 2016). To re-establish a tiger population in Sariska National Park 7 tigers from Ranthambore Tiger Reserve in 6 or 7 years from 2008. Successfully, in 2017, there were total 14 tigers present in Sariska National Park including re-located tigers and their offsprings. This re-location of tigers from Ranthambore Tiger Reserve to Sariska National Park was world's first tiger re-introduction. The bringing back of top most predators affects both cultural and ecological structure (Brownlow, 2000). This helped in restoration of ecosystem and human-wildlife interaction, and re-introduction of species (Jorgensen, 2015).

1.7.2 Case Study 2: Rise in Tiger Poaching in India for International Market

The illegal trafficking is still a threat to wild animals for various purposes which weakens the conservation efforts to protect wildlife. The illegal poaching of wild animals like tiger is also occur in India (Tensen, 2016). The main purpose of trafficking of body parts or derived products of tigers is their usage as decorative purposes, fashion articles and also their use in making of traditional medicines (Shepherd et al., 2020). In India, there are very strict laws against poaching and trafficking of wild animals but still Bengal tiger face trafficking threats. Being a apex predator, tiger has a great ecological importance. The number of Bengal tiger is very less in the world but despite of this demand of their body parts and derived products keep increasing in illegal international market (Mohan et al., 2021). Due to extreme poaching, the number of wild tigers reduced to half in last 20 years (Morgan et al., 2021). In the world, China is the largest exporter of body parts and derived products of wild animals (Jiao et al., 2021). In India despite of considering trade and poaching of tiger is illegal under Wildlife Protection Act, 1972 still there is a cases of illegal trade of tigers especially via Madhya Pradesh, Bihar, Maharashtra and others (Palita, 2007). The most of tiger poaching and smuggling cases happen in protected areas including various tiger reserves and national parks. Most of illegal trade occur via India–Nepal–China trade rout (Sharma et al., 2014). India spend a hefty amount of money and numerous efforts on protection and conservation of tigers. This also become successful as number of tigers is increasing in India but this doesn't give a privilege to minimize the conservation efforts or neglect the future threats of poaching and smuggling of tigers.

1.7.3 Case Study 3: Expansion of Tiger Conservation in Southern India

Tiger is a threatened species from long time due to habitat destruction, decrease in number of their prey, illegal poaching and smuggling (Joshi et al., 2016). To protect tigers, countries form a various protected areas to conserve tigers. India also has various tiger reserves for which funding is provided to maintain suitable conditions and provide essential availabilities for tigers (National Tiger Conservation Authority (NTCA), 2015). Designation of tiger reserves not merely enough to conserve tigers but also a careful protection of tigers are required. In recent times, Karnataka is one of the best examples of biggest protected area expansion. Cauvery Wildlife Sanctuary, Karnataka and Malai Mahadeshwara Hills Wildlife Sanctuary, Karnataka are a part of this extensive forest tract. This area has a great scope conservation of large and endangered species like tiger and Asian elephants (Wikramanayake et al., 2010). The increased in protection of these protected areas was done government officials and other conservation groups (Gubbi et al., 2016). Firstly, these forest areas were managed under 10-year plan which includes protection against forest fires, poaching and plantation of trees. The government also provide a fund for organizing a anti-poaching camps, protection infrastructure, patrolling vehicles and also an arrangement of physical barriers to stop entering of various animal species into agricultural fields to avoid human-animal conflict. After these 10 years, reserved forest areas converted to a protected area. In Tamil Nadu, almost similar type of expansion was seen between 2011 and 2014. This expansion helps in recovery and survival of endangered animal species like tigers. These protected areas consist of tigers, elephants, leopard, sloth bear, Nilgiri barb and others which are endemic to Southern India (Gubbi et al., 2014). With an expansion of these protected areas, there is a great chances of increasing in population of carnivores animals like tigers. This also helps in recovery of habitat and preys which is essential of the survival of tigers.

1.7.4 Case Study 4: Life Threatening Conflicts Between Humans and Elephants in South Bengal

There is always a report on human-elephant conflicts from various places. A study on human-elephant conflict in South Bengal showed a devasting consequences. Population of elephants largely shrinked till 1980s mainly due to deforestation. At that time, conflicts between humans and elephants were negligible. Only few cases of conflict between these two mammalian species were reported. During recent period of time, the co-existence of humans and elephants in various districts of South Bengal almost become impossible. In 3 years from 2017 to 2020, human-elephant conflict led to death of around 50 humans and around 35 humans got injured. Also, a death of 15 elephants suspected due to human-elephant conflict. The main reason of human-elephant conflict in South Bengal is deforestation, developmental projects and expansion of agricultural land. This leads to the reduction of food for elephants. To fulfil their feeding and nutritional needs, elephants started raiding agricultural fields which increases conflict between humans and elephants. This situation is still prevailing in various districts of South Bengal (Khanra et al., 2020).

1.7.5 Case Study 5: Train-Elephant Collisions in West Bengal

Most of animals have wide range of habitat to fulfil their essential needs on daily, annual or seasonal basis. Due to habitat connectivity, animals easily re-locate or disperse to minimize competition and risks of inbreeding within a population (Wilcove et al., 1998). During migration, many cases on collisions between animals and vehicles have been reported. This increases the mortality rate of animals either on roads or on railway tracks (Seiler & Helldin, 2015). The problem of collisions occurs in almost every country, mainly in U.S.A., U.K., Canada and India (Child et al., 1991). In India, train-wildlife collisions is a serious threat. The network of Indian Railway is one of the largest rail networks in the world (Indian Railways, 2014) which also extends through forests and negatively impact the natural habitat of wildlife. Large mammalian species like leopards, lions, rhinoceros, elephants and others have been killed due to train collisions (Dasgupta & Ghosh, 2015). Elephant is one of the most frequent victims of train collisions (Roy et al., 2009). Railway track passes through various habitats of elephants in many states which increases the risk of death of elephants due to train collision. Train-elephant collision commonly occurs in Siliguri–Alipurduar railway line in northern West Bengal, Haridwar–Dehradun railway line in Uttarakhand, Coimbatore–Thrissur railway line in Tamil Nadu and Kerala, Delhi–Howrah railway line in Jharkhand and Kharagpur–Adra railway line in southern West Bengal. The Siliguri–Alipurduar railway line in West Bengal reported highest number of mortalities of elephants due to elephant-train collisions. The death of 27 elephants occurs in between 1974–2002 and this increased to 65 deaths in between 2004–2015 (Roy & Sukumar, 2016). The Siliguri–Alipurduar track was built in 1910–1911 by clearing forests. This railway line passes through forests, crop lands, rivers, streams, tea gardens and elephant corridors (Ray, 2002). With time both passengers and speed of train increased. These collisions have significant impact in reduction of local elephant population in West Bengal (Government of West Bengal, 2015).

1.7.6 Case Study 6: Threat of Electrocution to Elephants in Assam

Direct contact between humans and wildlife increases due to rapid increase in population and destruction of natural habitats of wildlife. There are many cases of human-wildlife conflicts have been reported. In context of protection of themselves and their resources like agricultural lands, people started installing fences as barriers to restrict movement of animals (Lindsey et al., 2012). Electrified fences are usually installed to avoid direct human-wildlife conflict and mainly to protect crop field from raided by elephants, especially in India (Menon et al., 1997). The Asian Elephants already categorized as Endangered species in Red List of Threatened Species of IUCN. The main threat to Asian elephants is of growing human population and loss of their habitat. This reduces a access of resources to elephants, eventually they raid to crops which leads to conflict (Aziz et al., 2016). Electrocution is very common cause of death of elephants. 25% of total elephant deaths in India occur only in Assam due to electrocution. Assam is a home for more than 5000 elephants and practices like electrocution is prevailing in many districts of state. In Sonitpur district, electric fences are frequently installed. About 140 deaths of elephants occur between 2003–2016 in Sonitpur district out of which about 35 deaths of elephants occurred due to electrocution. Fences powered by using inverters, direct supply of power from household or illegally tap electricity from electric poles. The electrified fences are mainly installed around agricultural lands, human habitats or tea plantations. These electrified fences keep elephants out of human habitats and agricultural land. The death of elephants due to electrocution is seem to be natural death as there is no sign of unnatural death on elephant's body. Electrocution is very lethal for elephants as well as for humans. This also cause a hurdle in conservation efforts for protection of elephants (Kalam et al., 2018) (Table 1.1).

Table 1.1 Significant research/review articles explaining exploitation of key species like Tigers and Elephants

Article title	Authors	Year	Species focused	Exploitation type
"The tiger trade: a review of its extent, impacts, and conservation strategies"	K. L. MacDonald, P. W. L. de Souza	2013	Tiger	Poaching, illegal trade, habitat destruction
"Elephant population dynamics and the illegal ivory trade"	R. J. Smith, A. T. Johnson	2014	Elephant	Poaching, illegal trade, population decline
"Human-elephant conflict in Asia: A review of the literature"	S. S. Bhattacharya, R. K. Mishra	2015	Elephant	Human-wildlife conflict, habitat loss
"A review of elephant poaching and the illegal ivory trade"	S. G. Elliot, J. R. Miller	2016	Elephant	Poaching, ivory trade
"The effects of habitat loss and fragmentation on tiger populations"	L. T. Sharma, R. S. Pradhan	2017	Tiger	Habitat destruction, fragmentation
"Conservation of wild tigers: past, present, and future"	H. D. K. Sharma, A. S. Bhatt	2018	Tiger	Poaching, illegal wildlife trade
"The elephant ivory trade and its impact on African elephant populations"	M. D. Keane, A. M. Campbell	2019	Elephant	Illegal trade, poaching
"Tigers and human-wildlife conflict: A review of mitigation strategies"	J. T. Kruger, L. M. W. L. Allen	2020	Tiger	Human-wildlife conflict, poaching

References

Adams, R., & Mulligan, M. (2019). *Buddhism and conservation: Understanding the relationship between culture, religion, and biodiversity preservation*. Cambridge University Press.

Aggarwal, K. C. (2006). *Wildlife of India—Conservation and management.*

Alok, C. K. (2018). *Wildlife: An introduction.*

Aziz, M. A., Shamsuddoha, M., Maniruddin, M., Morshed, H. M., Sarker, R., & Islam, M. A. (2016). Elephants, border fence and human-elephant conflict in northern Bangladesh: Implications for bilateral collaboration towards elephant conservation. *Gajah, 45*, 12–19.

Brownlow, A. (2000). A wolf in the garden. In C. Philo & C. Wilbert (Eds.), *Animal spaces, beastly places: New geographies of human-animal relations* (Vol. 10, pp. 143–158). Psychology Press.

Child, K. N., Barry, S. P., & Aitken, D. A. (1991). Moose mortality on highways and railways in British Columbia. *Alces, 27*, 41–49.

Choudhary, S. (2024). *The Wildlife (Protection) Act, 1972: A comprehensive study on wildlife conservation in India*. SSRN, 1–44.

Dasgupta, S., & Ghosh, A. K. (2015). Elephant–railway conflict in a biodiversity hotspot: Determinants and perceptions of the conflict in northern West Bengal, India. *Human Dimensions of Wildlife, 20*, 81–94.

Government of West Bengal. (2015). *Synchronized population estimation of the Asian elephant in forest divisions of Northern West Bengal-2014*. West Bengal, Directorate of Forests, Office of Conservator of Forests.

Gubbi, S., Reddy, V., Nagashettihalli, H., Bhat, R., & Madhusudan, M. D. (2014). Photographic records of the ratel (*Mellivora capensis*) from the southern Indian state of Karnataka. *Small Carnivore Conservation, 50*, 42–44.

Gubbi, S., Mukherjee, K., Swaminath, M. H., & Poornesha, H. C. (2016). Providing more protected space for tigers (*Panthera tigris*): A landscape conservation approach in the Western Ghats, southern India. *Oryx, 50*(2), 336–343. https://doi.org/10.1017/S0030605315000072

Guha, R., & Gadgil, M. (1989). State forestry and social conflict in British India. *Indian Economic and Social History Review, 26*(2), 121–147.

Indian Railways. (2014). *Statistical summary—Indian railways*. Retrieved November 5, 2016.

Jiao, Y., Yeophantong, P., & Lee, T. M. (2021). Strengthening international legal cooperation to combat the illegal wildlife trade between Southeast Asia and China. *Frontiers in Ecology and Evolution, 9*, 105. https://doi.org/10.3389/fevo.2021.645427

Jorgensen, D. (2015). Rethinking rewilding. *Geoforum, 65*, 482–488.

Joshi, A. R., Dinerstein, E., Wikramanayake, E., Anderson, M. L., Olson, D., Jones, B. S., Seidensticker, J., Lumpkin, S., Hansen, M. C., Sizer, N. C., Davis, C. L., Palminteri, S., & Hahn, N. R. (2016). Tracking changes and preventing loss in critical tiger habitat. *Science Advances, 2*(4), e1501675. https://doi.org/10.1126/sciadv.1501675

Kalam, T., Nath, N. K., & Das, J. P. (2018). Lethal fence electrocution: A major threat to Asian elephants in Assam, India. *Tropical Conservation Science, 11*, 1–9. https://doi.org/10.1177/1940082918817283

Khanra, A., Ghosh, P., & Ghosh, S. (2020, March). *Human-elephant conflict: A case study in human-dominated landscapes of South Bengal*. Paper presented at Animal Research Today: Basic and Applied Arena. https://doi.org/10.13140/RG.2.2.14023.01447

Klappenbach, L. (2014). *Understanding the threats to animals and wildlife – Examining natural threats and man-made threats* [online].

Lema, B., Teressa, H., Sime, D., & Weyya, G. (2025). Ethnozoological study of traditional medicinal animals used by indigenous people in Ethiopia: A review. *Traditional Medicine Research, 10*(5), 28. https://doi.org/10.53388/TMR20240909002

Lindsey, P. A., Masterson, C. L., Beck, A. L., & Romanach, S. (2012). Ecological, social and financial issues related to fencing as a conservation tool in Africa. In *Fencing for conservation* (pp. 215–234). Springer.

Mehta, J., & Shah, R. (2019). *Hinduism and ecology: The role of religious beliefs in the protection of the environment*. Oxford University Press.

Mengistu, M., Kebebew, M., & Meyer-Rochow, V. B. (2024). Ethnozoological study of medicinal animals used by the inhabitants of the Kucha District, Gamo Zone, Southern Ethiopia. *Journal of Ethnobiology and Ethnomedicine, 20*, 72. https://doi.org/10.1186/s13002-024-00714-8

Menon, V., Sukumar, R., & Kumar, A. (1997). *A god in dis tress: Threats of poaching and the ivory trade to the Asian elephant in India*. Indian Institute of Science, Asian Elephant Conservation Centre.

Miller, J. R. (2010). Hindu rituals and wildlife conservation: The impact of sacrifice and worship. *Ethnobiology Journal, 35*(1), 50–64.

Mishra, C. (2021). The role of local communities in conservation of wildlife in India. *Biodiversity and Conservation, 30*(3), 621–625.

Mohan, G., Yogesh, J., Nittu, G., Shameer, T. T., Backer, S. J., Nandhini, S., Ramakrishan, B., Jyothi, M., & Sanil, R. (2021). Factors influencing survival of tiger and leopard in the high-altitude ecosystem of the Nilgiris, India. *Zoology and Ecology, 32*(2), 116–133. https://doi.org/10.35513/21658005.2021.2.6

Morgan, K. I., Ewart, K. M., Nguyen, T. Q., Sitam, F. T., Ouitavon, K., Lightson, A. L., Kotze, A., & McEwing, R. (2021). Avoiding common numts to provide reliable species identification for tiger parts. *Forensic Science International: Reports, 3*, 100166. https://doi.org/10.1016/j.fsir.2020.100166

National Tiger Conservation Authority (NTCA). (2015). *Ministry of Environment, Forests and Climate Change, Government of India*. Retrieved from https://projecttiger.nic.in

Palita, S. K. (2007). *Royal Bengal Indian Tiger: Past, present and future–An analysis in Orissa context.*

Rajak, S., & Chatruvedi, K. (2023a). The evolution of wildlife protection laws and policies in India: A historical analysis. *IJNRD, 8*, 354–363.

Rajak, S., & Chatruvedi, K. (2023b). Historical role of institutions in Forest and wildlife conservation in India. *JETIR, 10*(8), 426–445.

Ray, S. (2002). *Transformations on the Bengal frontier: Jalpaiguri 1765–1948.* Routledge.

Rohner, T. (2016). The release of captive animals in Buddhist communities: Implications for conservation. *Biodiversity and Conservation, 25*(9), 1921–1930.

Roy, M., & Sukumar, R. (2016, March). *Survey of elephant movement paths/corridors across rail tracks in northern West Bengal.* Final report to International Elephant Foundation.

Roy, M., Baskaran, N., & Sukumar, R. (2009). The death of jumbos on railway lines in northern West Bengal. *Gajah, 31*, 36–39.

Sebastian, S. (2016, July 24). A tale of two tiger reserves. *The Hindu.*

Seiler, A., & Helldin, J. O. (2015). Greener transport infrastructure—IENE 2014 international conference. In A. Seiler & J. O. Helldin (Eds.), *Proceedings of IENE 2014 international conference on ecology and transportation* (Vol. 11, pp. 5–12). Nature Conservation.

Shahabuddin, G. (2010). *Conservation at the crossroads: Science, society, and the future of India's wildlife.* Orient Blackswan Private Ltd.

Sharma, A., & Tripathi, R. (2024). Wildlife in India during the colonial period. *Educational Administration: Theory and Practice, 30*(5), 5771–5776.

Sharma, M., Shrivastav, A. B., & Pandey, G. (2011). *Wildlife management: Some tips.*

Sharma, K., Wright, B., Joseph, T., & Desai, N. (2014). Tiger poaching and trafficking in India: Estimating rates of occurrence and detection over four decades. *Biological Conservation, 179*, 33–39. https://doi.org/10.1016/j.biocon.2014.08.016

Shepherd, C. R., Kufnerova, J., Cajthaml, T., Frouzová, J., & Gomez, L. (2020). Bear trade in the Czech Republic: An analysis of legal and illegal international trade from 2005 to 2020. *European Journal of Wild life Research, 66*(6), 1–10. https://doi.org/10.1007/s10344-020-01425-7

Singh, A., & Singh, H. S. (2020). Human—Wildlife conflict and mitigation strategies in India: A review. *Biodiversity and Conservation, 29*(11), 3485–3509.

Tensen, L. (2016). Under what circumstances can wildlife farming benefit species conservation? *Global Ecology and Conservation, 6*, 286–298.

Thompson, R. (2004). The cow in Hindu culture and the protection of cattle in India. *Journal of South Asian Studies, 21*(1), 34–49.

Tian, M., Potter, G. R., & Phelps, J. (2023). What is "wildlife"? Legal definitions that matter to conservation. *Biological Conservation, 287*, 110339.

Usher, M. (1986). Wildlife conservation evaluation: Attributes, criteria and values. In M. Usher (Ed.), *Wildlife conservation evaluation* (pp. 3–44). Chapman and Hall Ltd.

Wikramanayake, E., Dinerstein, E., Seidensticker, J., Lumpkin, S., Pandav, B., Shrestha, M., Mishra, H., Ballou, J., Johnsingh, A. J. T., Chestin, I., Sunarto, S., Thinley, P., Thapa, K., Jiang, G., Elagupillay, S., Kafley, H., Pradhan, N. M. B., Jigme, K., Teak, S., Cutter, P., Aziz, M. A., & Than, U. (2010). A landscape-based conservation strategy to double the wild tiger population. *Conservation Letters, 4*(3), 219–227. https://doi.org/10.1111/j.1755-263X.2010.00162.x

Wilcove, D. S., Rothstrin, D., Dubow, J., Philips, A., & Locos, E. (1998). Quantifying threats to imperiled species in the United States. *Bioscience, 48*, 607–615.

Yarrow, G. (2009). *Wildlife and wildlife management.* [online] Dc.statelibrary.sc.gov. Available at https://dc.statelibrary.sc.gov/server/api/core/bitstreams/7070ca48-d63f-496e-aae4-58c35a537f20/content

Chapter 2
India's Role in Global Wildlife Trade Networks

Shrey Dandriyal, Kavita Bisht, and Bharti Chauhan

Contents

Abstract India occupies a crucial place in the global wildlife trade, functioning at once as a source, a transit corridor, and a consumer of numerous species and wildlife-derived products. While regulated trade in items such as medicinal plants, marine resources, and handcrafted goods sustains livelihoods, illegal trafficking

S. Dandriyal
Wildlife Institute of India, Chandrabani, Dehradun, Uttarakhand, India

K. Bisht (✉)
Laboratory for the Conservation of Endangered Species (LaCONES), CSIR-Centre for Cellular and Molecular Biology, Hyderabad, Telangana, India

B. Chauhan
Faculty of Life Science, Chaudhary Bansi Lal University, Bhiwani, Haryana, India

A. Poonia et al. (eds.), *Wildlife Trade in India*,
https://doi.org/10.1007/978-981-95-3694-8_2

continues to place severe pressure on tigers, pangolins, rhinos, red sanders, and a range of exotic birds and reptiles. This trade is fuelled by traditional practices, luxury demand, and a growing urban market, and is enabled by porous borders, corruption, and limited enforcement capacity. Trafficking networks connect India to Myanmar, Bangladesh, Nepal, and China, with recent years seeing a shift toward online platforms, cryptocurrency payments, and dark web markets. Through case studies and trade records, this chapter identifies key species, major routes, and enforcement shortcomings, and proposes targeted measures to enhance monitoring, strengthen interagency collaboration, and raise public awareness to safeguard biodiversity.

Keywords Wildlife trade · Illegal trafficking · Key species · Biodiversity · Case studies

2.1 Introduction

India takes part in a focal title role in the global wildlife trade, doing the work of source, transit hub, and consumer of numerous species and their derivatives. As one of the most biology and diversity countries, India has over 7% of the world's recorded species, including diverse endemic and endangered flora and fauna (Challender et al., 2020). However, this abundance has also made it a central point for both legal and illegal wildlife trade, with significant conservation, economic, and law enforcement ramification.

The legal wildlife trade in India is modulated by domestic laws, such as the Wildlife Protection Act of 1972, and international compliance like the Convention on International Trade in Endangered Species of Wild Fauna and Flora (CITES). Legal trade supports subsistence and contributes to economic growth, especially in sectors involving medicinal plants, marine products, and sustainably harvested timber (Dutta et al., 2020). Despite these regulations, challenges remain in imposition and observation.

Illegal wildlife trade, alternately, is one of the substantial black-market industries, frightening numerous species with extinction. Poaching of tigers, leopards, pangolins, and exotic birds, escort trafficking of scarce reptiles and marine life, has been documented in India (Nijman et al., 2019). Solicitude for wildlife products, consumed traditional medicine, luxury markets, and rituals, continues to fuel this underground trade (TRAFFIC, 2020). Improvements in digital platforms have facilitated trafficking, posing new provocations for law implementation (Shrestha & Bawa, 2018).

This chapter explores India's role in the global wildlife trade by examining the key species embroiled, the socio-economic drivers of trade, and the efficacy of authoritative frameworks scrutinizes. Additionally highlights the need for increased conservation strategies, cross-border cooperation, and technological intermediation to curb illegal wildlife trade and encourage sustainable practices.

2.1.1 *Cultural and Medicinal Drivers of Demand*

India's traditional medicine systems, including Ayurveda, Unani, and Siddha, have historically used wildlife-derived ingredients, some of which remain in demand despite legal restrictions (CITES, 2020). Similarly, wildlife is incorporated into cultural and religious practices, with products like peacock feathers, star tortoises, and certain owl species used in rituals and ornamentation (Nuwer, 2014). In addition, luxury goods such as shahtoosh wool and ivory carvings continue to hold status value, driving illicit markets (TRAFFIC, 2021).

2.1.2 *Economic and Urban Consumption Trends*

India's rapidly expanding middle class has fuelled demand for exotic pets, such as parrots, reptiles, and ornamental fish, creating a domestic market linked to international wildlife trafficking. The interface between domestic consumption and cross-border trade, as evidenced by Wildlife Crime Control Bureau (WCCB) seizure data between 2020 and 2024, demonstrates India's dual role as both a consumer and an exporter of wildlife products. This dynamic is closely tied to global Sustainable Development Goal 15, which emphasizes the need to combat illegal wildlife trade to preserve biodiversity and ensure sustainable ecosystems.

2.2 Routes, Mechanisms, and Hotspots of Wildlife Trafficking

India's geographic position and porous borders contribute to its vulnerability as a wildlife trafficking conduit. The country's shared borders with Nepal, Myanmar, Bangladesh, Bhutan, and China, as well as its extensive coastline, create numerous opportunities for illegal wildlife flows.

2.2.1 *Geographic Vulnerabilities*

Key wildlife trafficking routes include India's northeastern borders with Myanmar and Bangladesh, which serve as entry and exit points for species destined for Southeast Asian markets. Coastal smuggling routes through Sri Lanka and the Maldives also facilitate maritime wildlife trafficking. Additionally, wildlife products are frequently concealed in air cargo or postal shipments, with major seizures reported at hubs such as Delhi and Mumbai airports.

Myanmar

India shares its eastern border with Myanmar, with the four bordering Indian state being Arunachal Pradesh, Nagaland, Manipur, and Mizoram. There are International Tiger Trade Routes from India via Moreh in Manipur and from Myanmar to China through the Sino-Myanmar border and Ruili (Martin, 2014). The lack of adequate checkpoints at Tamu (Myanmar) and Moreh (India) have facilitated large-scale illegal trade between the countries. There is an alarming flow of illegal Wildlife Products out of India towards Myanmar, especially pangolin scales. Between 2009 and 2017, it is estimated that approximately 6000 pangolins were illegally hunted throughout India to meet increasing market demand for their meat and scales (Martin, 2014). They are smuggled to Myanmar via Manipur after being killed on-site or transported alive. Manipur accounts for 36 percent of seizures, followed by Tamil Nadu, which accounts for ten percent (Martin, 2014). Two of the active locations for wildlife trafficking in the Northeast are Behiang village of Churachandpur in Manipur and Dimapur town in Nagaland (Martin, 2014). Wild plants including orchids and medicinal plants like *Paris polyphylla* (Arunachal 24, 2018) are being smuggled to Myanmar from Manipur and Arunachal Pradesh respectively (Martin, 2014). To better understand these mechanisms, the following investigation is informative. A set of poachers were supplied tiger body parts to Myanmar via various routes and intermediaries in Mizoram and Meghalaya and from there to China through Ruili (Martin, 2014). The investigators have found that even though they are barely educated, they run bank accounts, make online payments, and book consignments and flight tickets. They book these consignments often to Guwahati with a transporter and storage facility instead of carrying the illegal wildlife products. They collect the goods after traveling by train or flight and drive to Shillong to deliver them (Martin, 2014).

Bangladesh

India shares its border with Bangladesh across five states: West Bengal, Assam, Meghalaya, Tripura, and Mizoram. West Bengal has the longest border and plays a crucial role in trade and cross-border movement (Das, 2008). The Sundarbans mangrove forest covers parts of the Indian states of West Bengal and Bangladesh and identified 'Calcutta—Sundarbans' as one of the hotspots for trade. Three tiger heads, four tiger skins, and twenty-four kg of bones were discovered in the custody of a suspect in 2011 on the Bangladesh side of Sundarbans. Trafficking between Bangladesh, India, and Myanmar occurs via land routes facilitated by private cars, freight trucks, and public transport such as publicly and privately owned buses and cars. The use of sleeping pills to keep the live felid calm and undetected is used during transportation by road. Between India and Bangladesh, 16 transit points are known, out of which 11 are involved in trafficking tiger parts and seven in other felids.

Nepal

Nepal acts as supply and transit hub for illegal wildlife trade, with trafficking networks targeting a wide range of endangered species including tigers, rhinos, snow leopards, leopards, red pandas, and pangolins (Paudel et al., 2023). The porous Himalayan border, particularly along the Araniko (also spelled Arniko) Trail, enables smugglers—often linked with Chinese networks—to transport wildlife parts into Tibet and beyond, complicating enforcement efforts (Oxpeckers, 2021; Koirala, 2025). The trade has evolved in sophistication and scale: while once driven primarily by economic incentives, with many perpetrators motivated by "extra income and the ease of IWT compared to other employment," recent studies show that poverty alone does not explain participation in illegal wildlife trade (Paudel, 2020). Nepal's law enforcement landscape involves its Department of National Parks and Wildlife Conservation as well as agencies like the Nepal Army and Police, though jurisdictional confusion, lack of expertise, and limited coordination—especially with units like the Armed Police Force—hinder effective action. Moreover, seizures of pangolin scales and rhino horns suggest an increasingly organized and persistent trafficking network despite national and international regulations under CITES and Nepal's 1973 wildlife protection law (NTNC, n.d.; Paudel et al., 2023).

China

China is a major consumer of the global wildlife trade (Heinrich et al., 2019). Records of wildlife trade going back 2000 years (Menon, 2019), and since then, wildlife trophies such as ivory, Rhinoceros horn, coral, and rare animals have been transported to China. India shares its border with China through the Ladakh (Union Territory) and the states of Himachal Pradesh, Uttarakhand, Arunachal Pradesh and Sikkim. Presently, there is not easy access to China from India, as borders across large extend of the Himalayas are disputed and are mostly remains closed. Areas of western and northwestern Nepal bordering China and India offer the shortest route to China from India. Tiger and Tibetan antelope are some of the documented species that are illegally traded across the Indo-Chinese border. China stands as a major hub for both consumption and complex trafficking networks of wildlife products. The country is the primary global market for pangolins—4-legged mammals having scales that are prized in ancient traditional Chinese medicine and their meat is considered a luxury food item and resulting in over 100,000 individuals trafficked annually into China and Vietna. Prosecution records from 2014 to 2019 reveal that Guangdong, Guangxi, and Yunnan provinces serve as key trade hotspots for pangolin parts, highlighting critical intervention points (Huang et al., 2021). Traditionally, ethnic ties and family networks have enabled clandestine wildlife trafficking, particularly in border towns, leveraging trust and shared identity to evade enforcement (Wong, 2019; van Uhm & Wong, 2019). Responding to public health and conservation imperatives, China permanently banned the trade and consumption of terrestrial wildlife in February 2020 and upgraded pangolins to the highest protection level under national law (Salisbury, 2021).

2.2.2 Corruption and Transport Networks

Traffickers often exploit weaknesses in transport networks, including document forgery and bribery at ports, to bypass inspection (INTERPOL, 2023). The misuse of legal trade permits to launder illicit shipments has been identified as another mechanism facilitating illegal trade.

2.3 Key Species in Legal and Illegal Trade

India is central to the trade of several high-value species, both through regulated and illicit markets.

2.3.1 Key Species in Legal Trade

1. **Red Sandalwood (*Pterocarpus santalinus*)**: This treasured hardwood species, endemic to India, is traded under strict regulations. Sandalwood is Used for furniture-making, traditional medicine, and cosmetics, it is listed under CITES Appendix II to ensure controlled trade and avert overharvesting.
2. **Himalayan Yarsa Gumba (*Ophiocordyceps sinensis*)**: A highly evaluated medicinal fungus found in the Himalayas, its accumulation is regulated to avert depletion (Siriwat & Nijman, 2020). Trade is essentially managed through cooperative systems that encourage sustainable harvesting.
3. **Marine Products (Shrimp, Ornamental Fish, Pearls)**: among the world's ranks India leading exporters of farmed shrimp, aquarium fish, and cultured pearls (Ghosh et al., 2021). The Marine Products Export Development Authority (MPEDA) enforces renewable harvesting and export regulations.
4. **Medicinal Plants (*Rauwolfia serpentina*, *Aconitum species*, *Withania somnifera*)**: this is commonly used in Ayurveda, these plants are cultured and traded under strict government policies (Lal & Singh, 2021). India is a key exporter of herbal medicines.
5. **Handicrafts and By-products**: hand-carved wooden items, leather from domesticated animals, and Pashmina wool, supporting rural artisans and herders are included in Ethical trade of wildlife-derived handicrafts (TRAFFIC, 2021).

To secure sustainability, authorities emphasize stringent monitoring, institutional capacity-building, and the integration of advanced surveillance technologies in wildlife trade regulation (WWF, 2020).

Illegal Wildlife Trade in India

In spite of strict laws, illegal wildlife trade give rise to a severe menace to conservation and wild biodiversity (Nijman et al., 2019). Factors such as high market demand, weak enforcement, and organized crime networks enable this illicit trade (United Nations Office on Drugs and Crime (UNODC), 2024).

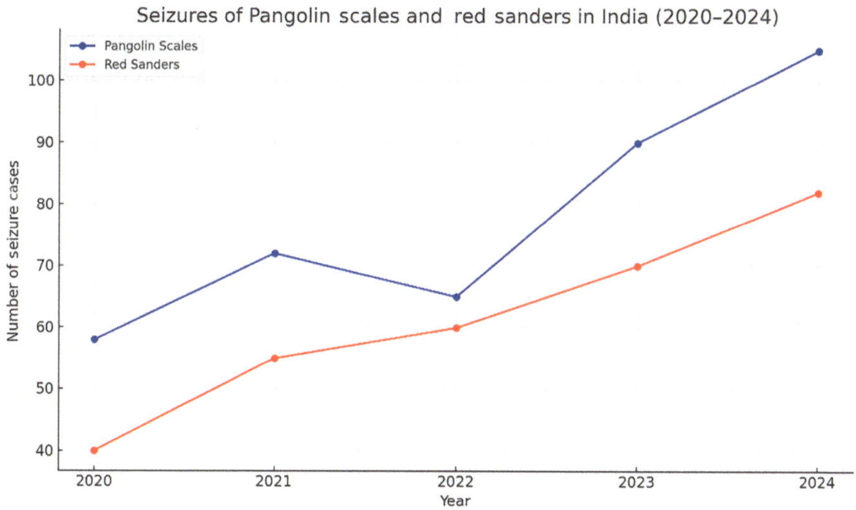

Fig. 2.1 Graph of seizures (2020–2024) for pangolin scales, red sanders

2.3.2 Key Species in Illegal Trade

6. **Tigers and Leopards**: They are poached for their bones, skins and many body parts, which are used for traditional medicine and as status symbols (WWF, 2021).
7. **Pangolins**: pangolin scales and meat are high in demand Among the most trafficked mammals, especially in traditional Asian medicine (Challender et al., 2020).
8. **Rhinos**: The One-Horned Rhinoceros faces poaching hazards due to the demand for its horn (Sharma & Patel, 2021).
9. **Exotic Birds and Reptiles**: Many of the native bird species, such as hornbills, parrots, and also with reptiles like star tortoises and monitor lizards, are illegally entrapped for the pet trade (D'Cruze et al., 2020).

2.4 The Digital Expansion of Wildlife Trade

Technological advancement has accelerated the reach of wildlife trafficking networks, shifting parts of the illegal trade into online spaces.

Internet and the Rise of Wildlife Trafficking
The Economic and Social Council (2013) highlights that illegal wildlife trade is a widespread environmental crime with far-reaching global consequences (Chaurasia, 2023). The rise of the Internet, which supports billions of daily transactions worldwide, has made such trade easier (Chaurasia, 2023; Harrison et al., 2016). Today, a

significant portion of wildlife—including live animals, plants, fungi, and their parts—is traded online through e-commerce platforms, private forums, and messaging applications (Arunachal, 2018). The Internet enables forms of wildlife trade that were previously impossible (Li, 1991; St-Cyr-Leroux, 2023), making online monitoring of this activity essential for both conservation and biosecurity efforts (Fukushima et al., 2021; Stringham et al., 2021).

Most of the wildlife trade on the Internet runs on open websites, which are accessible to the public (Hernandez-Castro & Roberts, 2015; Yu & Jai, 2015). However, wildlife trade is also becoming more prevalent on the deep web, which includes social media and private messaging apps (WCCB, 2021) (e.g., Facebook and WhatsApp) (Stringham et al., 2021; TRAFFIC, 2020). Wildlife trafficking thrives on other parts of the Internet, such as social media and e-commerce sites, but it is not as common on the dark web. Décary-Hétu noted that "People want to have as many customers and sales outlets as possible because trade is all about profit" (Stoner et al., 2016).

2.4.1 *Social Media and E-commerce Platforms: New Frontier for Wildlife Trafficking*

The Internet has expanded rapidly. But it has also benefited criminal activity, which has changed and evolved into a new type of crime known as "cybercrime" (Demeau et al., 2019). The marketing and sale of endangered wildlife online is illegal under both national and international laws, but growing research is showing that this activity occurs on social media platforms, e-commerce websites, and the dark web (Harrison et al., 2016; Di Minin et al., 2018). Facebook is one of the most widely used social media sites in the world, with 2.27 billion monthly users worldwide (Abbruzzese, 2018; Xu et al., 2020). By facilitating easier connections between buyers and sellers, the Internet has significantly expanded and globalized the potential of the wildlife trade (Lavorgna, 2014; Yu & Jia, 2015). Those who illegally trade endangered species are less likely to face the consequences because digital platforms can offer anonymity (Soriano-Redondo et al., 2023).

Claim that wildlife traffickers use well-known, legitimate channels such as classified ads, discussion boards, and websites for auctions (Demeau et al., 2019). To increase their extent of privacy, animal traffickers on Facebook use other digital platforms like WhatsApp or BlackBerry Messenger to carry out their transactions (Jain, 2001; Demeau et al., 2019). The first online representation of the wildlife market was created in 2005 by the International Fund for Animal Welfare (IFAW). The majority of online data on wildlife trafficking is restricted to tracking advertisements that are placed on websites within a specific time frame. In just 1 week, nearly 9000 advertisements for animals and products derived from them were reported by IFAW (2005) (Demeau et al., 2019).

However, live animals were also promoted; for examples, a tiger and a gorilla were listed for more than $70,000 and $9000, respectively. eBay auction site is the primary marketplace for illegal traders (IFAW, 2007, 2008; Demeau et al., 2019). Traffickers have, however, shifted the majority of their operations to social media in recent years. By 2015, approximately 80% of online wildlife trafficking was conducted through social media platforms like Facebook and Instagram (Jain, 2001; Demeau et al., 2019).

A Chinese e-commerce platform sold more than 70,000 individuals of 95 protected species of turtles and parrots in just 150 days (Ye et al., 2020). Similarly, it was found that Facebook and other social media forums had a large number of posts selling turtles, elephant ivory, rhino horn, and hawksbill shells (Sung et al., 2021; Xu et al., 2020; Soriano-Redondo et al., 2023). Using sophisticated automated data collection and filtering techniques, we track the illegal trade for these species on four online platforms—Twitter, YouTube, Flickr, and TikTok—as well as the search engine Google (Soriano-Redondo et al., 2023).

According to another report, a lot of tigers' products, like skin fragments or canine teeth are readily available on social media platform like Facebook and can be delivered by mail because of their small size, which makes it possible for them to be delivered undetected (Demeau et al., 2019).

Due to the rise in internet activities, illegal wildlife trade moves to online platforms, which changes the pattern of wildlife trafficking. Traffickers sell and distribute illegal wildlife products by taking advantage of a variety of digital platforms, such as social media, e-commerce sites, and messaging apps like WhatsApp (Global Initiative Against Transnational Organized Crime, 2024).

2.4.2 Cryptocurrencies and Darknet Markets

The dark web refers to that area of the internet that is hidden or not easily available through normal search engines like Google, Yahoo Search, Microsoft Bing, etc. For this, special software is required, such as Tor (The Onion Router), which can only be downloaded from the official Tor Project Website (Chaurasia, 2023). These anonymous networks, like those that use onion routing (like the Tor network, https://www.torproject.org), are usually made to provide users with anonymity. Through these networks, users can exchange money (such as cryptocurrencies) and communicate anonymously without revealing who they really are (Harrison et al., 2016).

Crypto-markets generally described as "online forums on which goods and services are exchanged between users by concealing their identities through digital encryption," are websites on the dark web that claimed to sell drugs, including heroin, cannabis, and synthetic drugs. Encryption methods are used in the crypto markets to secure transactions involving virtual currencies like Bitcoin (Demeau et al., 2019).

These cryptocurrencies, mainly Bitcoin, facilitate end-to-end encrypted transactions, especially for activities on the dark web, including wildlife contraband trade. It is more difficult to track financial flows because of the extra layer of anonymity this provides (Chaurasia, 2023). In some transactions, certain animals and drugs are also used as currencies, like parrots trading for heroin (Demeau et al., 2019). That's why wildlife traffickers use the dark web fearlessly, as it provides anonymity, and all the conversations and transactions on the dark web are encrypted (Chaurasia, 2023).

Unlike standard email service providers like Gmail or Yahoo Mail, Proton Mail is used by wildlife traffickers to communicate; it offers end-to-end encrypted email service and does not store user information. Social media, e-commerce, and other websites can be readily accessed on the surface web by entering their name into standard search engines like Google, Bing, and others. Posts that involve the illegal wildlife trade can then be found with ease by using keywords. Darknet marketplaces, in contrast to surface websites, are only accessible via the Tor browser and a specially designed URL. Alphanumeric codes that end in "onion" make up these URL (Chaurasia, 2023) (Fig. 2.2, Table 2.1).

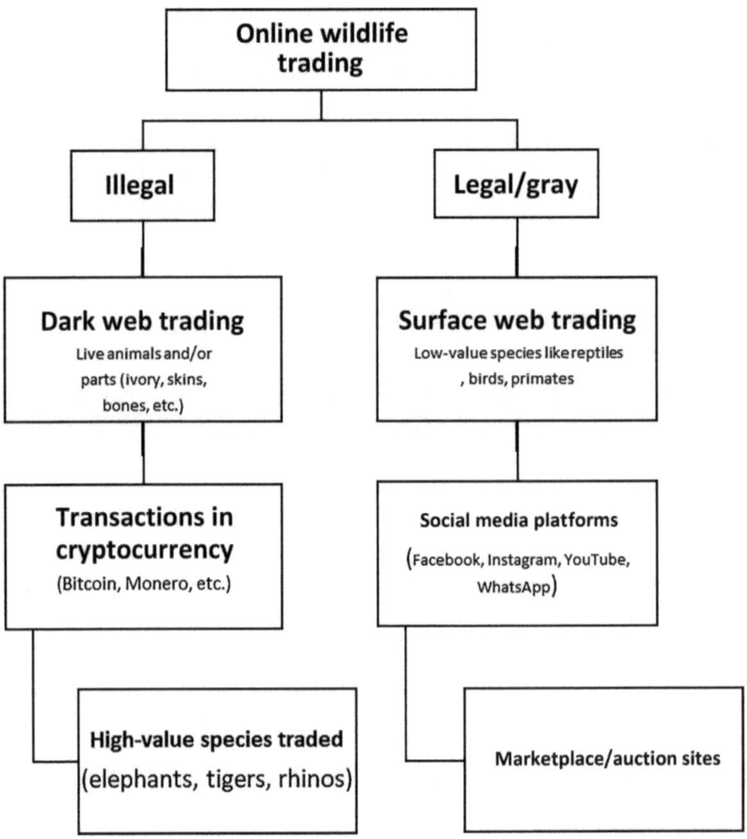

Fig. 2.2 Online wildlife trade: legal, illegal, and gray pathways

Table 2.1 Documented cases of wildlife trade on the dark web: a structured overview

S. no.	Animal used	Name of product	Quantity	Name of website/platform	Country of origin	Country of sale	References	Comments
1	Pangolin	Pangolin Scales	423 kg seized	Darknet marketplace (noted in Interpol "Golden Strike")	Africa → China	China	INTERPOL (2022a, b), UNODC (2024)	Seizure linked to darknet platform and crypto payments
2	Ivory/ tiger	Ivory pieces & tiger parts	Thousands of pieces (ivory & tiger)	INTERPOL darknet ads (2015–17)	Asia/ Africa	Darknet buyers	INTERPOL–IFAW Report (2017)	Ads for highly endangered species parts noted between 2015–17
3	Rhino horn	Rhino horn products	Estimated dozens (ads count)	Same darknet platforms	Africa	Global	INTERPOL–IFAW Report (2017)	Emerging use of dark web due to anonymity and crypto

2.4.3 *Trends, Evidence, and Challenges in Wildlife Trade on Dark Web*

Illegal trade activity is rapidly rising on the surface web, so it has been suggested that it could shift to anonymous networks like the darknet or dark web (Harrison et al., 2016). One well-known example is the "Silk Road", which is a marketplace for the trading of illicit goods, mostly drugs. It makes sense that people who want to engage in the illegal wildlife trade might move to these networks, as traditional illegal trades, like those involving drugs and firearms, have been moving there (Harrison et al., 2016). Due to the increase in interest in the illegal wildlife trade, the conservation community has become aware of the dark web (Roberts & Hernandez-Castro, 2017).

After a year, bycatch—the illegal wildlife trade—is the main source of revenue. Cacti traded for their hallucinogenic qualities and counterfeit luxury goods, particularly Chanel handbags, that contain reptile skin are the two main forms of illegal wildlife trade we have discovered (Roberts & Hernandez-Castro, 2017). After looking through roughly 2 million advertisements over 5 years, an Australian research team looking into the dark web trade in wild animals and plants discovered that almost 3500 of them were for wildlife. From 150 species that were for sale, approximately 90% were meant for recreational drug use. Some studies claim that the illegal trade in wild animals is challenging in a number of ways. There are chances that these species might be harmful or carry pathogens, endangering human health and biosecurity. Because some of these species are invasive and others are endangered, it also poses a threat to biodiversity (St-Cyr-Leroux, 2023).

Using a list of keywords associated with illegal online wildlife trade on the surface web, archived markets were systematically searched. These keywords represented 49 taxa (e.g., bear, gorilla, python) and 89 products (e.g., fur, skin, taxidermy), as defined by IFAW (2014b), along with terms like "cactus" and "orchid" (Hughes, 2021). Additionally, 30 keywords commonly used in this context, such as "faux ivory" and "ox bone," were employed to specifically target elephant ivory trade (Hughes, 2021). For comparison, terms related to illicit drugs such as "heroin," "cocaine," and "cannabis" were also queried. Out of 9852 items searched, only four of the 121 wildlife-related keywords returned results. Harrison et al. (2016) report that three of these—"bone," "faux," and "ivory"—serve as codenames for elephant ivory trade (Harrison et al., 2016) (Table 2.2).

2.4.4 *Major Species Trafficked Online*

Animals (n = 18), fungi (n = 28), and plants (n = 101 species) made up the majority of the traded species. Many plants were marketed for their medicinal qualities as well as their potential as drugs. Animals were exchanged for a variety of goods, such as food, medicine, pets, decorative items, furs, and skins, as well as clothing. Among the 18 mammal species observed in trade, the Sonoran Desert toad (*Incilius*

Table 2.2 Wildlife products sold on the darknet: a factual overview

S. no.	Species group	Product	Quantity/ listings	Websites	Notes
1	Rhino, elephant, tiger	Horns, ivory, parts	21 ads (2015–17)	Darknet markets	"Very limited amounts" of listings (ifaw.org, awf.org)
2	Cacti (San Pedro)	Echinopsis pachanoi	1 listing	Darknet	Only one non-CITES plant listed
3	Various plants/ fungi	Psychoactive wildlife materials	~3332 ads (~600/ year)	Multiple darknet sites	90% were traded for drug use
4	Wildlife bycatch	Reptile-skin bags, etc.	Few known cases	AlphaBay etc.	Bycatch; not core wildlife trade

alvarius), valued for its psychoactive (psychedelic) secretions, and the raccoon (*Procyon lotor*), traded primarily for its fur, were the most frequently traded (Stringham et al., 2023). Seventeen species were recorded as being harvested directly from the wild, appearing in 52 advertisements by sellers. According to the IUCN, three of these wild-harvested species are at risk of extinction: the Japanese spiky sea cucumber (*Apostichopus japonicus*; endangered), peyote (*Lophophora williamsii*; vulnerable), and caterpillar fungus (*Ophiocordyceps sinensis*; vulnerable). Additionally, the trade included elephant tusks (ivory), rhinoceros horns, and teeth and skins of tigers and lions (Stringham et al., 2023).

The two species most impacted by the illegal wildlife trade worldwide between 2015 and 2021 are the rhino (an animal) and the cedar (a plant). Elephant ivory accounts for 15% of the illegal animal trade, pangolin scales for 28%, and rhino horn for 29%. Other animals that are illegally traded include carnivores, turtles and tortoises, snakes, seahorses, eels (5%), crocodilians (5%), parrots and cockatoos (2%), and others (Drishti, 2024) (Fig. 2.3).

The largest market, comprising 47% of illegally traded plants, was made up of cedars and other Sapindales, such as mahogany, holy wood, and guaiacum; rosewoods came in second at 35%, and agarwood and other Myrtales at 13%. Due to overexploitation, wildlife crime contributes to biodiversity degradation, which leads to population declines and extinction threats. Long-term ecological issues, comprising sex-ratio imbalances and lower reproduction rates can result from overexploiting wildlife (Drishti, 2024).

Multiple enforcement operations have resulted in the seizure of thousands of wildlife products and the identification of around 100 suspects across 23 countries, leading to international arrests and further investigations into wildlife trafficking (INTERPOL, 2022a). Seized items included 1202 pieces of ivory totaling over four tonnes, 50 rhino horns weighing 72 kg, 46 kg of totoaba bladders, 423 kg and 78 pangolin scales, 52 kg and more than 3785 pieces of molluscs, 33 red corals,

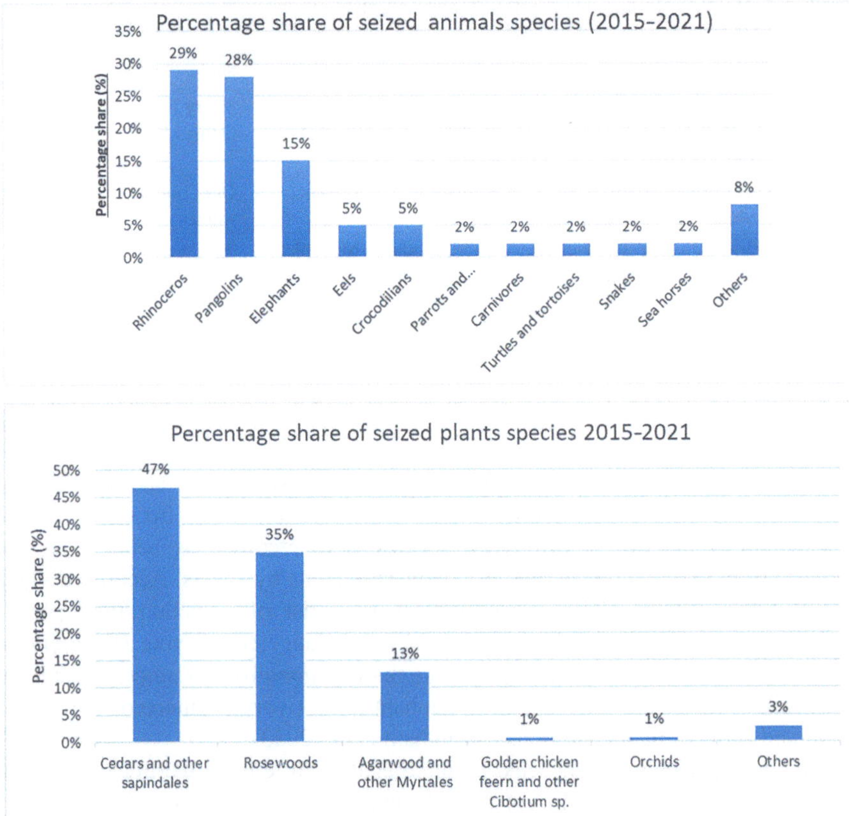

Fig. 2.3 (**a**, **b**) Distribution of major wildlife species groups seized between 2015 and 2021, categorized into plants and animals. The data reflects the percentage share based on the standardized seizure index (Drishti, 2024)

42 shark teeth, and 1336 additional CITES-protected species. Live specimens included over 120 birds, as well as three turtles and tortoises (INTERPOL, 2022a).

Wild cat species are also extensively traded. Live animals are kept as pets, housed in zoos or private wildlife collections, or consumed as luxury food. Their skulls and skins are used for decorative purposes, claws and teeth are fashioned into jewelry, and bones and other parts are utilized in traditional medicine or witchcraft. Leopard skins, in particular, are sought for ornamental uses, and some communities believe they can treat rheumatism (Nijman et al., 2019). The legal and illegal trade in live specimens and their products, including hunting trophies, food, clothing, ornaments, pets, and traditional remedies, produces billions of dollars every year (Dutta et al., 2020; Hughes, 2021; Soriano-Redondo et al., 2023).

2.5 Discussion: Enforcement and Policy Challenges

Wildlife trafficking in India continues despite a strong legal framework, largely due to enforcement limitations and evolving criminal networks.

2.5.1 India's Enforcement Gaps

Challenges include understaffed wildlife crime units, a shortage of trained personnel, and inadequate coordination across agencies when confronting sophisticated trafficking syndicates (WCCB, 2023). Additionally, recent amendments to the Wildlife Protection Act in 2022 have left some loopholes that traffickers can exploit, requiring further legal review and harmonization.

2.5.2 Recommendations for Disrupting Trade Networks

To address these challenges, India could deploy artificial intelligence-based surveillance tools, such as automated image recognition for detecting wildlife products in online marketplaces. Strengthening interagency task forces with representatives from customs, forest departments, and cybercrime agencies would enhance enforcement. Lastly, public awareness campaigns directed at urban consumers could help reduce demand for illegal wildlife products and build a more informed citizen base that supports conservation objectives.

2.6 Discussion & Conclusion

India plays an important role in the global wildlife trade, stabilizing economic interests with conservation responsibilities. While legal trade holds up livelihoods and biodiversity conservation, illegal trafficking remains a crucial challenge (Shepherd, 2019). The imposing of existing laws needs to be strengthened, with considerable coordination between domestic agencies and international organizations. Besides, public awareness campaigns can play a remarkable role in reducing demand for illegal wildlife products. Human errors, such as weak execution of policies, insufficient monitoring, and corruption, additionally complicate enforcement efforts. Addressing these gaps requires a varied approach, including technological advancements, stricter penalties, and increased global cooperation. Eventually, the conservation of India's rich biodiversity depends on a concerted effort from governments, law enforcement agencies, conservationists, and the public to ensure sustainable wildlife trade and curb illicit activities (WCCB, 2021).

Wildlife plays a crucial role in human societies, serving as a source of food, medicine, and ornamentation, leading to a significant international market for the trade of wildlife parts (Hughes, 2021). With the continuous rise in human population, the demand for wildlife products has surged, causing both legal and illegal wildlife trade to flourish (Hale & Hale, 2003). While many taxa are legally traded under strict regulations, the illegal wildlife trade has become a pressing conservation concern globally, affecting thousands of species and resulting in far-reaching consequences on the environment, local livelihoods, food security, national security, and sustainable development (IUCN (International Union for Conservation of Nature), 2012; Biggs et al., 2017; Dickson, 2008; Brashares et al., 2014).

Illegal wildlife trade, often classified as green crime by criminologists, poses one of the greatest threats to the survival of endangered species (Patel & Iyer, 2021). It ranks among the largest sources of criminal income alongside trafficking of arms, drugs, and people (Hughes, 2021). The issue has garnered increasing societal, media, and political attention, resulting in substantial financial investments in response to surging poaching incidents, particularly those involving elephants and rhinoceroses. In Asia, illegal wildlife trade remains a major conservation challenge, threatening various taxa, including birds (Ghose & Kaul, 2005; Aiyadurai, 2011), mammals (Dexel & Deutschland, 2002; Li & Lu, 2014; Abbas et al., 2015; Nijman et al., 2019), fishes, and plants. The primary driver of this trade is consumer demand (Hale & Hale, 2003; Shepherd, 2019; Nijman et al., 2019), while poverty serves as a significant contributing factor in regions such as the Himalayas (Rao & Bhat, 2023; Duffy et al., 2016; Ji et al. 2024). Despite existing legislation to regulate wildlife trade, the low risk-to-reward ratio for illegal traders sustains the trade (Rao & Bhat, 2023). Further complicating efforts to curb this trade is its transboundary nature, making enforcement more challenging (Nijman et al., 2019; Shepherd, 2019; Patel & Iyer, 2021).

While enforcement efforts have significantly curbed illegal trade, the black market continues to threaten conservation progress (Bhattarai et al., 2017; Dhakal & Baral, 2015; Martin, 2014). A significant portion of illegally-traded wildlife is exported to China, primarily for use in traditional medicine and luxury goods (Chow et al., 2014; Gao et al., 2016).

The collection of turtles for trade is among the most severe threats to Asian species, and in China too, many freshwater turtles and tortoises are actively traded and some of them are in threatened category (Ji et al., 2024; Gong et al., 2006). Subsistence hunting has evolved into a commercial wildlife trade to supply markets in China, Thailand, and Vietnam. Over the past two decades, illegal trade in animal parts has emerged as the most significant threat to endangered species such as the tiger (*Panthera tigris*), rhinoceros (*Diceros bicornis* and *Rhinoceros unicornis*), sloth bear (*Ursus ursinus*), and snow leopard (*Panthera uncia*). The decline in tiger populations is particularly alarming; three of the eight tiger species were declared extinct in the 1900s, and India's tiger population has plummeted from 40,000 in the early twentieth century to approximately 1400 in 2006–2007 (Jain, 2001).

The illegal ivory trade continues to fuel elephant poaching, with India seizing an average of 270 kg of raw ivory each year between 1996 and 2001. Despite the 1989 global ivory trade ban enforced by the CITES, African elephant (*Loxodonta africana*) populations declined by approximately 60% between 1979 and 1987 due to poaching. Poaching pressure extends beyond elephants, impacting other wildlife such as leopards (*Panthera pardus*), sambar deer (*Cervus unicolor*), spotted deer (*Axis axis*), blackbuck (*Antilope cervicapra*), and Indian star tortoises (*Geochelone elegans*). Additionally, illegal trade extends to plant species such as red sanders (*Pterocarpus santalinus*), agarwood (*Aquilaria spp.*), and kuth root (*Saussurea lappa*).

Economic globalization has intensified the demand for wildlife products, reinforcing market-driven exploitation (Freese & Trauger, 2000). Market forces have historically contributed to wildlife declines, as seen in the cases of bison (Bos bison), passenger pigeon (Ectopistes migratorius), and whale populations (Bolen & Robinson, 1999). While economic growth is often measured by indicators such as GDP and per capita income (Hall et al., 2000), conservationists must consider the impacts of rapid development on wildlife trade (Gossling, 1999).

The role of policy is critical in mitigating illegal wildlife trade. Despite CITES regulations protecting over 32,000 species, conservation remains a secondary concern to economic interests (Cooney, 2001). The effectiveness of CITES implementation in countries like India is debated, especially concerning flagship species such as tigers, elephants, leopards, and rhinoceroses. While strategies such as ecotourism have proven successful in countries like Kenya, their viability in mitigating illegal trade in India remains uncertain. To ensure sustainable conservation, policymakers must integrate international frameworks such as the Convention on Biological Diversity (CBD) with national legislation while addressing socio-economic factors that drive illegal wildlife trade (Bennett & Robinson, 2000).

References

Abbas, F. I., Rooney, T. P., Mian, A., Bhatti, Z. I., & Haider, J. (2015). The distribution, population status, and wildlife product trade of Himalayan Musk deer in Gilgit-Baltistan, Pakistan. *Journal of Bioresource Management, 2*(3), 38–47.

Abbruzzese, J. (2018). *Facebook hits 2.27 billion monthly active users as earnings stabilize* [www document]. URL https://www.nbcnews.com/tech/tech-news/facebook-hits-2-27-billion-monthly-active-users-earnings-stabilize-n926391

Aiyadurai, A. (2011). Wildlife hunting and conservation in Northeast India: A need for an interdisciplinary understanding. *International Journal of Galliformes Conservation, 2*, 61–73.

Arunachal. (2018, August 30). Jamoh raises Smuggling of rare medicinal plant issue. *Arunachal 24*.

Bennett, L. E., & Robinson, G. J. (2000). *Hunting of wildlife in tropical forests* (Environment Department papers. Biodiversity series–Impact studies 76). World Bank.

Bhattarai, B. R., Wright, W., Poudel, B. S., Aryal, A., Yadav, B. P., & Wagle, R. (2017). Shifting paradigms for Nepal's protected areas: History, challenges and relationships. *Journal of Mountain Science, 14*(5), 964–979.

Bhupathy, S., Choudhury, B. C., Hanfee, F., Kalyar, S. M., Khan, M. H., Platt, S. G., & Rashid, S. M. A. (2000). Turtle trade in South Asia: Regional summary (Bangladesh, India and Myanmar). In P. P. Van Dijk, B. L. Stuart, & A. G. J. Rhodin (Eds.), *Asian Turtle Trade: Proceedings of a workshop on conservation and trade of freshwater turtles and tortoises in Asia, Phnom Penh, Cambodia, 1–4 December 1999* (Chelonian Research Monographs 2) (pp. 101–105). Chelonian Research Foundation.

Bichel, N., & Hart, A. (2023). A history of hunting and hunting perceptions. In *Trophy hunting*. Springer.

Biggs, D., Cooney, R., Roe, D., Dublin, H. T., Allan, J. R., Challender, D. W., & Skinner, D. (2017). Developing a theory of change for a community-based response to illegal wildlife trade. *Conservation Biology, 31*(1), 5–12.

Bolen, E. G., & Robinson, W. L. (1999). *Wildlife ecology and management* (4th ed.). Prentice Hall.

Bradley, E. M., & Phipps, M. (1996). A review of the wild animal trade in Cambodia. *Traffic Bulletin, 16*, 45–60.

Brashares, J., Abrahams, B., Fiorella, K. J., et al. (2014). Wildlife decline and social conflict. *Science, 345*, 376–378.

Chakrabarty, K., Kumar, A., & Menon, V. (1994). *Trade in agarwood.* WWF/TRAFFIC India Report, New Delhi.

Challender, D., Harrop, S. R., & MacMillan, D. C. (2020). *Global Wildlife Trade and Conservation.* Cambridge University Press.

Chaurasia, A. K. (2023). *Unmasking the dark web: The hidden world of illegal wildlife trade.* Wild Hub.

Choudhury, B. C., & Bhupathy, S. (1993). *Turtle trade in India: A study of tortoises and freshwater turtles.* TRAFFIC/WWF.

Chow, A. T., Cheung, S., & Yip, P. K. (2014). Wildlife markets in South China. *Human-Wildlife Interactions, 8*(1), 108.

CITES. (2020). *Convention on international trade in endangered species of wild fauna and flora annual report.* CITES Secretariat.

Cooney, R. (2001). CITES and CBD: Tensions and synergies. *RECIEL, 10*, 259–267.

D'Cruze, N., Singh, B., & Carder, G. (2020). The illegal trade of star tortoises in India: Market trends and conservation implications. *Herpetological Journal, 30*(2), 45–61.

Das, P. (2008). India–Bangladesh border management: A review of government's response. *Strategic Analysis, 32*, 367–388.

Demeau, E., Monroy, M. E. V., & Karolan, J. (2019). Wildlife trafficking on the internet: A virtual market similar to drug trafficking? *Revista Criminalidad, 61*(2), 101–112.

Dexel, B., & Deutschland, N. (2002). *The illegal trade in snow leopards: A global perspective.* German Society for Nature Conservation.

Dhakal, M., & Baral, H. (2015). *Tiger conservation in south Asia: Lessons from terai arc landscape, Nepal.* Paper presented in 2nd international conference on tropical biology ecological restoration in Southeast Asia: Challenges, gains, and future directions, SEAMEO BIOTROP, Bogor, Indonesia, 12–13 October 2015.

Di Minin, E., Fink, C., Tenkanen, H., & Hiippala, T. (2018). Machine learning for tracking illegal wildlife trade on social media. Nature Ecology & Evolution, 2(3), 406–407.

Dickson, B. (2008). CITES and the livelihoods of the poor. *Oryx, 42*, 548–553.

Drishti, I. A. S. (2024). *World wildlife crime report 2024.*

Dublin, H. T., Mulliken, T., & Barnes, R. F. W. (1995). Four years after the CITES ban: Illegal killing of elephants. Ivory and stockpiles. *TRAFFIC Bulletin, 15*, 61–62.

Duffy, R., St John, F. A., Büscher, B., & Brockington, D. (2016). Toward a new understanding of the links between poverty and illegal wildlife hunting. *Conservation Biology, 30*, 14–22.

Dutta, S., Banerjee, A., & Sharma, P. (2020). Sustainable use of wildlife resources in India. *Journal of Wildlife Policy, 45*(3), 223–245.

Economic and Social Council. (2013). *Crime prevention and criminal justice responses to illicit trafficking in protected species of wild fauna and flora.* Resolution 2013/40. Available from http://www.unodc.org/documents/commissions/CCPCJ/Crime_Resolutions/2010 2019/2013/ECOSOC/Resolution_2013-40.pdf. Accessed April 2015.

Environmental Investigation Agency. (2022). *Transnational wildlife crime: India's role and responses*. EIA Reports.

FAO. (2022). *State of the world's forests 2022: India's role in biodiversity conservation*. Food and Agriculture Organization.

Freese, C. H., & Trauger, D. L. (2000). Wildlife markets and biodiversity conservation in North America. *Wildlife Society Bulletin, 28*, 42–51.

Fukushima, C. S., Tricorache, P., Toomes, A., Stringham, O. C., Rivera-Téllez, E., Ripple, W. J., Peters, G., Orenstein, R. I., Morcatty, T. Q., Longhorn, S. J., Lee, C., Kumschick, S., de Freitas, M. A., Duffy, R. V., Davies, A., Cheung, H., Cheyne, S. M., Bouhuys, J., Barreiros, J. P., et al. (2021). Challenges and perspectives on tackling illegal or unsustainable wildlife trade. *Biological Conservation, 263*(109), 342.

Gao, Y., Stoner, K. J., Lee, A. T., & Clark, S. G. (2016). Rhino horn trade in China: An analysis of the art and antiques market. *Biological Conservation, 201*, 343–347.

Ghose, D., & Kaul, R. (2005). Conservation implications of wild animal biomass extractions in Northeast India. *Animal Biodiversity and Conservation, 28*(2), 169–179.

Ghosh, A. (2018). The white hunters and the Indian wilderness in an era of modern technology: An exploration of the white hunting narratives of the Raj. *International Journal of Research in Engineering, IT and Social Sciences, 8*(7), 315–319.

Ghosh, P., Sen, A., & Das, R. (2021). Marine trade and conservation in India. *Marine Biology Research, 67*(2), 119–132.

Global Initiative Against Transnational Organized Crime. (2024, June). *Monitoring online illegal wildlife trade.*

Gong, S., Wang, J., Shi, H., Song, R., & Xu, R. (2006). Illegal trade and conservation requirements of freshwater turtles in Nanmao, Hainan Province, China. *Oryx, 40*(3), 331–336. https://doi.org/10.1017/S003060530600094

Gopal, S., Devi, R., & Menon, V. (2021). Smuggling networks and illegal wildlife trade in South Asia. *Crime & Conservation Journal, 12*(3), 78–96.

Gossling, S. (1999). Ecotourism: A means to safeguard biodiversity and ecosystem function? *Ecological Economics, 29*, 303–320.

Gupta, A., & Sharma, K. (2022). Bamboo trade and conservation challenges in India. *Journal of Forest Economics, 16*(1), 55–72.

Hale, D., & Hale, L. H. (2003). China takes off. *Foreign Affairs, 82*, 36–53.

Hall, A. S. C., Jones, W. P., Donavan, T. M., & Gibbs, J. P. (2000). The implications of mainstream economics for wildlife conservation. *Wildlife Society Bulletin, 28*(1), 16–25.

Harrison, J. R., Roberts, D. L., & Castro, J. H. (2016). Assessing the extent and nature of wildlife trade on the dark web. *Conservation Biology, 30*(4), 900–904.

Heinen, J. T., Shukurov, A., & Sadykova, M. (2001). Wildlife trade and conservation in Central Asia. *Environmental Policy and Law, 31*(2), 84–89.

Heinrich, S., Ross, J. V., & Cassey, P. (2019). Of cowboys, fish, and pangolins: US trade in exotic leather. *Conservation Science and Practice, 1*, e75.

Hernandez-Castro, J., & Roberts, D. (2015). Automatic detection of potentially illegal online sales of elephant ivory via data mining. *PeerJ Computer Science, 1*, 1–11.

Huang, X.-Q., Li, J.-H., Zhang, L., & Zhang, M.-L (2021). Prosecution records reveal pangolin trading networks in China, 2014-2019. *Zoological Research, 42*, 666–670.

Hughes, A. C. (2021). Wildlife trade. *Current Biology, 31*, R1218–R1224.

IFAW. (2005). *Caught in the web: Wildlife trade on the internet*. Available at https://www.ifaw.org/united-states/resource-centre/caught-web

IFAW. (2007). *Bidding for extinction, a snapshot survey of illegal trade in elephant ivory on eBayTM sites in: Australia – Canada – China – France – Germany – Netherlands – UK – USA*. Available at http://www.ifaw.org/sites/default/files/Report2007BiddingforExtinction.pdf

IFAW. (2008). *Killing with keystrokes: An investigation of the illegal wildlife trade on the World Wide Web*. Available at http://www.ifaw.org/sites/default/files/Killing with Keystrokes.pdf

IFAW. (2014a). *Wanted dead or alive, exposing online wildlife trade*. Available at http://www.ifaw.org/sites/default/files/IFAW-Wanted-Dead-or Alive-Exposing-Online-Wildlife-Trade-2014.Pdf

IFAW. (2014b). *Preventing illegal trade in restricted wildlife and improving the effectiveness of eBay's ivory ban*. Unpublished.

Indian Express. (2024, March 23). 3 years, 100 tigers killed & counting: New-age poaching mafia taps tech, digital payments, hawala networks. *The Indian Express*.

Indian Ministry of Environment, Forest & Climate Change. (2025). *India's national biodiversity action plan*. Government of India.

INTERPOL. (2022a). *Ivory, rhino horns, pangolin and tiger parts seized in transit from Africa to Asia*. Available at https://www.interpol.int/en/News-and-Events/News/2022/Ivory-rhino-horns-pangolin-and-tiger-parts-seized-intransit-from-Africa-to-Asia

INTERPOL. (2022b). *Operation Golden Strike results in seizure of 423 kg of pangolin scales*.

Interpol–IFAW Report. (2017). *21 Darknet wildlife ads selling ivory, rhino, tiger parts*.

IUCN. (2012). *The IUCN red list of threatened species* (Version 2015-4).

IUCN. (2021). *Red list of threatened species: India report*. International Union for Conservation of Nature.

Jain, P. (2001). *Project tiger status report*. Ministry of Environment and Forests.

Ji, X., Wang, Z., Zhang, T., Chen, D., Yang, W., Li, M., Shi, H., & Liu, L. (2024). Current status and trends of illegal trade in hawksbill turtle products in Hainan Island, China. *Frontiers in Marine Science*. https://doi.org/10.3389/fmars.2024.1498647 Frontiers+1

Katuwal, H. B., Parajuli, K., & Sharma, S. (2016). Money overweighed the traditional beliefs for hunting of Chinese Pangolins in Nepal. *Journal of Biodiversity & Endangered Species, 4*, 3.

Koirala, K. R. (2025, March). *Nepal collaborates with neighbors to gain wildlife crime intel but struggles with enforcement*. Shahi, sources cited in Mongabay.

Kumar, A., Mehta, S., & Verma, R. (2022). Medicinal plant trade and regulatory mechanisms. *Indian Journal of Herbal Science, 12*(1), 45–61.

Lal, R., & Singh, P. (2021). Challenges in combating wildlife crime: A case study of India. *Crime Studies Review, 18*(4), 101–120.

Lavorgna, A. (2014). Wildlife trafficking in the internet age. *Crime Science, 3*, 5.

Li, M. W. (1991). *A study of "silk road" trade history*. Gansu People Press.

Li, J., & Lu, Z. (2014). (Indian Express, 2024). Snow leopard poaching and trade in China 2000–2013. *Biological Conservation, 176*, 207–211.

Martin, J. (2014). *Drugs on the dark net: How crypto markets are transforming the global trade in illicit drugs*. Palgrave McMillan.

Menon, V. (2019). *The illegal wildlife trade in India*. Oxford University Press.

Nijman, V., Morcatty, T., Smith, J. H., Atoussi, S., Shepherd, C. R., Siriwat, P., & Bergin, D. (2019). Illegal wildlife trade–surveying open animal markets and online platforms to understand the poaching of wild cats. *Biodiversity, 20*(1), 58–61.

NTNC. (n.d.) *Wildlife crime*. The National Trust for Nature Conservation. ntnc.org.np

Nuwer, R. (2014, October 30). The blackmarket trade for endangered animal flourishes on the web. *Newsweek*. Available from http://www.newsweek.com/2014/11/07/extinctcom-280884.html. Accessed November 2015.

Oxpeckers. (2021, November 12). *Nepal's cross-border snow leopard smuggling networks*. oxpeckers.org

Patel, J., & Iyer, M. (2021). Wildlife forensics and the role of DNA profiling in combating illegal trade. *Journal of Forensic Science, 28*(1), 33–49.

Paudel, K. (2020, January 18). Poverty alone not a leading cause for involvement in illegal wildlife trade, new study says. *The Kathmandu Post*.

Paudel, K., Acharya, R., Baral, N., Heinen, J. T., & Jnawali, S. R. (2023). Pangolin seizures in Nepal indicate priority areas for conservation interventions. *Oryx, 57*(6), 727–734. Cambridge University Press & Assessment.

Paudel, K., Potter, C., & Phelps, J. (n.d.). *Monitoring illegal wildlife trade and strengthening public awareness in Araniko-Trail, Nepal-China border region.* The Rufford Foundation (project period 2016–2017). rufford.org

Rao, R., & Bhat, N. (2023). Aquaculture and sustainable marine trade: The case of India. *Journal of Marine Studies, 20*(1), 102–119.

Raza, H., & Kapoor, S. (2018). Rhino horn poaching in India: Law enforcement and conservation measures. *Wildlife Conservation Studies, 15*(2), 75–89.

Roberts, D. L., & Hernandez-Castro, J. (2017). Bycatch and illegal wildlife trade on the dark web. *Oryx, 51*(3), 393–394.

Salisbury, S. (2021). *Pangolins and a pandemic: The evolving Chinese Wildlife Trade.* US-China Today.

Sharma, R., & Patel, A. (2021). Digital platforms and the wildlife trade: Emerging challenges. *Conservation Technology Journal, 29*(1), 55–73.

Shepherd, C. R. (2019). The illegal reptile trade in India: Emerging trends and law enforcement strategies. *Herpetological Conservation & Biology, 14*(2), 199–215.

Shrestha, U., & Bawa, K. (2018). Harvesting and trade of Yarsa Gumba in the Himalayas. *Himalayan Ecology Studies, 10*(2), 33–50.

Siriwat, P., & Nijman, V. (2020). Wildlife trade shifts from brick-and-mortar markets to virtual marketplaces: A case study of birds of prey trade in Thailand. *Journal of Asia-Pacific Biodiversity, 13*, 454–461.

Soriano-Redondo, A., Alwasiti, H., Kulkarni, R., Correia, R. A., Bryukhova, S., Lita, N. M., et al. (2023). Online wildlife trade in species of conservation concern. *Conservation Letters, 16*(6), e12985.

St-Cyr-Leroux, B. (2023). *A wildlife market on the dark web.* UdeMNouvelles. Available at https://nouvelles.umontreal.ca/en/article/2023/07/05/a-wildlife-market-on-the-dark-web/

Stoner, S., Krishnasam, K., Wittmann, T., Delean, S., & Cassey, P. (2016). *Reduced to skin and bones re-examined: Full analysis. An analysis of Tiger seizures from 13 range countries from 2000 to 2015.* Available at http://www.traffic.org/publications/reduced-to-skin-and-bones-reexamined-full-analysis-an-analy.html

Stringham, O. C., Toomes, A., Kanishka, A. M., Mitchell, L., Heinrich, S., Ross, J. V., & Cassey, P. (2021). A guide to using the internet to monitor and quantify the wildlife trade. *Conservation Biology, 35*, 1130–1139.

Stringham, O. C., Maher, J., Lassaline, C. R., Wood, L., Moncayo, S., Toomes, A., et al. (2023). The dark web trades wildlife, but mostly for use as drugs. *People and Nature, 5*(3), 999–1009.

Stuart, C., & Stuart, T. (1996). *Africa's vanishing wildlife.* Smithsonian Institution Press.

Sung, Y.-H., Lee, W.-H., Leung, F. K.-W., Fong, J. J.-W., & Fong, J. J. (2021). Prevalence of illegal turtle trade on social media and implications for wildlife trade monitoring. *Biological Conservation, 261*(109), 245.

TRAFFIC. (2020). *India's wildlife trade report.* TRAFFIC International.

TRAFFIC. (2021). *The role of India in global wildlife crime networks.* TRAFFIC International.

United Nations Office on Drugs and Crime (UNODC). (2024). *World wildlife crime report 2024: Linkages of pangolin and ivory trade.*

van Uhm, D. P., & Wong, R. W. Y. (2019). Establishing trust in the illegal wildlife trade in China. *Asian Journal of Criminology, 14*, 23–40.

WCCB. (2021). *Wildlife crime control bureau annual report 2021.* Government of India.

WCCB. (2023).https://www.dailypioneer.com/2023/pioneer-exclusive/staff-crunch-hinders-work-of-wildlife-crime-controlbureau.html

Wong, R. W. Y. (2019) *The illegal wildlife trade in China: Understanding the distribution networks.* Palgrave Studies in Green Criminology (SpringerLink).

WWF. (2020). *India's big cat crisis: Poaching and conservation efforts.* World Wildlife Fund.

WWF. (2021). *Wildlife trade and conservation in Asia: A regional perspective*. World Wildlife Fund.

Xu, Q., Cai, M., & Mackey, T. K. (2020). The illegal wildlife digital market: An analysis of Chinese wildlife marketing and sale on Facebook. *Environmental conservation, 47*(3), 206–212.

Ye, Y.-C., Yu, W.-H., Newman, C., Buesching, C. D., Xu, Y.-L., Xiao, X., Macdonald, D. W., & Zhou, Z.-M. (2020). Effects of regional economics on the online sale of protected parrots and turtles in China. *Conservation Science and Practice, 2*, e161.

Yu, X., & Jai, W. (2015). *Moving targets: Tracking online sales of illegal wildlife products in China*. Available at http://static1.1.sqspcdn.com/static/f/157301/26245505/1432122394320/China-monitoring-report.pdf?token=Q8AuzEmc9lBrlrqSNkYIn6E%2FseI%3D

Yu, X., & Jia, W. (2015). *Moving targets: Tracking online sales of illegal wildlife products in China*. TRAFFIC.

Part II
Legal Frameworks and Enforcement Challenges

Chapter 3
India's Wildlife Protections Law: A Critical Review

Sharda Kalra, Kapish Samota, Devinder Singh, Asha Poonia, and Ravikant Verma

Contents

S. Kalra · K. Samota
Department of Zoology, Chaudhary Bansi Lal University, Bhiwani, Haryana, India

D. Singh
Department of Laws, Panjab University, Chandigarh, India

A. Poonia
Faculty of Life Sciences, Chaudhary Bansi Lal University, Bhiwani, Haryana, India
e-mail: asha.poonia@cblu.ac.in

R. Verma (✉)
Department of Zoology, CCS Haryana Agricultural University, Hisar, Haryana, India

A. Poonia et al. (eds.), *Wildlife Trade in India*,
https://doi.org/10.1007/978-981-95-3694-8_3

Abstract India's Wildlife Protection Act (WPA), enacted in 1972, is one of the most comprehensive legislative frameworks for wildlife conservation in the world. This chapter provides a critical review of the Act, examining its evolution, effectiveness, and limitations in protecting India's rich biodiversity. It begins by tracing the historical context that led to the enactment of the WPA, highlighting its primary objectives and key provisions, such as the establishment of conservation areas, hunting regulations, and trade of wildlife species. The analysis also explores the Act's alignment with international agreements like CITES and its role in mitigating wildlife trade and habitat loss. Despite its successes, the WPA has faced challenges in enforcement, legal ambiguities, and balancing conservation with the rights and livelihoods of local communities. This chapter delves into case studies illustrating these issues, such as conflicts arising from human-wildlife interactions and the effectiveness of penalties and deterrents for wildlife crime. Additionally, it assesses recent amendments to the Act, addressing their implications for conservation policy and governance in India. The chapter concludes by proposing pathways for reform, emphasizing the need for participatory approaches that integrate local and indigenous knowledge, adaptive management strategies, and stronger institutional frameworks. By critically evaluating the WPA's strengths and weaknesses, this chapter aims to contribute to a nuanced understanding of India's conservation paradigm and its potential for addressing emerging threats to biodiversity in the twenty-first century.

Keywords Wildlife Protection Act (WPA) · Wildlife conservations · Amendments · Protected areas · Hunting · Poaching · CITES

3.1 Introduction

Wildlife refers to any wild or undomesticated animal and plants lives in their natural habitat in terrestrial and aquatic bodies such as Mammals, reptiles, Amphibians and birds etc. The Wildlife Protection Act, 1972, sec 2(37) define "wild life" as any animal, aquatic or land vegetation which forms part of any habitat (Section 2 (37) Wild Life (Protection) Act, 1972). The Sec 2(36) define "wild animal" as any animal specified in Schedules I–IV and found wild in nature (Section 2 (37) Wild Life (Protection) Act, 1972). The exploitation of wildlife for financial profit, cultural practices, and international trade has reached unprecedented levels globally. This unsustainable exploitation is a major driver of biodiversity loss, pushing numerous species to the brink of extinction (Bhupathy et al., 2013; Willmott et al., 2022; Bulte & Horan, 2003). Key activities such as illegal poaching, overharvesting, and unregulated trade have led to dramatic population declines, particularly among species targeted for their high economic or cultural value (Willmott et al., 2022;

Rajak & Chaturvedi, 2023). This has not only impacted individual species but has also disrupted ecosystem integrity, diminishing essential ecological services (Bulte & Horan, 2003; Rajak & Chaturvedi, 2023; Choudhary, 2024; Karanth et al., 2018). Immediate, evidence-based interventions are required to curtail these pressures and promote sustainable use to ensure the survival of threatened species. A recent study on bird trade in India highlighted the significant scale of both legal and illegal trade, revealing that a large number of birds are being smuggled out of the country through porous borders. This illegal activity poses a serious threat to avian biodiversity and undermines conservation efforts. To curb such smuggling and protect India's rich biodiversity, there is an urgent need for stricter regulations (Table 3.1), enhanced enforcement mechanisms, and better cross-border coordination to prevent the illegal trade of birds (Jayakrishnan, 2023; Mitra & Bhaumik, 2013; Wildlife Protection Act, 1972; Kalra et al., 2024).

3.2 Historical Background of WPA 1972 Act

Wildlife conservation has very long ancient history in India, and it is intricately linked to the country's historical, religious, and cultural beliefs. India's tremendous biodiversity and the fundamental importance put on living things appear in the many ways that wildlife has been honored and protected over the ages. The ancient, medieval, colonial, and post-independence eras are the several historical periods that can be used to trace the development of wildlife conservation in India (Rajak & Chaturvedi, 2023; Choudhary, 2024; Karanth et al., 2018; Jayakrishnan, 2023; Mitra & Bhaumik, 2013).

3.2.1 Ancient India: Honoring the Natural World

Wildlife was cherished and considered as crucial to the environment in ancient India. The significance of living in harmony with environment is emphasized in the Vedic texts, such as the Rigveda and Atharvaveda. These Vedas includes hymns that honors and praise animals. The Indian religions of Jainism, Buddhism, and Hinduism placed a strong emphasis on ahimsa, or non-violence, toward all living things. In Sanatan Dharma, wildlife was encouraged by the frequent association of animals with spirits. For example: The fish (Matsya) and tortoise (Kurma), two of Lord Vishnu's avatars, serve to emphasize the sacred nature of these animals. They worshipped the snake, which was associated with Lord Shiva, and the peacock, which was connected to Lord Kartikeya, Lioness is associated with goddess Durga and Swan with the goddess Saraswati (Rajak & Chaturvedi, 2023; Choudhary, 2024). These association render the protection of the wildlife. Yajur Veda, Bridha Samiti, Yajnavalkya Smriti, and Vishnu Samhita emphasize harmonial co-existence of human and wildlife (Rajak & Chaturvedi, 2023; Choudhary, 2024; Velmurugan, 2017).

Table 3.1 Evolution of Wildlife Protection Laws in India which includes a timeline of the key legislations and amendments (Wildlife Protection Act, 1972; https://www.indiacode.nic.ir/bitstream/123456789/6198/1/the_wild_life_%28protection%29_act%2C_1972.pdf)

Year	Law/amendment	Key provision	Impact	Challenge
1972	Wildlife Protection Acts (1972)	This law provided comprehensive legal protection to wildlife, habitats, and ecosystems. It established six schedules for species protection, with Schedule I & II providing the highest protection. It created provisions for national parks, conservation reserves and wildlife sanctuaries. The law introduced hunting regulations and penalties for poaching and illegal trade.	The law helps in reducing the legal hunting. The establishment of protected areas increases conservation. It created the legal framework for wildlife crime prosecution.	Weak enforcement, lack of awareness, inadequate funding for conservation and loopholes in legal interpretation are main conservation challenge with respect to this law.
1982	Amendment	This amendment strengthened penalties for illegal wildlife trade and poaching. It extended protection to include more plant species. It also empowered forest officers for stricter enforcement.	It provides enforcement actions against poaching. It promotes conservation of endangered species. It also increased funding and spread awareness.	Habitat destruction and human-wildlife conflict were emerged as challenges. The illegal trade persisted due to high demand.
1986	Amendment	The amendment included new provisions for the protection of the environment. It addressed pollution-related concerns affecting wildlife habitats. Strengthened protection measures for wildlife habitats and ecosystems.	This amendment emphasis on habitat conservation and integrated wildlife protection with environmental policies.	Lack of coordination between environmental and wildlife laws, continued habitat degradation due to deforestation and industrial expansion.
1991	Amendments	This amendment strengthened penalties for illegal wildlife trade and poaching. It extended protection to include more plant species and prohibited hunting of all species, removing previous exceptions. It empowered forest officers for stricter enforcement.	It enhanced conservation of endangered species by increasing funding and spreading awareness. More rigorous enforcement were implemented against poaching.	Continued habitat destruction, human-wildlife conflict and increasing illegal trade due to high demand were main conservation challenges.

2002	Amendments	This amendment established the National Board for Wildlife (NBWL) for policy guidance. The Central Zoo Authority was created to regulate zoos and for stricter regulations on trade of scheduled species.	The amendment increased accountability in zoos and wildlife trade regulations by improving wildlife governance and strengthening conservation policies.	NBWL faced bureaucratic inefficiencies in this amendmentand continued illegal trade and habitat encroachment were observed prominent challenges
2006	Forest Rights Act, 2006	This act was recognized as rights of forest-dwelling communities that allowed traditional livelihoods in conservation areas and balanced conservation with human rights.	It empowered indigenous communities in conservation efforts. Improved community-led conservation initiatives and increased conflicts between authorities and local communities.	Increasing conflicts between conservationists and indigenous communities for exploitation of forest products. Lack of proper implementation guidelines and risk of misuse of rights for commercial gain are main challenges faced in this act.
2013	Amendments	It strengthened the laws on poaching and wildlife crime. It also imposed higher penalties for offenses related to endangered species and established provisions for more conservation reserves.	More cases of wildlife crime detected and prosecuted. It expanded conservation areas. The national and international cooperation on wildlife crime are strengthened through this amendments.	The rise in organized wildlife crime networks, lack of trained enforcement personnel which are legal loopholes exploited by traffickers.
2022	Wildlife Protection (Amendment) Act, 2022	This amendment rationalized schedules from six to four for better management. It also increased penalties for violations and improved provisions for captive breeding and conservation. It also strengthened CITES implementation.	Through compliance with CITES, increasing penalties and improved conservation policies for lesser-known species.	Enforcement, concerns over relaxed provisions for certain species trade and potential conflicts over exemptions for certain species remains challenge.

Indian sacred literature, including the Gita, Ramayana, and Mahabharata, emphasize the importance of compassion and kindness towards all living beings, advocating against cruelty. Wildlife in India holds significant cultural & spiritual value, and various communities continue to protect species for these importance (Velmurugan, 2017).

In ancient India, wildlife Conservation was a fundamental component of dharma (obligation). This philosophy was best demonstrated by Emperor Ashoka, who ruled in the third century BCE, by implementing laws that protected wildlife. With an emphasis on biodiversity preservation, his decrees specifically prohibited the hunting of particular wild animal, including some birds, squirrels, and rhinoceroses. Ashoka also encouraged the planting of medicinal plant gardens and afforestation to improve ecological balance (Rajak & Chaturvedi, 2023; Choudhary, 2024).

3.2.2 Medieval India: Hunting and Protection

In medieval India between 2000 to 5000 BC at the time of Harappa and Mohenjo-daro civilization, bone traces of swamp deer, Chital, hare jackal with many other animals have been found as a result of overhunting and exploitation (Rajak & Chaturvedi, 2023; Choudhary, 2024). Emperor Ashoka warned the tribes that lived in forests that his tolerance had its limitations since he understood how difficult it was to maintain authority over them. Although people in forests and even farmers probably hunted animals on a regular basis, it was not impossible to enforce laws against hunting, fishing, and forest fires. There were infractions of these regulations, as evidenced by the 100 panas (coins) penalties levied for poaching deer in royal forest reserves. "Our king killed very few animals," Ashoka wrote in his fifth pillar edict. He also highlighted the importance of non-violence by saying, "I have enforced the law against killing certain animals and many others, but the greatest righteousness among men comes from exhortation in favor of non-injury to life and abstention from killing living beings" (Velmurugan, 2017; Dixit & Singh, 2024).

The introduction of Islamic rule and the rise of feudal institutions during the middle Ages significantly altered people's perspective towards wildlife. Hunting become associated with royal pleasure and power, although other rulers understood the social and ecological value of conservation. The Mughal kings, particularly Akbar, had a strong interest in wildlife and worked to record and control its use. The Ain-i-Akbari, Akbar's administrative handbook, provided ethical hunting principles with a focus on moderation to avoid overexploitation. In keeping with a dual mindset of respect and practical application, the text also discussed the domestication and care of wild animals. Furthermore, forest areas were designated as hunting reserves by regional rulers throughout the Indian subcontinent. Despite being primarily used for recreational purposes, these reserves frequently functioned as unofficial sanctuaries, unintentionally aiding in the preservation of habitat and the protection of animals (Velmurugan, 2017; Dixit & Singh, 2024).

3.2.3 Exploitation and Regulation in the British Era

Wildlife management underwent a dramatic change throughout the British colonial era, mostly due to commercial and cultural interests. The colonial elite made hunting a formal sport, which led to the extinction of species. It became very common to hunt animals like rhinoceroses, elephants, and tigers as trophy (Rangarajan, 1996; Sekar & Srivastav, 2024; Gautam & Kaur, 2023; Thawrani & Banait, 2022). During this time, The Indian Forest Act of 1865 and its later changes (such as 1878 and 1927) were implemented by the British with the primary goal of regulating forest resources for timber (Rajak & Chaturvedi, 2023; Choudhary, 2024; Rangarajan, 1996). These laws had no provisions for protecting animals, but they unintentionally affected wildlife ecosystems. Game regulations were put in place by the colonial administration to control hunting, although its purpose was not to protect wildlife but rather to preserve game for elite hunters. This era marked the establishment of the first protected areas, like Corbett National Park (formerly Hailey National Park) in 1936, which indicated the beginning modern conservation initiatives in India. In spite of these steps, colonial policies gave resource exploitation precedence over conservation, which led to widespread habitat loss and a sharp decrease in wildlife animals (Rajak & Chaturvedi, 2023; Choudhary, 2024; Velmurugan, 2017; Sekar & Srivastav, 2024). Apart from these, many legislation were enacted in British era to protect the wildlife such as Shore Nuisance (Bombay and Kolaba) Act, 1853 which limited the contamination of seawater, The Merchant Shipping Act, 1858 which addressed the prevention of oil-related marine contamination, The Fisheries Act, 1897, The Bengal Smoke Nuisance Act, 1905, Bombay Smoke Nuisance Act, 1912, and Wild Birds and Animals Protection Act, 1912 (Meena & Maikhuri, 2023).

3.2.4 After Independence: A Transition to Conservation

India's strategy for protecting wildlife changed substantially after gaining independence. It became clear that a strong legislative framework was required to prevent habitat loss and wildlife exploitation. Initially, Indian Board for Wildlife was founded in 1952, functioned as a consultative organization to promote the conservation of wildlife. The government established a number of national parks and wildlife sanctuaries, but these initiatives were initially dispersed and lacked a unified legal framework. Strong legislation was implemented in consequences of concerning decline of species such as tiger and increased awareness of environmental conservation. Finally, in 1972, the wildlife Protection Act was passed, establishing the framework for the protection of animals (Rajak & Chaturvedi, 2023; Choudhary, 2024; Saravanan, 2023).

Throughout these periods, India's cultural traditions continued to play a role in wildlife protection. Sacred groves, maintained by local communities, served as unofficial conservation areas. These grooves, often associated with lords or spirits, provided shelter for many species, illustrating the enduring bond between Indian society and its natural heritage. Both domestic issues and international commitments served as catalysts for the Act. India's policymakers were influenced by the 1972 Stockholm Conference on the Human Environment, which was crucial in emphasizing the need for protecting the environment on a global scale. Poaching, unregulated commerce, and habitat destruction have been identified as the main threats to biodiversity (Choudhary, 2024; Saravanan, 2023; Soni, 2020). The Act was created to ensure that wildlife protection became a shared responsibility by offering a consistent framework throughout jurisdictions. Based on their conservation status, it divided species into schedules and the most protected schedules had the highest penalties for hunting or trading animals. The Act also empowered the creation of national parks, wildlife sanctuaries, and conservation reserves, providing legal sanctity to protected areas. The concept of eco-sensitive zones was also introduced around these areas to mitigate human-wildlife conflicts (Rajak & Chaturvedi, 2023; Choudhary, 2024; Saravanan, 2023).

3.3 The Wildlife Protection Act, 1972 and Its Evolution

Wildlife protection acts are the core legislation to protect the wildlife from poaching, hunting and trade local to global. It came into force in 1972 but amended many time to keep adding of new species in schedule I to schedule IV to prevent the species from the threat of extinction. It has six schedule from I to VI providing varying degree of protection to wild animals. The schedule I and II have most threatened species and their trade is highly punishable than schedule III and IV with lower penalties. Schedule V includes the Vermin species such as fruit bats, Crow, rats, mice which may be hunted freely. Schedule VI has the endemic plant species and prohibited from farming and cultivation. Each schedule was divided into Parts which included all the threatened species from mammals, reptiles, birds, amphibians, fishes, invertebrates and plants (Rajak & Chaturvedi, 2023; Wildlife Protection Act, 1972; Soni, 2020).

There was an urgent need for strong central laws because of the increasing devastation of natural resources and the loss of wild animals. However, as "Wild animals and birds" were included in the "State List" of Schedule VII in the Indian constitution, the state legislature was the sole body with the authority to enact laws. After sometimes, a resolution were passed in support of article 252 by the state legislature of Andhra Pradesh, Bihar, Gujarat, Haryana, Himachal Pradesh, Madhya Pradesh, Manipur, Punjab, Rajasthan, Uttar Pradesh, and West Bengal giving authority to the Parliament to pass necessary legislation to conserve the wildlife. The Central government recognized the need for comprehensive legislation to conserve birds and wild animals for all the issues related to this, because the laws currently in place in various states were inadequate (Karmakar, 2013). 'Article 17-A and 17-B' were introduced in List-III by "42nd amendment in 1976" to empowering authority to the Parliament to enact comprehensive laws throughout the nation

(Ramesh, 1999). The Wildlife Protection Act of 1972 was extended to the entire nation, with the exception of Jammu and Kashmir, by the Wildlife Protection Amendment Act of 1991 (Rajak & Chaturvedi, 2023; Wildlife Protection Act, 1972; Tyler, 1973). So this is the critical step towards the biodiversity conservation and establishing protected areas.

Human encroachment has emerged as a major driver of habitat destruction, threatening wildlife sanctuaries across the country. The inadequacies in earlier legislation, coupled with the rapid loss of natural habitats, underscored the urgent need for comprehensive and effective legal provisions at the national level. This necessity culminated in the formulation of the Wildlife Protection Act of 1972. The primary goals of this legislation were to standardize wildlife conservation laws across India, establish a robust network of protected areas including National Parks and Wildlife Sanctuaries, and regulate the illegal trade of wildlife and their derivatives (Tyler, 1973; Roy & Roy, 2017).

After the independence in 1947, the wildlife conservation is started with Rhinoceros preservation act and the elephant preservation act in 1950 to protect some threatened species from exploitation (Thawrani & Banait, 2022; Singh & Sharma, 2024). After that a meeting if advisory committee were called to related issue of wildlife protection and to create wildlife sanctuaries and national parks. Consequently in1952, the Indian board of wildlife were inaugurated in Mysore. However, It was challenging to change the public perception towards wildlife conservation because many middle class Indians, former princes and forester were still involved in hunting, poaching and wildlife trade either to make financial profit and for the entertainment. Many wildlife including monkey were exported to united states in huge quantity between 1950 to 1960 and skins of most threatened species such as tiger, leopard and snakes were sold for foreign currency (Rajak & Chaturvedi, 2023; Choudhary, 2024; Tyler, 1973; Roy & Roy, 2017).

The Government attempt to issue gun to farmer in safeguard of their crops from wild animals. Consequently, a free roaming animal tiger (Cheetah), went extinct in 1951.However, Indian government starts to change policies to shift towards wildlife conservation and 8 tiger were imported from Namibia on 17 September, 2022 (Rajak & Chaturvedi, 2023).

3.3.1 Amendment of 1982: Creation of "Closed Areas"

Through the introduction of the idea of "closed areas," this amendment established a legal foundation to bans hunting and ensure wildlife protection in certain places. Wildlife wardens were granted enhanced powers to enforce these regulations effectively (Rangarajan, 1996). This was a significant step toward limiting human intervention in vital ecosystems and establishing protected habitats. Apart from these, comprehensive law was made enacted on both animals and Plant. Strengthening enforcement against the export of medicinal plants and timber species, such as **Sandalwood (Santalum album)**, widely targeted for its aromatic wood and essential oil. The protection of **Red Sandalwood (Pterocarpus santalinus)** in India highlights the focus on high-value plant species. Stricter monitoring and export controls were implemented to address smuggling (Johnson, 1984; Pascal et al., 2025; Narayanan et al., 2014; Rangarajan, 1996; Sekar & Srivastav, 2024).

3.3.2 1986 Amendment: Strengthened Enforcement Measures and Addressed Habitat Destruction

New enforcement agencies were formed to regulate and monitor the unsustainable trade, poaching and hunting of wildlife. Policies to regulate land use change that led to habitat destruction such as deforestation and agriculture were also introduced. Project Tiger was expanded in India during this time to include new reserves, ensuring the preservation of habitat. The Bengal Tiger (Panthera tigris tigris), whose habitat was constantly disturbed, benefited directly from the addition of anti-deforestation measures (Chhikara & Chaudhary, 2021; Rangarajan, 1996).

3.3.3 1991 Amendment: Extending Conservation's Purview

The term "wildlife" was broadened to include plant species in recognition of the ecological interdependence of plants and animals. Punishments for poaching, illegal trade, and habitat destruction were intensified to deter violations. By eliminating delays in the legal system, the creation of special courts offered a quicker resolution of charges pertaining to wildlife crime. Such as special court in Jodhpur, Rajasthan play a vital role in hearing the cases against people involved in wildlife crime and violating the wildlife laws. The establishment and creation of wildlife corridor to connect fragmented habitats facilitating movement of migratory species and gene flow. For instance, the Kaziranga-Karbi Anglong corridor in Assam, India, was created to allow elephants and other animals to move safely (Chhikara & Chaudhary, 2021; Mitra & Bhaumik, 2013; Rangarajan, 1996).

3.3.4 2002 Amendment: Emphasizing Community Participation

The concept of community reserve was first time introduced empowering local community to play an active role in conservation of wildlife and their habitat. These reserves provided legal protection to community managed areas which are outside of protected region such as National Park and sanctuaries but have ecological significance. Communities can contribute in biodiversity conservation if their livelihood and traditions were sustained (Tyler, 1973; Rangarajan, 1996; Biggs et al., 2017). The amendment introduced compensation strategies for victims of incidents involving wildlife, such as crop damage or harms by wild animals to address the rising problem of conflicts between humans and wildlife. This clause intended to reduce intolerance toward wildlife by encouraging collaboration between local residents and conservation officials. The conservation reserves were protected through stringent regulation with severe penalties for the hunting and poaching. Following this amendment, many community reserve were established by local community such as Fisherfolk is a community which was dependent on the Chilka lake, Orissa for their livelihood starts participating actively to maintain ecological health and illegal prawn farming and wildlife crime. Similarly, Sirohi community reserve, Meghalaya located in Khasi Hills were protected and maintained by people of Khasi tribes. They tried to protect the habitat of rare Blyth's Tragopan, a vulnerable pheasants (Chhikara & Chaudhary, 2021; Rangarajan, 1996).

3.3.5 2006 Amendment: Integrating Habitat and People

The amendment aligned with the Scheduled Tribes and Other Traditional Forest Dwellers (Recognition of Forest Rights) Act, 2006 (commonly known as the Forest Rights Act). As part of their cultural traditions, indigenous communities have the right to access, use, and manage forest resources in a sustainable manner. The amendment emphasized the protection of critical wildlife habitats, ensuring that conservation efforts accounted for both ecological and human needs. It provided a framework for identifying Critical Wildlife Habitats (CWHs) within protected areas, where human activities could be regulated to preserve biodiversity. For example, Orissa's Simlipal National Park, a tiger reserve which was recognized as Critical Wildlife Habitat by the 2006 regulations. Local communities surrounding the park participated in eco-development initiatives to help conserve habitat and ensure the preservation of their traditional rights (Biggs et al., 2017). Indian wildlife laws were brought into compliance with international agreements, including the Convention on Biological Diversity (CBD), which emphasizes the value of traditional knowledge in conservation initiatives (Narayanan et al., 2014; Chhikara & Chaudhary, 2021).

3.3.6 2022 Amendment: Complying with International Agreements

This amendment is the important step in the evolution of wildlife protection act, 1972, because wildlife conservation legislation were align with the international agreement such as CITES (Convention on international trade on endangered species of fauna and flora). The Indian government is fully committed to maintaining and regulating the trade its side of the agreement.In accordance with treaties like the Convention on International Trade in Endangered Species of Wild Fauna and Flora (CITES), the government is fully committed to maintaining its side of the agreement (Gautam & Kaur, 2023; Singh & Sharma, 2024). An analysis of CITES trade data revealed that a significant number of wild animals and birds have been exported from India. A species-specific assessment indicated that birds accounted for the majority of exports, both in terms of the number of individuals and the diversity of species involved. This highlights the need for targeted measures to monitor and regulate the trade of avian species to ensure compliance with conservation priorities and international trade regulations (Poonia et al., 2022a). The main aim of this amendment is to regulate the trade of endangered species and addressing the challenge associated with invasive species introduction and stringent mechanism to conserve the biodiversity.

Important feature of the 2022 Amendment's CITES Framework Integration

1. The amendment introduced specific provisions to streamline the implementation of CITES regulations in India. It ensured adherence to global standards by establishing precise rules for the transportation, possession, and trade of species listed in the CITES Appendices (Table 3.2). A Clear mandate was provided for maintaining a national inventory of CITES-listed species to monitor trade activities more efficiently (Boora, 2023).
2. The Amendment included measures to regulate and monitor the threats of invasive species to native biodiversity. It required the identification, listing, and regulation of invasive species to prevent their spread and mitigate ecological harm. To prevent the illegal wildlife trade, government of India issued an advisory to dealing with imported exotic species in India and declaration of stock of wildlife in 2020 (Ghanghas & Poonia, 2021; Kalra et al., 2024).
3. The amendment revised the schedules under the Wildlife Protection Act, aligning them with global conservation priorities (Table 3.2). New species were added to the schedules, reflecting updated scientific assessments of their conservation status.
4. Penalties for violations, particularly those involving the illegal trade of CITES-listed species, were enhanced (Table 3.2 and 3.3). Stricter enforcement mechanisms were introduced to deter wildlife crimes and ensure accountability.
5. The schedules were reduced from six to four comprised of schedule I with most threatened species and schedule II with least threatened species, schedule III with vermin species such as fruit bats, crows etc. and schedule IV with the species protected by CITES (Table 3.2). The major differences between WPA, 1972 and the Wild Life (Protection) Amendment Act, 2022 is given below Table 3.2.

3.4 Rationale for the Wild Life (Protection) Amendment Act, 2022

Despite the Wildlife (Protection) Act's comprehensive framework, the increasing pressures of biodiversity loss, habitat degradation, and illegal wildlife trade necessitated updates to enhance its efficacy (Kalra et al., 2024; Poonia et al., 2022a; Chowdhury, 2014). To address these emerging challenges, contemporary conservation needs and international standards including implementation of the provisions of the Convention on International Trade in Endangered Species of Wild Flora and Fauna (CITES), of which India is a nearly member since 1976, the Wild Life (Protection) Amendment Bill, 2021 (the bill) was introduced in the winter session of Parliament i.e. Lok Sabha on 17 December 2021, and was passed in the Rajya Sabha on 8 December 2022 (Kalra et al., 2024; Poonia et al., 2022a). This amendment aims to strengthen the "conservation, protection and management of wild life" by strengthening legal provisions such as enhancing the penalties for wildlife crimes, implementation of CITES regulations, to address gaps in previous legislation (Table 3.3).

Global connectivity due to advancement in technology has made the connections between the countries like never before. This advancement has provided countless new ways to do businesses including wildlife trading whose approximate worth falls between 91 and 258 billion dollars annually (Poonia et al., 2022b; Nijman, 2010; Hughes, 2021). Various studies have reported Asia as a major source and consumer of wildlife, whereas Europe is a major consumer (Hughes, 2021; Krishnasamy & Zavagli, 2020). Present evolving dynamics of globalization, wildlife trade and climate change while balancing the needs of local people for livelihood generation for access and benefit-sharing, makes it necessary to update the existing laws to foster species diversity, ecosystem health and climate stability. The Act also reflects India's commitment to adhering to global conventions such as CITES, to strengthen its claim as leading country in wildlife conservation and tackling climate change (Table 3.4). Many other organization participate actively to conserve the biodiversity such a IUCN,IWC, Interpol etc. given in (Table 3.4).

3.5 The Forest (Conservation) Amendment Bill, 2023

The 2023 amendments to the Forest Conservation Act (FCA) of 1980 introduce provisions that exempt specific forest lands from its jurisdiction and exclude certain activities from being classified as non-forest purposes. These changes are anticipated to reduce overall forest cover and potentially undermine the rights granted to forest-dwelling communities under existing central legislation (PRS, 2023; Saxena, 2024).

Table 3.2 Comparative placement of species in different Schedules in The Wild Life (Protection) Act, 1972 and The Wild Life (Protection) Amendment Act, 2022 (Wildlife Protection Act, 1972; https://www.indiacode.nic.in/bitstream/123456789/6198/1/the_wild_life_%28protection%29_act%2C_1972.pdf)

The Wild Life (Protection) Act, 1972						The Wild Life (Protection) Amendment Act, 2022			
Schedule I Part I Mammals (77 species) Part II Amphibians and Reptiles (24 species) Part IIA Fishes (12 species) Part III Birds (46 species) Part IV Crustacea and Insects: Butterflies and Moths (129 species belonging to 6 different families), Coconut or Robber Crab (1 specie), Dragon Fly (1 specie) Part IV A Coelenterates (5 species) Part IV B Mollusca (9 species)	Schedule II Part I (14 species) Part II Beetals (341 species belonging to 13 families), 24 species	Schedule III 8 species	Schedule IV 10 species, [a]Birds (80 species), [b]Snakes (14 species), [c]Butterflies and Moths (19 species belonging to 5 different families), Mollusca (15 species) Horseshoe crabs (1 specie)	Schedule V Vermin (4 species)	Schedule VI Plants (6 plant species)	Schedule I: 597 species Part A Mammals (183 species), Part B Birds (208 species), Part C Reptiles (62 species), Part D Amphibians (4 species), Part E Fishes (20 species), Part F Echinodermata (1 species), Part G Mollusks (9 species), Part H Anthropods (1 species), Part I Butterflies (96 species), Part J Dragonfly (1 species), and Part K corals (12 species)	Schedule II 1553 species Part A Mammals (34 species), Part B Birds (1134 species), Part C Reptiles (22 species), Part D Amphibians (32 species), Part E Fishes (10 species), Part F Mollusks (15 species), Part G Anthropods (10 species), Part H Butterflies (295 species), Part I porifera/ sponges (1 species)	Schedule III (Specified Plants, 19 entries in place of 6 entries originally	Schedule IV for species listed in CITES Appendix I (627 species); Appendix II (497 species), and Appendix III (217 species)

Part IV C Echinodermata (1 specie)									
Same degree of protection to included wild animals listed in Schedule I & Part II of Schedule II i.e. characterized by highest protection with severe penalties.		Same degree of protection to included wild animals listed in Schedule III, IV & Part I of Schedule II i.e. characterized by moderate to low protection with lesser penalties for violations.		Allows hunting of animals categorized as vermin.	Hunting is prohibited. However, enforcement is less stringent compared to higher schedules.	Highest level of protection is offered to species listed in Schedule I with severe penalties for offences against animals (like poaching, hunting, or trade).	Species listed in Schedule II are given lower protection than species listed in Schedule I but still enjoy strong safeguards with moderate penalties.	Species listed in Schedule III receive even less protection compared to those in Schedules I & II. The protection is mostly against hunting and capture, but penalties for offenses are milder.	Species listed in Schedule IV are also protected, but the level of protection and penalties for offenses are significantly lower than those in the first three schedules of the Act.

[a]Transferred to Part II of Schedule II as entry 2B by notification No. S.O. 859(E), dated 24-11-1986 (Wildlife Protection act, 1972)
[b]Transferred to Part II of Schedule II as entry 2C by notification No. S.O. 859(E), dated 24-11-1986 (Wildlife Protection act, 1972)
[c]Transferred to Part II of Schedule II as entry 2E by notification No. S.O. 859(E), dated 24-11-1986 (Wildlife Protection act, 1972)

Table 3.3 Comparative analysis of the Wild Life (Protection) Act, 1972 and the Wild Life (Protection) Amendment Act, 2022: key provisions and enhancements for biodiversity protection (Wildlife Protection Act, 1972)

Feature	Wild Life (Protection) Act, 1972	Wild Life (Protection) Amendment Act, 2022
Protected Species Lists	**Sections 6–9 & Schedules I–V**: Listed species under Schedule I (highest protection), Schedule II (regulated hunting), and others, with fewer species protected.	**Amendment of Sections 6 & 7**: Expanded schedules (now covering more species under Schedules I & II) by adding numerous species to the protection lists to address growing threats.
Regulation of Invasive Alien Species	No specific provision on invasive alien species in the original Act.	**New Chapter VB (Sections 62A–62C)**: Introduces regulation of invasive alien species. Section 62A empowers the central government to prohibit or regulate the import, trade, and possession of invasive species that threaten biodiversity.
Human-Wildlife Conflict	**Sections 11 & 12**: Limited provisions to address human-wildlife conflict (primarily focused on killing or capturing wild animals in self-defense or to protect property).	**Amendment of Section 11 & Addition of Section 11B**: Expands the scope for addressing human-wildlife conflict. Allows states to designate conflict-prone areas as **no-go zones**, empowering local governments to take preventive measures. Section 11B gives authorities power to control such zones and reduce fatalities.
CITES Compliance	**Section 40(1)**: General provisions on wildlife trade without alignment to international conventions like CITES.	**Amendment of Sections 49E & 49F**: Introduces a streamlined process to comply with CITES regulations. The amendment enhances regulation on the import/export of listed species under the CITES framework, ensuring that India meets its international wildlife trade obligations.
Increased Penalties for Violations	**Section 51**: Penalties for poaching, illegal trade, and related violations existed but were generally moderate, with a fine of up to ₹25,000 or imprisonment up to 3 years.	**Amendment of Section 51**: Significant increase in penalties. Imprisonment now ranges from 3 to 7 years, with fines raised to up to ₹1,00,000 for offenses involving Schedule I species, acting as a strong deterrent against wildlife crimes.
Community Involvement in Conservation	**Section 36C**: Community reserves could be created, but local involvement in conservation was minimal.	**Amendment of Section 36A, New Section 36E**: Strengthens provisions for **community reserves** and introduces eco-sensitive zones under Section 36A. Encourages greater **community participation** in wildlife conservation and habitat management. Local communities now have a more prominent role in managing eco-sensitive areas.

Focus on Sustainable Development	**Sections 29–30**: Provisions for protected areas (e.g., national parks, sanctuaries) with limited consideration for the balance between development and conservation.	**Amendment of Sections 30A & 33**: Integrates a more sustainable approach to wildlife conservation, emphasizing the balance between human development and environmental protection. Explicit guidelines on managing habitats without compromising biodiversity.
Wildlife Habitat Protection	**Section 33–35**: General provisions on the protection of national parks and sanctuaries with no mention of eco-sensitive zones.	**New Section 36E & Amendment to Section 35**: Focuses on **eco-sensitive zones** and emphasizes the reduction of human activities in critical habitats. The introduction of these zones ensures that buffer areas around national parks are also protected.
Wildlife Conservation Infrastructure	**Section 4**: Provided for the appointment of Chief Wildlife Wardens and Wildlife Advisory Boards but did not significantly address capacity-building for enforcement agencies.	**Amendment of Section 4 & Section 6(1)**: Enhanced focus on strengthening **enforcement agencies**. The amendment provides for the training and equipping of wildlife officials with better resources to manage and protect biodiversity.
Scientific and Technological Advances	No mention of the use of technology or scientific research for wildlife conservation.	**New Sections 61B & 61C**: Introduces provisions for leveraging **modern scientific tools** such as geospatial mapping, environmental DNA monitoring, and population data collection to monitor species and habitats more effectively.

Table 3.4 Wild Life (Protection) Act, 1972 vs. Wild Life (Protection) Amendment Act, 2022 in Light of International Agreements

International agreement/feature	Wild Life (Protection) Act, 1972	Wild Life (Protection) Amendment Act, 2022	Need for improvement
International Whaling Commission (IWC) (IWC, 1946)	No mention of IWC regulations or provisions for the conservation of marine mammals.	Still missing: No reference to IWC or regulations for whale and marine mammal conservation.	Marine conservation inclusion: The Act should incorporate IWC provisions for protecting marine mammals, particularly whales, and support global conservation efforts.
International Union for Conservation of Nature (IUCN) (IUCN, 1948)	No reference to IUCN Red List or specific guidelines for assessing species' threat status.	Schedules I & II amended: Aligns the lists of protected species with the IUCN Red List categories, ensuring better protection of endangered and critically endangered species.	Dynamic updates required: A system for regularly updating the species lists based on changes in IUCN status and threats to biodiversity is necessary.
Asia-Pacific Forestry Commission (FAO) (APFC, 1949)	No reference to sustainable forest management in line with APFC recommendations.	Still missing: The Act does not integrate sustainable forestry practices under FAO's Asia-Pacific Forestry Commission recommendations.	Sustainable forestry inclusion: Incorporating APFC principles on forest management would improve forest biodiversity conservation efforts in the Asia-Pacific region.
INTERPOL's Wildlife Crime Working Group (INTERPOL, 1923)	The 1972 Act did not engage with INTERPOL's wildlife crime initiatives or establish any legal framework for cross-border cooperation with INTERPOL's Environmental Crime Program.	The 2022 Amendment enhances cooperation with international enforcement agencies, including INTERPOL, by aligning India's wildlife crime legislation with international crime protocols.	Further strengthening joint enforcement operations with INTERPOL, particularly in sharing data on organized wildlife trafficking networks, could improve law enforcement effectiveness. Capacity building for wildlife law enforcement agencies in collaboration with INTERPOL is also needed.
International Plant Protection Convention (IPPC) (IPPC, 1952)	No provisions for plant protection from pests or invasive species management in line with IPPC.	Still missing: No specific mention of plant protection from pests or regulation of invasive species in accordance with IPPC guidelines.	Strengthen plant biodiversity protection: The Act needs specific mechanisms for managing plant pests and invasive species threats to ecosystems, as per IPPC objectives.

WCO (World Customs Organization) (WCO, 1952)	The **1972 Act** did not address cooperation with **customs authorities** globally in the context of **illegal wildlife trade**. There was no mention of collaboration with **WCO** on wildlife crime detection at borders.	The **2022 Amendment** indirectly supports international trade enforcement by strengthening wildlife trade regulations in **Sections 49A–49F**, but lacks explicit mention of **WCO collaboration**.	The Act should include explicit provisions for cooperation with **customs authorities** globally through **WCO** protocols to better detect and prevent the illegal trade of wildlife at international borders.
International Consortium on Combating Wildlife Crime (ICCWC) (ICCWC)	The 1972 Act did not acknowledge or integrate cooperation with ICCWC partners. The Act lacked provisions related to global cooperation on wildlife crime, particularly with organizations like INTERPOL, UNODC, World Bank, and WCO, which are part of ICCWC.	The 2022 Amendment aligns partially with ICCWC's international enforcement strategies by strengthening wildlife crime penalties under Section 51 and enhancing protections for endangered species. Provisions in Sections 49A-49F establish authorities to regulate and prevent illegal wildlife trade. However, specific reference to ICCWC's tools is still lacking.	The Act should formally integrate ICCWC's Wildlife and Forest Crime Toolkit to align Indian wildlife crime prevention mechanisms with international best practices. Strengthen cooperation with ICCWC partners such as INTERPOL, UNODC, and the World Bank to enhance data-sharing, intelligence, and enforcement operations.
ICCWC Wildlife and Forest Crime Toolkit (ICCWC Toolkit)	The **1972 Act** did not adopt the **ICCWC Toolkit** for assessing and combating wildlife crime or developing data-driven enforcement mechanisms.	The 2022 Amendment has improved data collection and enhanced wildlife crime penalties, but does not adopt the ICCWC Toolkit, which provides a comprehensive framework for evaluating national responses to wildlife and forest crime.	The Act should incorporate the ICCWC Toolkit, focusing on wildlife crime analysis, intelligence gathering, and capacity building to align with international standards. This will allow for improved crime detection, prosecution, and prevention strategies based on global best practices.
ICCWC Indicator Framework for Combating Wildlife Crime (ICCWC Indicators)	The 1972 Act did not use ICCWC's Indicator Framework for assessing the effectiveness of wildlife crime enforcement and policy implementation.	The 2022 Amendment improves enforcement and provides for better wildlife crime tracking but does not adopt ICCWC's Indicator Framework for evaluating the effectiveness	The Act should integrate the ICCWC Indicator Framework, which provides measurable indicators to evaluate wildlife crime enforcement, policy effectiveness,

(continued)

Table 3.4 (continued)

International agreement/feature	Wild Life (Protection) Act, 1972	Wild Life (Protection) Amendment Act, 2022	Need for improvement
	No focus on international wildlife crime indicators was present.	of law enforcement and policy measures.	and law enforcement capacity. This would enhance monitoring and accountability in wildlife crime prevention.
Capacity Building with ICCWC Partners (INTERPOL, UNODC, WCO)	No mention of capacity-building programs or cooperation with ICCWC partners (e.g., INTERPOL, UNODC, World Customs Organization). The Act lacked legal mechanisms to facilitate training programs or technical assistance with international partners.	The 2022 Amendment introduces better legal enforcement mechanisms for wildlife crimes, but it does not include provisions for capacity building with ICCWC partners. Section 49C provides for scientific authorities, but no specific training with ICCWC is mentioned.	The Act should include formal agreements with ICCWC partners to facilitate training and capacity-building for law enforcement officials, customs agents, and judiciary in handling wildlife crimes. Collaborative technical assistance and knowledge-sharing programs should be added to the Act.
Financial Investigations and Anti-Money Laundering (ICCWC's Financial Guide)	The 1972 Act did not address the financial aspects of wildlife crime, such as money laundering or financial investigations linked to illegal wildlife trade.	The 2022 Amendment increases penalties but does not tackle the financial dimensions of wildlife crime (e.g., money laundering, asset forfeiture). There is no integration with ICCWC's Financial Investigation Guide.	The Act should align with ICCWC's Financial Investigation Guide and include provisions for financial investigations targeting wildlife crime syndicates, asset seizures, and money laundering. This will disrupt the financial networks supporting organized wildlife crime.
Ramsar Convention on Wetlands (Ramsar, 1971)	No direct reference to Ramsar or wetland-specific provisions. Wetlands were not given special status.	Sections 26A & 26B: Aligns with Ramsar by introducing specific legal provisions for the protection and restoration of wetlands, including the designation of wetlands as protected areas.	Clearer ecological management: More structured frameworks are needed for monitoring the ecological character of Ramsar sites, including restoration programs for coral reefs, along with funding mechanisms to

			ensure proper implementation of Ramsar's goals.
CITES (Convention on International Trade in Endangered Species of Wild Fauna and Flora) (CITES, 1973) Management Authority for CITES Scientific Authority under CITES	No explicit reference to CITES. Limited general provisions for trade in wildlife under Sections 40 & 44, Section 40: Covered general trade in wildlife, with limited emphasis on international trade. Section 44 mentioned international trade but did not specify obligations under multilateral agreements. Lacked structured compliance mechanisms for international obligations under CITES. No specific provision for the establishment of a national authority or scientific authority responsible for implementing CITES obligations	New Chapter VB (Sections 49A-49F): Introduces specific provisions and comprehensive mechanism for regulating international trade to regulate trade in species covered under CITES. Section 49A establishes procedures for dealing with the import, export, and re-export of wildlife in accordance with CITES standards, ensuring India's legal framework is aligned with international conventions. Introduces a CITES Management Authority (responsible for implementing and overseeing India's compliance with CITES. This includes granting permits, tracking trade, and ensuring regulations are enforced in line with international commitments) and Scientific Authority (New Section 49C tasked with offering guidance on the trade of CITES-listed species. The Authority ensures that international trade does not harm species survival and offers scientific assessments on sustainable trade quotas) to regulate international trade of species listed under CITES. It mandates that no person can import or export any specimen listed under CITES without	Need for enforcement enhancement: Greater focus on capacity-building and cross-border enforcement mechanisms is required to ensure compliance with CITES regulations. Wildlife Crime control Bureau (WCCB) needs transparency in public availability of wildlife crimes for better understanding and disposal of wildlife crimes.

(continued)

Table 3.4 (continued)

International agreement/feature	Wild Life (Protection) Act, 1972	Wild Life (Protection) Amendment Act, 2022	Need for improvement
		prior permission from the CITES Management Authority, ensuring stricter compliance.	
Protection of Migratory Species under CMS (Convention on the Conservation of Migratory Species of Wild Animals) (CMS) (CMS, 1979)	No specific reference to **CMS** or specific provisions to protect migratory species covered under international agreements.	**New Section 62B**: Introduces a specific provision for the protection of migratory species listed under CMS in accordance with CMS, giving special legal status and protection measures to migratory species passing through India as part of their natural migration routes.	**Regional coordination gaps**: More cooperation with neighboring countries is needed to ensure the protection of migratory species and their habitats.
United Nations Convention on the Law of the Sea (UNCLOS) (UNCLOS, 1982)	The **1972 Act** had no reference to **marine biodiversity protection in exclusive economic zones (EEZs)** or ocean conservation areas under **UNCLOS**.	The **2022 Amendment** provides a framework for protecting **marine biodiversity** in Indian waters, including the **EEZ**, in line with **UNCLOS** obligations. **Section 18B** recognizes the need to regulate marine resources and protect ecosystems like coral reefs.	Clearer **legal mechanisms for managing marine resources** and enforcing laws in **EEZs** need to be established. The Act should provide more detailed provisions for **fishing restrictions and coral reef management** within EEZs.
International Tropical Timber Organization (ITTO) (ITTO, 1986)	No reference to sustainable forest management or timber conservation.	Still missing: The Act does not address the objectives of ITTO, such as sustainable management of tropical forests and timber resources.	Need for inclusion: Incorporating ITTO's principles of sustainable forest management would strengthen conservation efforts in tropical forest regions.
Financial Action Task Force (FATF) (FATF, 1989)	The **1972 Act** had no specific provisions addressing the **financial aspects of wildlife crime** or money	The **2022 Amendment** increases penalties for wildlife crime but does not explicitly target the **financial**	The Act should align with **FATF recommendations** on addressing the **financial dimensions of wildlife crime**, such as money laundering

	laundering linked to illegal wildlife trade.	**flows** linked to wildlife trafficking networks.	and the illicit financial networks supporting wildlife trafficking. **Asset seizure** provisions for wildlife crime profits should be introduced.
International Cooperation on Wildlife Crime	Section 58: General provisions existed for cooperation with foreign governments, but no specific mention of multilateral environmental agreements (MEAs).	Amendment of Section 58: Enhances international cooperation in combating wildlife crime, especially in the context of CITES enforcement. This includes information sharing with foreign governments, collaboration on investigations, and support for cross-border enforcement.	
United Nations Convention on Biological Diversity (CBD) (CBD, 1992)	No alignment with **CBD** objectives for sustainable use or biodiversity protection. The Act provided general wildlife protection, but lacked a comprehensive biodiversity focus. No specific mention of **marine biodiversity** or **coral reef ecosystems** in the Act. **Section 18** vaguely mentioned "territorial waters," but there was no focus on marine ecosystems.	**Sections 36D–36F**: Aligns with **CBD** goals, incorporating provisions for **sustainable use of biodiversity**, **ecosystem restoration**, and habitat protection. Integrates the **Kunming-Montreal Global Biodiversity Framework** targets. **Section 18B** expands the protection of marine species and ecosystems, aligning with **CBD Aichi Biodiversity Targets**. Focus is placed on the protection of **endangered marine species** and **critical habitats**, including coral reefs.	**Further alignment with CBD targets**: A more structured approach for meeting specific **CBD Post-2020 Biodiversity Framework** goals, particularly ecosystem restoration and species recovery by 2030. **Stronger monitoring of coral reef health** and protection against climate impacts and pollution is required to fully align with **CBD goals**. The Act should implement **marine protected area networks** in line with **Post-2020 Global Biodiversity Framework**.
Rio Convention (1992)—Convention on Biological Diversity (CBD) (CBD, 1992)	The 1972 Act primarily focuses on species conservation and wildlife habitats in isolated areas (parks, sanctuaries) but lacks comprehensive biodiversity approach toward	Sections 36D–F in the 2022 Amendment reflect efforts to aligns somewhat with biodiversity objectives of the Rio Convention.	The Act needs to better integrate ecosystem services, sustainable use, genetic diversity and biodiversity monitoring to fully align with the CBD's Aichi Biodiversity Targets

(continued)

Table 3.4 (continued)

International agreement/feature	Wild Life (Protection) Act, 1972	Wild Life (Protection) Amendment Act, 2022	Need for improvement
	biodiversity, as emphasized in the Rio Convention.		and the Post-2020 Global Biodiversity Framework. Integration of climate change considerations into wildlife conservation is required.
UN Framework Convention on Climate Change (UNFCCC) (UNFCCC, 1992)	No reference to climate change adaptation in wildlife conservation.	Still missing: No explicit mention of UNFCCC objectives or provisions for adapting wildlife conservation strategies to climate change.	Climate change integration: Incorporating climate adaptation strategies for ecosystems and species protection in line with UNFCCC goals would strengthen conservation efforts. Additionally, reef restoration and resilience building programs, should be integrated. Legal frameworks addressing ocean acidification and its impact on coral ecosystems also needs to be incorporated.
UN Convention to Combat Desertification (UNCCD) (UNCCD, 1994)	No focus on combating desertification or restoring degraded land.	Still missing: No specific provisions for addressing land degradation and desertification.	Land restoration needed: Future amendments should integrate UNCCD goals for restoring degraded lands and protecting ecosystems vulnerable to desertification.
Kyoto Protocol (1997) (Kyoto, 1997)	Not directly aligned, as the 1972 Act does not focus on climate change mitigation or impacts of greenhouse	While the **2022 Amendment** acknowledges the importance of habitat restoration (**Section 36D**), it indirectly contributes to **climate**	The Act should include specific provisions for **climate change adaptation** and mitigation, especially for ecosystems vulnerable to

	gas emissions on ecosystems and biodiversity	**change mitigation** or carbon sequestration efforts by protecting forests and ecosystems that act as carbon sinks.	climate impacts (e.g., coral reefs, wetlands). Greater emphasis is needed on **carbon sinks** and **nature-based solutions** as outlined in the **Kyoto Protocol**.
International Network for Bamboo and Rattan (INBAR) (INBAR, 1997)	No mention of bamboo and rattan conservation under INBAR.	Still missing: The Act does not cover the conservation or sustainable use of bamboo and rattan resources.	Incorporate INBAR objectives: Provisions should be added to support the conservation and sustainable management of bamboo and rattan, aligning with INBAR's mission.
UNODC (United Nations Office on Drugs and Crime) (UNODC, 1997)	The **1972 Act** did not formally recognize or integrate **UNODC frameworks** on addressing **organized wildlife crime** as part of the broader **transnational organized crime** efforts.	The **2022 Amendment** aligns with **UNODC's** international crime prevention strategies, enhancing provisions for combatting **organized wildlife crime**. **Section 51** increases penalties for wildlife offenses, promoting stronger deterrence.	A **formalized partnership with UNODC** could be included to enhance **cross-border cooperation**, intelligence sharing, and joint operations in combating transnational wildlife crime. More focus on addressing the **financial flows** supporting wildlife crime is needed.
Cartagena Protocol on Biosafety (CBD) (Cartagena, 2000)	No mention of biosafety or regulations regarding genetically modified organisms (GMOs).	Still missing: No provisions aligned with the Cartagena Protocol for regulating biosafety and managing GMOs that may impact biodiversity.	Biosafety inclusion: The Act should address risks posed by GMOs to biodiversity, in compliance with the Cartagena Protocol on Biosafety under the CBD.
FAO International Treaty on Plant Genetic Resources for Food and Agriculture (ITPGRFA) (FAO, 2001)	No mention of ITPGRFA or the need to conserve plant genetic resources for food security.	Still missing: The Act does not address the conservation of plant genetic resources for food and agriculture.	Need for inclusion: Provisions should be added for the protection and sustainable use of plant genetic resources in line with ITPGRFA for food security and biodiversity.

(continued)

Table 3.4 (continued)

International agreement/feature	Wild Life (Protection) Act, 1972	Wild Life (Protection) Amendment Act, 2022	Need for improvement
Asia-Pacific Forest Invasive Species Network (APFISN) (APFISN, 2004)	No provisions for managing forest invasive species in accordance with APFISN guidelines.	Still missing: The Act does not address forest invasive species threats under the Asia-Pacific Forest Invasive Species Network framework.	Forest invasive species management: The Act should introduce provisions for managing forest invasive species threats, aligned with APFISN principles.
UNODC Wildlife and Forest Crime Analytic Toolkit (UNODC Toolkit, 2008)	The 1972 Act did not incorporate data collection, analysis, or reporting mechanisms based on international standards like the UNODC Wildlife Crime Toolkit.	The 2022 Amendment strengthens data collection and reporting mechanisms on wildlife crime, though it does not formally adopt the UNODC Toolkit framework.	The Act should integrate the UNODC Toolkit for wildlife crime analysis and reporting to enhance wildlife crime investigation, prosecution, and international reporting based on globally accepted standards.
Copenhagen Accord (2009) (Copenhagen, 2009)	No specific reference to climate change resilience or adaptation strategies for wildlife in the face of climate change, which the Accord highlights as critical for ecosystems.	The 2022 amendment focuses on species-specific protection, but does not incorporate specific climate resilience measures for ecosystems or acknowledge the importance of reducing emissions for biodiversity conservation	The Wild Life Amendment Act should integrate ecosystem resilience measures to climate change, particularly to protect vulnerable species and habitats in line with the Copenhagen Accord's climate goals. Emphasis on habitat restoration in response to climate impacts is needed.
Doha Amendment to the Kyoto Protocol (2012) (Doha, 2012)	The 1972 Act does not engage with the international mechanisms for climate change mitigation, greenhouse gas reductions, clean development, or adaptation strategies outlined in the Doha Amendment.	The 2022 Amendment does not make direct references to climate mitigation targets, focusing more on wildlife conservation but not on climate-related commitments.	The Act should be updated to incorporate climate mitigation strategies relevant to wildlife and habitat conservation, aligning with the Doha Amendment's focus on reducing emissions through land-use changes and forest conservation.

International Cooperation on Ecosystem Services (IPBES) and incorporation of IPBES Findings (IPBES, 2012)	No mention of the **IPBES** (Intergovernmental Science-Policy Platform on Biodiversity and Ecosystem Services) and its global biodiversity findings. General provisions existed for cooperation with foreign entities, without focusing on ecosystem services or biodiversity assessments.	**New Section 36C**: Acknowledges the importance of scientific and policy inputs from international bodies like **IPBES**. The Act emphasizes utilizing global biodiversity assessments from IPBES to inform national policies, ensuring that conservation practices are science-based and globally aligned. **New Sections 36D–36E**: Strengthens international cooperation on biodiversity and ecosystem services by utilizing **IPBES** reports and scientific assessments. This includes better alignment of India's biodiversity protection efforts with international scientific research and policy guidance from IPBES.	There is need for strengthening biodiversity assessments, ecosystem service protection, climate adaptation strategies based upon recommendations of IPBES.
Paris Agreement (2015) (Paris, 2015)	The **1972 Act** did not address **climate change** or the impacts of global temperature rise on biodiversity and ecosystems.	The **2022 Amendment** focuses on **species and** habitat conservation but does not incorporate provisions to address climate adaptation or mitigation in the context of global warming, as required by the Paris Agreement.	The Act should establish **climate resilience** frameworks for wildlife conservation and include mechanisms to assess and mitigate the impacts of climate change on protected areas. Specific reference to temperature rise, habitat shifts, and biodiversity loss due to global warming is needed. The Wildlife Protection Act needs to integrate concrete action plans for wildlife adaptation in line with the Paris Agreement's climate goals, particularly in relation to temperature-sensitive species and habitats.

(continued)

Table 3.4 (continued)

International agreement/feature	Wild Life (Protection) Act, 1972	Wild Life (Protection) Amendment Act, 2022	Need for improvement
United Nations Sustainable Development Goals (SDGs) (UN, 2015)	No specific mention of **SDG 15 (Life on Land)** or integration of biodiversity conservation into sustainable development frameworks. No reference to **SDG 14 (Life Below Water)**, which focuses on **marine ecosystem conservation** and reducing marine pollution.	**Section 36G**: Integrates the protection of biodiversity into broader sustainable development frameworks, aligning with **SDG Goal 15** (Life on Land) to prevent biodiversity loss. **Section 18B** aligns more closely with **SDG 14** by focusing on protecting **marine ecosystems** and addressing **illegal fishing**, coral mining, and pollution threats to coral reefs.	**Incorporate SDG indicators**: Specific indicators should be provided in the Act to measure progress toward achieving the SDG 15 and **SDG 14** goals related to life on land and forest protection along with **ocean pollution control**, sustainable fishing practices, and reducing the impact of coastal development.
Global Coral Reef Monitoring Network (GCRMN) (GCRMN, 2020)	The **1972 Act** did not mandate any coral reef monitoring activities or integrate **GCRMN standards**.	The **2022 Amendment** acknowledges the need for **marine ecosystem monitoring**, including **coral reefs**, and aligns with global standards, though **Section 61B** still lacks specifics on coral monitoring.	**Dedicated coral reef monitoring programs** in line with **GCRMN standards** should be explicitly mentioned. A **national coral reef database** should be established to track changes in reef health and biodiversity.
Kunming-Montreal Global Biodiversity Framework (2022) (CBD, COP15)	No provisions for compliance with the **Post-2020 Global Biodiversity Framework** or its 2030 and 2050 targets.	**Section 36F**: Includes legal obligations to meet the **2030 and 2050 targets** for species recovery, ecosystem restoration, and halting biodiversity loss.	**Operational framework needed**: The Act needs specific operational guidelines for implementing and monitoring progress towards these global biodiversity goals.
South Asia Wildlife Enforcement Network (SAWEN) (SAWEN, 2016)	The 1972 Act has limited provisions for cross-border wildlife crime prevention, though it recognizes the need for international cooperation.	**The 2022 Amendment** includes stronger provisions for regulating wildlife trade and enhancing enforcement, particularly under **Chapter VB (Sections 49A–49F)**, which addresses illegal wildlife trade	The Act does not formally recognize SAWEN as a key partner in wildlife crime enforcement. An explicit reference to **regional cooperation under SAWEN** would enhance India's commitment to joint

		and implements stricter penalties for violations. The Act also aligns with SAWEN's objective of **combating wildlife crime** through enhanced protection of endangered species, aligned with international obligations such as **CITES**. **Section 51** increases penalties for offenders, supporting deterrence in line with SAWEN's regional efforts to curb poaching and illegal wildlife trafficking.	enforcement efforts. Act lacks provisions for **real-time intelligence-sharing** with SAWEN member countries, which is critical for tracking transboundary wildlife crimes. Legal amendments should include formal mechanisms for **data exchange, joint investigations**, and **cross-border cooperation** in collaboration with SAWEN. The Act should mandate **capacity-building initiatives** with SAWEN to train enforcement officials in dealing with sophisticated wildlife crime networks operating across South Asia. These initiatives could involve **joint training programs** and the **sharing of best practices** in wildlife crime enforcement. The Act currently lacks provisions to combat the **financial aspects of wildlife crime**, such as money laundering. Enhanced cooperation with SAWEN and other regional bodies could support investigations into the financial underpinnings of wildlife trafficking.

The amendments to the Forest Conservation Act (FCA) exclude two categories of land from its scope: (1) land recorded as forest prior to October 25, 1980, but not formally notified as such, and (2) land converted from forest use to non-forest use before December 12, 1996. This provision may conflict with the 1996 Supreme Court judgment aimed at preventing deforestation (Saxena, 2024; PRS India, 2023). The exemption of land in border areas for national security projects poses a significant risk to forest cover and wildlife, particularly in the ecologically sensitive northeastern states. Additionally, granting broad exemptions for activities such as the establishment of zoos, eco-tourism facilities, and reconnaissance surveys may result in adverse impacts on forest ecosystems and biodiversity.

However apart from these amendments, the evolution of wildlife protection laws in India has been significantly influenced by landmark cases and judicial interventions (Table 3.5). These cases have shaped the interpretation, implementation, and strengthening of wildlife conservation laws, particularly the **Wildlife Protection Act (WPA), 1972**, and its subsequent amendments (Table 3.5).

3.6 Trend in Wildlife Crime in Last Two Decades

Analyzing CITES data for India from last two decades 2001 to 2020 reveals a significant number of wildlife crime including trafficking of species such as Mammals, birds, reptiles, fishes and invertebrates. However there is a absence of records in the exported category for amphibians over the past two decades (Fig. 3.1). In contrast, only two transactions were documented under the imported category for amphibians during this period. These trends underscore the persistent challenges despite legal protections. The Wildlife Protection Act aligns with CITES commitments by restricting international trade in endangered species through its Schedule I and Schedule II listings. Additionally, the amendments to the Act, particularly in 2003 and 2022, introduced stricter penalties for violations, reflecting India's evolving legislative response to combat the growing sophistication of wildlife crime networks. The correlation between CITES data and Indian legislation highlights the effectiveness and gaps in enforcement, emphasizing the need for stronger inter-agency coordination and enhanced border security. The discontinuity observed in the line graph represents the absence of recorded trade data for that years, highlighting a potential gap in reporting or limited international trade from India. This discrepancy underscores the need for further investigation into whether the lack of records reflects actual trade patterns, underreporting, or gaps in enforcement and monitoring mechanisms (Fig. 3.1).

Table 3.5 Some major cases related to wildlife protection law's evolution

Sr. no.	Key judgements	Detail	Significance
1.	*T.N. Godavarman Thirumulpad v. Union of India* (1996) (Chowdhury, 2014; Bean, 1977; WCCB, 2008)	This case began as a public interest litigation (PIL) highlighting the unregulated deforestation in the Nilgiris. The definition of forests to include all areas recorded as forests, regardless of ownership were also expanded. The Supreme Court directed states to regulate deforestation and establish Forest Advisory Committees, ensuring stricter habitat protection under the Wildlife Protection Act (WPA), 1972.	Comprehensive interpretation of the term "forest" and protection of forest ecosystems. This case led to stricter enforcement of the WPA, ensuring better habitat protection for wildlife.
2.	*Animal Welfare Board of India v. A. Nagaraja and Others* (2014)	Details: The case challenged the practice of jallikattu, a traditional bull-taming sport in Tamil Nadu, as cruel to animals. Outcome: The Supreme Court ruled that such practices violated Sections 3 and 11 of the Prevention of Cruelty to Animals Act, 1960, and were contrary to the constitutional duty under Article 51A(g) to protect wildlife and show compassion to living creatures.	Strengthened animal rights and highlighted cruelty as a violation of constitutional values. The case influenced wildlife laws by emphasizing the ethical treatment of animals and the need for conservation aligned with moral principles.
3.	*Centre for Environmental Law, WWF v. Union of India* (1996)	Details: The case dealt with the illegal diversion of forest land for non-forest purposes. Outcome: The Supreme Court mandated that all protected areas under the WPA, 1972, including National Parks and Sanctuaries, should be strictly managed for conservation purposes. It reinforced the role of buffer zones and eco-sensitive areas.	Strengthened the establishment of protected areas like National Parks and Wildlife Sanctuaries. It highlighted the importance of maintaining ecological balance through habitat preservation.

(continued)

Table 3.5 (continued)

Sr. no.	Key judgements	Detail	Significance
4.	*Bombay Environmental Action Group v. State of Maharashtra* (1997)	Details: The case was filed to prevent the destruction of mangroves in Mumbai. Outcome: The Bombay High Court declared mangroves as "protected forests" under the WPA and directed their preservation. It also prohibited construction within 50 m of mangrove areas.	Protection of mangroves and coastal ecosystems as critical wildlife habitats. This case underscored the role of ecosystems like mangroves in sustaining biodiversity and wildlife.
5.	*M.C. Mehta v. Kamal Nath and Others* (1997)	Details: The case involved the diversion of the Beas River in Himachal Pradesh for private purposes. Outcome: The Supreme Court held that natural resources, including forests and wildlife, are held in trust by the state for public use. Any activity harming these resources violates this trust.	Reinforced the state's duty under the WPA to conserve wildlife and their habitats as a public trust. Established the "public trust doctrine" in environmental law.
6.	*Wildlife Rescue and Rehabilitation Centre v. Union of India* (2008)	Details: The petition highlighted the cruelty faced by wildlife used in circuses. Outcome: The Supreme Court banned the use of wildlife species in circuses, citing violations of the WPA and the Prevention of Cruelty to Animals Act.	Strengthened enforcement against exploitation of wildlife in captivity. Prohibited the use of animals in circuses and entertainment.
7.	*Goa Foundation v. Union of India* (2014)	Details: The case challenged illegal mining activities in Goa, which were degrading wildlife habitats. Outcome: The Supreme Court banned mining in eco-sensitive zones around National Parks and Wildlife Sanctuaries.	This case enhanced habitat protection under the WPA. Prohibited mining in and around protected areas.
8.	*Narmada Bachao Andolan v. Union of India* (2000)	Details: The construction of the Sardar Sarovar Dam led to submergence of forest areas and displacement of wildlife.	Balancing development with ecological concerns and highlighted the importance of

		Outcome: While the dam was allowed, the Court directed stringent compliance with environmental and wildlife protection measures.	Environmental Impact Assessments (EIAs) in protecting wildlife habitats.
9.	*Shivaji Rao Holkar v. Union of India* (2013)	Details: This case dealt with the protection of tiger habitats from human activities. Outcome: The Supreme Court imposed a temporary ban on tourism in the core areas of tiger reserves under Project Tiger. It later allowed regulated tourism with strict guidelines.	Strengthened the implementation of tiger conservation efforts under the WPA
10.	Supreme Court on Vermin Species (2022) *Centre for Environmental Law, WWF India v. Union of India* (2022)	Details: The Court reviewed petitions challenging the classification of certain species, such as nilgai and wild boar, as vermin for culling purposes. Outcome: It called for a balance between controlling wildlife-human conflict and conservation needs, stressing scientific assessments.	It Influenced the WPA's provision for vermin classification under Schedule V. As a result, Controversies over classification and culling of wildlife as vermin.
11.	*The State of Himachal Pradesh v. The State of Punjab* (2003)	This case deal with the protection of the forest ecosystem in Himachal Pradesh and Punjab. The Supreme Court ruled on the conservation of forest land and its utilization for developmental purposes.	It emphasized that the forest cover was crucial for maintaining biodiversity and required special protection from exploitation. The court ordered a halt on all non-forest activities in forest areas until proper clearances were obtained.
12	*State of Tamil Nadu v. K. K. Verma* (2009)	This case dealt with the protection of the lion population in the Gir Forest, Gujarat. The Supreme Court upheld the state's role in protecting endangered species, directing that no industrial or developmental activities be allowed within the Gir Sanctuary to safeguard the Asiatic lions' habitat.	This judgment reinforced the principle that wildlife protection must take precedence over developmental activities.
13	*Centre for Environmental Law, WWF v. Union of India* (2004)	This case dealt with illegal deforestation and the management of forest resources. The Supreme Court ruled in favor of stricter monitoring of forest areas, including preventing illegal deforestation and exploitation of forest resources.	It called for an audit of all mining leases and the imposition of guidelines to prevent illegal activities affecting biodiversity.

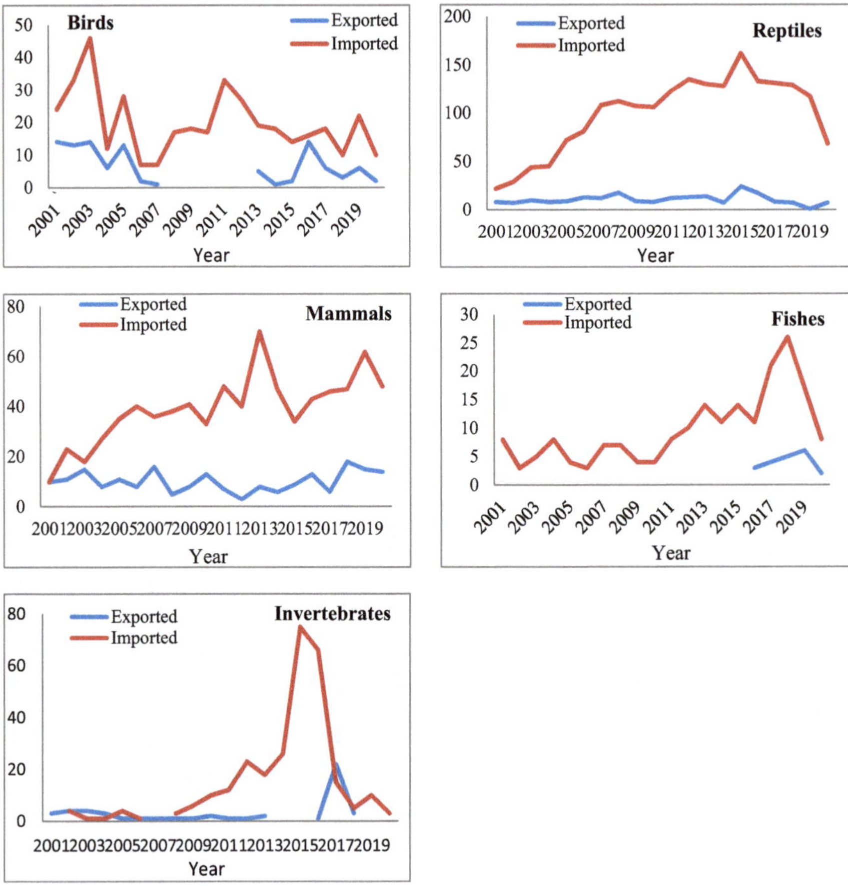

Fig. 3.1 Last two decades trends of Wildlife Crime reported from CITES

3.7 Challenges in Wildlife Conservation Under the Wildlife (Protection) Act, 1972

The Act has played a crucial role in conserving biodiversity; its implementation faces several scientific, legal, and administrative challenges that hinder its effectiveness.

3.7.1 Habitat Fragmentation and Loss

Habitat destruction and fragmentation is the biggest problem driven by human activities such as urbanization, deforestation, infrastructural development and agricultural expansion facing wildlife conservation (Velmurugan, 2017; Poonia et al., 2022a; Chartier et al., 2011). The wildlife protection acts has led to creation of protected areas such as national park, wildlife sanctuaries and conservation reserve, however linear infrastructure projects (roads, railways and transmission lines often intersect critical wildlife habitats leading to human wildlife conflict and genetic isolation of population diversity (Biggs et al., 2017; Kalra et al., 2024; Kumar & Verma, 2017; Kumar et al., 2002). For example, The Chilla-Motichur corridor in Rajaji National Park, Uttarakhand, is a significant elephant corridor, but rapid urbanization and infrastructure projects such as highways and railway lines have severely fragmented the habitat. This has led to an increase in human-elephant conflict, with elephants straying into human settlements in search of food, leading to casualties on both sides.

3.7.2 Human-Wildlife Conflict (HWC)

The encroachment of human settlements into forested landscapes has intensified conflicts between humans and wildlife, particularly involving species such as elephants, leopards, and tigers (Rajak & Chaturvedi, 2023; Poonia et al., 2022a; Chartier et al., 2011). Despite the WPA's stringent penalties for harming wildlife, retaliatory killings of animals due to crop depredation, livestock predation, and human casualties remain prevalent (Manral et al., 2016). The Act must address compensation mechanisms or alternative conflict-mitigation strategies, such as habitat restoration and community-based conservation programs. Many species such as jackal, elephant, tiger, antilopes etc. has been reported most common in human-wildlife conflicts.

3.7.3 Wildlife Trafficking and Illegal Trade

Despite the provisions of the WPA, illegal wildlife trade remains a persistent challenge due to high demand for animal parts, live birds, and reptiles in both domestic and international markets. India is a major source and transit hub for wildlife trafficking, particularly for species like pangolins, star tortoises, parakeets, and tigers. Inadequate surveillance, weak border controls, and lack of coordination with international law enforcement agencies (such as INTERPOL and CITES) hinder the effectiveness of anti-trafficking efforts (Kalra et al., 2024; Chartier et al., 2011).

3.7.4 Invasive Alien Species

Invasive Alien Species (IAS) pose a severe threat to biodiversity by disrupting native ecosystems, altering habitat structures, and outcompeting indigenous flora and fauna. These species, introduced either intentionally or accidentally, proliferate rapidly due to the absence of natural predators or regulatory ecological constraints in their non-native environments (Thomaz et al., 2015; Singh, 2005; Sundaram et al., 2012). IAS impact biodiversity at multiple levels, including genetic, species, and ecosystem diversity, leading to irreversible ecological imbalances. The introduction of invasive alien species (IAS), such as *Lantana camara*, *Prosopis juliflora*, and water hyacinth, has led to the degradation of native habitats and the displacement of indigenous species (Singh, 2005). *Eichhornia crassipes* (water hyacinth), originally introduced as an ornamental aquatic plant, has now become a major threat in India's freshwater systems, covering lakes, rivers, and wetlands such as the Chilika Lake (Odisha) and Loktak Lake (Manipur) (Sundaram et al., 2012). This dense floating vegetation reduces oxygen levels in water, blocks sunlight penetration, and disrupts fish breeding habitats, leading to declining populations of native fish species such as Indian major carps (*Catlacatla*, *Labeorohita*, *Cirrhinusmrigala*) (Soni, 2020; Dogra et al., 2009).

3.7.5 Political and Economic Pressures on Conservation Policies

Policy decisions regarding wildlife conservation are often influenced by economic interests, political lobbying, and short-term developmental goals (Saberwal, 2000). Despite the protective provisions of the Wildlife (Protection) Act, 1972 (WPA), large-scale infrastructure projects—including highways, dams, and mining—continue to be approved in or near Protected Areas (PAs), leading to habitat fragmentation, loss of biodiversity, and increased human-wildlife conflict (Archi Rastogi et al., 2014). The inadequacies in Environmental Impact Assessments (EIAs) and the dilution of regulatory frameworks have further weakened conservation efforts, allowing unsustainable development to take precedence over ecological integrity (Karmakar, 2013; Sen, 2021). A balance must be struck between economic growth and conservation priorities by adopting science-based decision-making, enforcing stringent EIAs, and enhancing legal mechanisms to protect wildlife and their habitats effectively.

3.7.6 *Limited Scientific Research and Management Strategies in Wildlife Conservation*

Limited Scientific Research and Adaptive Management Conservation efforts under the WPA often rely on outdated data, inadequate monitoring, and traditional management approaches (Distefano, 2005; Mawdsley et al., 2009). The Act does not emphasize scientific research, habitat modeling, or evidence-based policymaking. A lack of real-time ecological monitoring, genetic studies, and conservation technology (such as camera trapping and satellite telemetry) limits the effectiveness of wildlife management programs (Kadykalo et al., 2021; Hughes, 2021). Lack of updated data of species distribution often results in misinformed conservation policies, such as ineffective species recovery programs or misplaced protected area boundaries. Conservation programs under the WPA follow static, one-size-fits-all approaches rather than adaptive management frameworks that respond to changing ecological conditions. For example, protected area management plans are often revised every decade without incorporating new research findings, reducing their effectiveness in addressing emerging threats such as climate change, habitat fragmentation, and species migration patterns (Mawdsley et al., 2009; Singh, 2023; Poonia et al., 2022b; Nijman, 2010).

3.8 Conclusion

Wildlife protection laws are significant step towards wildlife conservation. India has shown a tremendous paradigm shift toward putting conservation first, starting with the comprehensive Wildlife (Protection) Act of 1972 and continuing through the early colonial-era laws that mainly attempted to control hunting and forest exploitation. The revision and amendment has created strong legislative framework for protecting threatened species, establishing protected areas and regulation of wildlife trade in association with international agreement like CITES. Incorporating scientific ideas into conservation initiatives, such as habitat restoration, species recovery plans, and biodiversity corridor creation, demonstrates the nation's dedication to conservation of biodiversity. Judiciary interpretation of constitutional provisions, specifically Article 48A and Article 51A(g), has also strengthened the state's and citizens' joint responsibility for protecting the environment. However there are many challenges such as habitat loss, human wildlife conflicts and illegal trade of wild animal and their products. To maintain the coexistence of people and wildlife, India's wildlife conservation strategy must integrate socio-economic factors, increase enforcement mechanisms, and promote collaborative governance. India may set an example by continuously modifying its legislative and regulatory frameworks to address new conservation demands. This evolutionary study emphasizes that wildlife conservation is essential to preserving ecological balance and ensuring the sustainability of natural resources for coming generations. Wildlife conservation is a social responsibility as well as a legislative undertaking.

References

Animal Welfare Board of India v. A. Nagaraja & Others, 7 SCC 547 (2014).

Archi Rastogi, A. R., Hickey, G. M., Ruchi Badola, R. B., & Hussain, S. A. (2014). Understanding the local socio-political processes affecting conservation management outcomes in Corbett Tiger Reserve, India. *Environmental Management, 53*(5), 913–929.

Bean, M. J. (1977). *The evolution of national wildlife law: Report to the council on environmental quality*. The Institute.

Bhupathy, S., Kumar, S. R., Thirumalainathan, P., Paramanandham, J., & Lemba, C. (2013). Wildlife exploitation: A market survey in Nagaland, North-Eastern India. *Tropical Conservation Science, 6*(2), 241–253.

Biggs, D., Cooney, R., Roe, D., Dublin, H. T., Allan, J. R., Challender, D. W., & Skinner, D. (2017). Developing a theory of change for a community-based response to illegal wildlife trade. *Conservation Biology, 31*(1), 5–12.

Bombay Environmental Action Group v. State of Maharashtra, SCC OnLineBom 1003 (1997).

Boora, S. (2023). Contradictions in conservation: India's forest (conservation) amendment bill, 2023. *International Journal of Environmental Studies, 80*(5), 1212–1215.

Bulte, E. H., & Horan, R. D. (2003). Habitat conservation, wildlife extraction and agricultural expansion. *Journal of Environmental Economics and Management, 45*(1), 109–127.

Centre for Environmental Law, WWF India v. Union of India, 10 SCC 1 (2022).

Centre for Environmental Law, WWF v. Union of India, 5 SCC 281 (1996).

Centre for Environmental Law, WWF v. Union of India, 6 SCC 548 (2004).

Chartier, L., Zimmermann, A., & Ladle, R. J. (2011). Habitat loss and human–elephant conflict in Assam, India: Does a critical threshold exist? *Oryx, 45*(4), 528–533.

Chhikara, U., & Chaudhary, P. S. (2021). Examining wildlife protection laws in India: Hiatus, gaps, flaws and amendments. *Shodh Sanchar Bulletin, 11*(41), 1–6.

Choudhary, S. (2024, May 1). *The Wildlife (Protection) Act, 1972: A comprehensive study on wildlife conservation in India*. Available at SSRN 4848859.

Chowdhury, N. (2014). Environmental risk regulation and the Indian Supreme Court: An exercise in de-formalization of the law? *Journal of Risk Research, 17*(1), 61–80.

Distefano, E. (2005). *Human-Wildlife Conflict worldwide: Collection of case studies, analysis of management strategies and good practices*. Food and Agricultural Organization of the United Nations (FAO), Sustainable Agriculture and Rural Development Initiative (SARDI), Rome, Italy. Available from: FAO Corporate Document repository http://www.fao.org/documents

Dixit, S., & Singh, K. (2024). Biodiversity conservation in ancient India with special reference to faunal diversity. *International Journal for Multidisciplinary Research, 6*(6), 1–5. https://doi.org/10.36948/ijfmr.2024.v06i06.31693

Dogra, K. S., Kohli, R. K., & Sood, S. K. (2009). An assessment and impact of three invasive species in the Shivalik Hills of Himachal Pradesh, India. *International Journal of Biodiversity and Conservation, 1*(1), 4–10.

Gautam, A., & Kaur, S. (2023). Management and conservation of wildlife in India: With special reference to Wildlife Protection (Amendment) Act, 2022. *Indian Journal of Integrated Research in Law, 3*, 1.

Ghanghas, A., & Poonia, A. (2021). Advisory for dealing with import of exotic live species in India and declaration of stock: A boom or bane for illegal wildlife trade in India. *Environmental Sciences and Ecology: Current Research, 2*, 1026.

Goa Foundation v. Union of India, 6 SCC 590 (2014).

https://www.indiacode.nic.in/bitstream/123456789/6198/1/the_wild_life_%28protection%29_act%2C_1972.pdf

Hughes, A. C. (2021). Wildlife trade. *Current Biology, 31*(19), R1218–R1224.

Jayakrishnan, R. (2023). An analysis of criminal laws against poaching animals and wildlife trafficking. *Indian Journal of Law and Legal Research, 5*(2), 1.

Johnson, M. K. (1984). Endangered Species Act amendments 1979 and 1982. *Wildlife Society Bulletin (1973–2006), 12*(3), 304–311.

Kadykalo, A. N., Cooke, S. J., & Young, N. (2021). The role of western-based scientific, indigenous and local knowledge in wildlife management and conservation. *People and Nature, 3*(3), 610–626.

Kalra, S., Davies, A., Martin, R. O., & Poonia, A. (2024). Insights from the media into the bird trade in India: An analysis of reported seizures. *Oryx, 58*(1), 69–77.

Karanth, K. K., Gupta, S., & Vanamamalai, A. (2018). Compensation payments, procedures and policies towards human-wildlife conflict management: Insights from India. *Biological Conservation, 227*, 383–389.

Karmakar, A. S. (2013). *The lacunae in wildlife protection laws in India* [Internet].

Krishnasamy, K., & Zavagli, M. (2020). *Southeast Asia at the heart of wildlife trade*. TRAFFIC, Southeast Asia Regional Office.

Kumar, A., & Verma, A. K. (2017). Biodiversity loss and its ecological impact in India. *International Journal on Biological Sciences, 8*(2), 156–160.

Kumar, A., Chellam, R., Choudhury, B. C., Mudappa, D., Vasudevan, K., Ishwar, N. M., & Noon, B. R. (2002). *Impact of rainforest fragmentation on small mammals and herpetofauna in the Western Ghats, South India*. WII-USFWS Collaborative Project Final Report, Wildlife Institute of India, Dehradun, 146 pp.

M.C. Mehta v. Kamal Nath and Others, 1 SCC 388 (1997).

Manral, U., Sengupta, S., Hussain, S. A., Rana, S. A., & Badola, R. U. (2016). Human wildlife conflict in India: A review of economic implication of loss and preventive measures. *Indian Forester, 142*(10), 928–940.

Mawdsley, J. R., O'Malley, R., & Ojima, D. S. (2009). A review of climate-change adaptation strategies for wildlife management and biodiversity conservation. *Conservation Biology, 23*(5), 1080–1089.

Meena, A. K., & Maikhuri, R. K. (2023). Evolution of environmental laws and policies in India. *A Global Journal on Humanity, VI*, 110–116.

Mitra, P. P., & Bhaumik, S. (2013). Laws for conservation of wildlife under the Indian federal structure. *Environmental, Law and Society Journal, 1*, 85.

Narayanan, S., de Wit, J., & Gupta, A. D. (2014). *Saving the Indian Tiger: A study on the effects of Wildlife (Protection) Amendment Act 2006 and National Tiger Conservation Authority* [Thesis].

Narmada Bachao Andolan v. Union of India, 10 SCC 664 (2000).

Nijman, V. (2010). An overview of international wildlife trade from Southeast Asia. *Biodiversity and Conservation, 19*(4), 1101–1114.

Pascal, A., Banerjee, D., Poojitha, D., Kalita, H., Rao, K. V., Ahmad, M., Priya, G., Chouhan, S. K., & Zeal, J. (2025). A walk-through the Wildlife (Protection) Act, 1972: Journey till now and the way ahead. *Zoo's Print, 40*(1), 69–83.

Poonia, A., Ghanghas, A., Sharma, S., Jangra, M., & Sindhu, V. (2022a). Trends in CITES listed bird's trade in South Asian countries in view of evolution of Indian laws during last four decades. *International Journal of Ecological Economics and Statistics, 43*(9), 115.

Poonia, S., et al. (2022b). Policy reform and wildlife conservation in India: An analysis of the Wild Life (Protection) Amendment Act, 2022. *Conservation and Society, 20*(3), 112–118.

PRS India. (2023). *The Forest (Conservation) Amendment Act 2023*. https://prsindia.org/billtrack/the-forest-conservation-amendment-bill-2023

Rajak, S., & Chaturvedi, K. (2023). The evolution of wildlife protection laws and policies in India: A historical analysis. *International Journal of Novel Research and Development (IJNRD), 8*(5), a354–a363.

Ramesh, M. K. (1999). The wildlife protection act, 1972 of India: An agenda for reform. *Asia Pacific Journal of Environmental Law, 4*, 271.

Rangarajan, M. (1996). The politics of ecology: The debate on wildlife and people in India, 1970–95. *Economic and Political Weekly, 31*, 2391–2409.

Roy, A., & Roy, A. (2017). Environmental conservation in ancient India. *International Journal of Sanskrit Research, 3*(4), 139–142.
Saberwal, V. K. (2000). Conservation as politics: Wildlife conservation and resource management in India. *Journal of International Wildlife Law and Policy, 3*(2), 166–173.
Saravanan, V. (2023). *Colonialism and wildlife: An environmental history of modern India*. Routledge India.
Saxena, K. B. (2024). The Forest (Conservation) Amendment Act 2023: A threat to forest lands and forest dwellers. *Social Change, 54*(1), 108–117.
Section 2 (37) Wild Life (Protection) Act. (1972).
Sekar, T., & Srivastav, A. (2024). Conserving wildlife: A vanishing asset. In *Forest policies, laws, and governance in India: Conservation challenges in the face of climate change* (pp. 281–353). Springer Nature Singapore.
Sen, A. (2021). *A political ecology of forest conservation in India: Communities, wildlife and the state*. Routledge.
Shivaji Rao Holkar v. Union of India, 4 SCC 605 (2013).
Singh, K. P. (2005). Invasive alien species and biodiversity in India. *Current Science, 88*(4), 539.
Singh, R. (2023). Legislative reform in India's wildlife laws: Bridging the gap between policy and practice. *Environmental Policy and Law, 52*(3), 121–129.
Singh, D., & Sharma, A. (2024, September 5). *Implementation of the Wildlife Protection Act, 1972: A focus on the protection of elephants*. Available at SSRN 4976716.
Soni, V. K. (2020). Analysis and evaluation of wildlife conservation laws and policies in India. *IJRAR, 7*, 4.
State of Himachal Pradesh v. State of Punjab, 4 SCC 442 (2003).
State of Tamil Nadu v. K. K. Verma, 9 SCC 515 (2009).
Sundaram, B., Krishnan, S., Hiremath, A. J., & Joseph, G. (2012). Ecology and impacts of the invasive species, Lantana camara, in a social-ecological system in South India: Perspectives from local knowledge. *Human Ecology, 40*, 931–942.
Thawrani, R., & Banait, J. (2022). An analysis of the Indian Wildlife Protection Act, 1972. *Indian Journal of Law and Legal Research, 4*(1), 1.
Thomaz, S. M., Kovalenko, K. E., Havel, J. E., & Kats, L. B. (2015). Aquatic invasive species: General trends in the literature and introduction to the special issue. *Hydrobiologia, 746*, 1–2.
Tyler, S. L. (1973). *A history of Indian policy*. Bureau of Indian Affairs.
Velmurugan, M. (2017). Historical development of wildlife protection in India. *International Journal of Current Research and Modern Education, 2*(2), 386–390.
WCCB. (2008). https://wccb.gov.in/WriteReadData/userfiles/file/Judgements/Judgement%20008.pdf
Wildlife Protection Act, 1972, No. 53, Acts of Parliament (India) (1972).
Wildlife Rescue and Rehabilitation Centre v. Union of India, 10 SCC 474 (2008).
Willmott, N. J., Wong, B. B., Lowe, E. C., McNamara, K. B., & Jones, T. M. (2022). Wildlife exploitation of anthropogenic change: Interactions and consequences. *The Quarterly Review of Biology, 97*(1), 15–35.

Chapter 4
Challenges and Opportunities in Wildlife Law Enforcement in India: Integrating Scientific Rigor for Enhanced Conservation

Asha Poonia, Priya Battu, Anil Ghanghas, Devinder Singh, and Ekashvi Thakur

Contents

A. Poonia
Faculty of Life Sciences, Chaudhary Bansi Lal University, Bhiwani, Haryana, India

P. Battu (✉)
Regeneration Biology Lab, IISER, Mohali, Punjab, India
e-mail: priyabattu@iisermohali.ac.in

A. Ghanghas
Department of Laws, Chaudhary Devi Lal University, Sirsa, Haryana, India

D. Singh
Department of Laws, Panjab University, Chandigarh, India

E. Thakur
Formerly at Department of Zoology, Panjab University, Chandigarh, India

A. Poonia et al. (eds.), *Wildlife Trade in India*,
https://doi.org/10.1007/978-981-95-3694-8_4

Abstract The Wild Life (Protection) Amendment Act, 2022 (WALP) marks a significant improvement in India's legislative framework for wildlife conservation, replacing the 1972 Act's narrow focus on the protection of individual species with a more holistic, ecosystem-based approach to biodiversity conservation, addressing critical issues such as the illegal wildlife trade, species protection, ecosystem restoration, sustainable management, zoonosis prevention, and international cooperation in wildlife conservation. The present work comprehensively evaluates the challenges and opportunities in wildlife law enforcement in India. The study undertakes a comprehensive literature review encompassing scientific research, relevant Indian legal documents, and reports from various organizations involved in wildlife conservation in India. The study also highlights the challenges in implementing wildlife law enforcement in India despite the Wildlife (Protection) Amendment Act 2022, strengthening biodiversity protection. It faces challenges in implementation due to limited institutional capacity, lack of financial resources, and procedural inefficiencies. Furthermore, the gaps in scientific rigor, particularly in the classification of invasive species and vermin, highlight the need for the adoption of internationally accepted risk assessment protocols. The study recommends harmonizing legal definitions with international standards, strengthening research and data-driven decision-making, enhancing international cooperation, and increasing investment in technological innovations. These steps would align the Act with global best practices, ensuring long-term biodiversity resilience and contributing to India's international conservation commitments.

Keywords Wildlife Law · India · Wild Life (Protection) Amendment Act, 2022 · Biodiversity conservation · CITES · Invasive species management · Wildlife crime · Ecosystem restoration · International conservation treaties · Genetic research, India · Adaptive management strategies · Vermin · Wildlife seizure · Legislative reform

4.1 Background of the Wildlife (Protection) Act, 1972

India is endowed with a plethora of wildlife but still faces striking challenges in its conservation due to habitat loss, hunting, smuggling and global warming (Rana & Kumar, 2023). Indian laws are considered very uncompromising when it comes to wildlife trade majorly due to The Wildlife (Protection) Act, 1972 which has been amended over time to meet the demands of protection of wildlife to establish a structured approach for the protection of wildlife and their habitats, addressing issues of poaching, trade, and habitat destruction (Poonia et al., 2022). Recent amendments in the act led to the creation of special areas protected by government like the wildlife sanctuaries and national parks and established a regulatory framework to govern the management of these areas (The Wild Life (Protection) Amendment Act, 2002; The Wild Life (Protection) Amendment Act, 2006; The Wild Life (Protection) Amendment Act, 2013). The act's main aims include but are not limited to providing a legislative framework aimed at safeguarding India's rich biodiversity and addressing the threats to wildlife and their habitats. The act also had provisions for protecting endangered species, setting up a system of wildlife crime control and promoting conservation through legal and institutional measures. Various acts like Fisheries act (1897), Indian Forests Act 1927, Mining and Mineral Development Regulation Act (1957), Prevention of Cruelty to Animals (1960), Wildlife Protection Act (1972), Water (Prevention and Control of Pollution) Act (1974), Forest Conservation Act (1980), Air (Prevention and Control of Pollution) Act (1981) etc. have been passed in favour of conservation of natural resources and wildlife (Srikanth et al., 2023). Similarly, Wetland (Conservation and Management) Rules, 2010 have been drafted to protect wetlands in India. Likewise, there are organisations which are set up by the Government of India for promoting research in area of wildlife conservation. Some of these institutes are the Wildlife Institute of India (WII, Uttarakhand), the Bombay Natural History Society (Mumbai, Maharashtra), etc. Despite immense conservation efforts and projects launched by the government of India, the challenges hamper the progress and success of these conservation projects and the illicit trade and abuse of wildlife continues (Srikanth et al., 2023). Hence, there is a need for collaborative efforts between communities and the government for the effective and sustainable conservation of wildlife. The Forest Rights Act (FRA), passed by the government of India in 2006, acknowledges the rights of people living in the buffer zones of forests. FRA can help in the transformation of the forest areas of India by including communities dependent on forests in conservation efforts. This can also help in achieving the goal of sustainable development (Kumar et al., 2017).

4.2 Rationale for the Wild Life (Protection) Amendment Act, 2022

The Wildlife (Protection) Act, 1972 of India is not an exhaustive act and only forbids the trade of over a few hundred species of flora and fauna and their derivatives. The act's applicability only to the trading of native animals generally excludes exotic wild animals. As a result, non-native animals are readily available for sale in pet shops or at live animal markets (Sinha, 2010). Despite the wildlife laws, wildlife resources are still abused, and many cases are still not reported. The killing of wild animals for sport, recreation, and profit, along with deforestation, has threatened our ecosystem and has resulted in extinction and loss of biodiversity in the country. The Wildlife Protection Act, Customs Act, and Import-Export policies in India have provisions for regulating the conservational measures and trade of wildlife species, especially the endangered species, however, illegal hunting and poaching activities and trade are still flourishing and these endangered species are still exploited. The elaborate conservational provisions of wild life protection act, have failed to protect the endangered/vulnerable species from exploitation. The customs acts and import-export policies have similarly been ineffective in stopping the illegal trade of wildlife (Kothari, 2006). Despite the Wildlife (Protection) Act's comprehensive framework, the increasing pressures of biodiversity loss, habitat degradation, and illegal wildlife trade necessitated updates to enhance its efficacy (Poonia et al., 2022; Wadhwa, 2022). To address these emerging challenges, contemporary conservation needs and international standards including implementation of the provisions of the Convention on International Trade in Endangered Species of Wild Flora and Fauna (CITES), of which India is an early member since 1976, the Wild Life (Protection) Amendment Bill, 2021 (the bill) was introduced in the winter session of Parliament i.e. Lok Sabha on 17 December 2021, and was passed in the Rajya Sabha on 8 December 2022 (Poonia et al., 2022). This amendment aims to strengthen the "conservation, protection and management of wildlife" by strengthening legal provisions such as enhancing the penalties for wildlife crimes and implementing CITES regulations to address gaps in previous legislation.

Global connectivity due to advancements in technology has made connections between countries like never before. This advancement has provided countless new ways to do business, including wildlife trading, whose approximate worth is between 91 and 258 billion dollars annually (Nijman, 2010; Hughes, 2021). Various studies have reported Asia as a major source and consumer of wildlife, whereas Europe is a major consumer (Hughes, 2021; Krishnasamy & Zavagli, 2020). The present evolving dynamics of globalisation, wildlife trade and climate change, while balancing the needs of local people for livelihood generation for access and benefit-sharing, make it necessary to update the existing laws to foster species diversity, ecosystem health and climate stability. The Act also reflects India's commitment to adhering to global conventions such as CITES to strengthen its claim as the leading country in wildlife conservation and in tackling climate change.

In view of the above, the present study was undertaken on the Wild Life (Protection) Amendment Act, 2022 to critically evaluate its alignment with International Standards, to evaluate its scientific rigor and impact on conservation outcomes, to identify gaps and provide recommendations for enhancing the legislative framework and its implementation based on scientific evidence and international best practices to enhancing overall biodiversity conservation.

4.3 The Wild Life (Protection) Amendment Act, 2022: Key Provisions and Changes

The Wildlife (Protection) Act, 2022, represents a significant revision of its predecessor, starting from its long title i.e. "The Wild Life (Protection) Act, 1972 for Protection of wild animals, birds and Plants" to "The Wild Life (Protection)

Amendment Act, 2022 for conservation, protection and management of wildlife". Tian et al. suggested that the name of an act plays a critical role in shaping the scope, interpretation, and implementation of conservation efforts. It also impacts how the law is understood by policymakers, conservation practitioners, and the public (Tian et al., 2023). As its title suggests, this recent legislative update is a shift in India's legal and policy framework towards a more comprehensive, ecosystem-based approach to wildlife governance. This change aligns with evolving global conservation trends, reflecting the need for sustainable management, zoonosis prevention, and international cooperation in wildlife conservation efforts by addressing gaps identified in previous legislation in alignment with contemporary international legislation. The act introduces several new key provisions and updates many existing provisions, showing substantial advancements. The Wild Life (Protection) Amendment Act of 2022 has several significant implications for wildlife conservation. Most important among them include an updated species list strengthened penalties for wildlife crimes, and enhanced Protection for Biodiversity, an updated list of protected areas and wildlife sanctuaries, first-time introduction of laws for the regulation of invasive alien species, community involvement, updated regulations for human-wildlife conflict and advocating the use of latest scientific infrastructure.

4.3.1 Update in the List of Protected Species

The Wild Life (Protection) Amendment Act, 2022 has updated the list of protected species as per the latest Red List (IUCN, 2024) data and habitats list as per the latest scientific data. Various species such as *Puntius denisonii* or Red Line Torpedo Barb (Endemic to Western Ghats; previously unprotected, now endangered) (Jain et al., 2022) *Semiplotus semiplotus* or Assamese Kingfish (Freshwater fish vulnerable due to habitat loss; Previously unprotected, now listed as endangered as per CITES list) (Jain et al., 2022). *Osteobrama belangerior* Manipur *Osteobrama* (Freshwater fish vulnerable due to habitat loss and overfishing; previously unprotected, now listed as endangered as per CITES list and IUCN status) (IUCN Red List of Threatened Species. 2010). *Ardeotis nigricepsor* Great Indian Bustard (Important bird for grassland ecosystem health; population severely declining in Rann of Kutch; previously critically endangered but not fully protected, now listed as critically endangered under stronger protection) (Srinivas et al., 2022), *Ursus arctos isabellinus* or Himalayan Brown Bear (important for Himalayan ecosystem ecology; and population declining due to habitat fragmentation and poaching; previously listed as endangered, now listed as critically endangered with strengthened protection) (Thakur et al., 2023). *Manis crassicaudata* or Indian Pangolin (Population declining due to extensive illegal trade and poaching; Important for controlling insect populations and maintaining forest ecosystem health; previously vulnerable, now listed as endangered with enhanced protection) (Gayen et al., 2024). *Platanista gangetica* or Ganges River Dolphin (fresh water mammal crucial for ecosystem health; population declining due to pollution, dam construction, and poaching;

previously listed as endangered with lesser protection, now listed as endangered under stronger protection) (Joshi, 2022) *Batagur kachugaor* or Red-Crowned Roofed Turtle (Important for riverine ecosystem health, population declining due to illegal trade and habitat degradation; previously unprotected, now under strengthened protection) (Talukdar et al., 2021) *Nilssonia leithiior* Leith's Softshell Turtle (Population decline due to habitat destruction and poaching for meat and shells; previously unprotected, now listed as critically endangered) (Gerard et al., 2023) *Heteroglaux blewittior* Forest Owlet (Population declining due to habitat shrinking; Previously critically endangered, now under Schedule I protection; mentioned in Wild Life (Protection) Act, 1972 as *Athene blewitti*) (Kulkarni et al., 2022). This expansion of the list of protected species and inclusion of CITES-listed species in different schedules depending on their level of threat and protection status is a wildly appreciated step for biological conservation.

4.3.2 Involvement of Local People in Wildlife Conservation

The Wildlife (Protection) Amendment Act, 2022, has greatly enhanced the preservation of Conservation Reserves and Community Reserves, territories where local people are directly engaged in wildlife and its habitats' protection. The Act also places considerable focus on raising community involvement in managing buffer areas around the protected areas; this not only enhances the local conservation arrangements but also contributes to the improved goodwill on wildlife management. The act recognizes the role of indigenous and local knowledge in managing protected areas and mitigating human- wildlife conflicts (Pascal et al., 2025). The Amendment also strengthened measures regarding Eco sensitive Zones around protected areas, including National Parks/Sanctuaries. In compliance with the provision of the CITES, the Act brought new legal measures for the control of wildlife trafficking (Nakamura & Kuemlangan, 2023).

4.3.3 Recognizing the Role of Science and Technology in Improving Conservation Efforts

It also improved the positions of Management and Scientific Authorities that are charged with the responsibility of making sure that trade in products does not affect protected species in conservation areas. It advocates the inclusion of the latest technology into conservation, ensuring that conservation strategies are informed by the latest scientific knowledge (Wadhwa, 2022) and strengthening the management and protection plan for the protection of threatened species and habitat restoration.

4.3.4 Greater Penalties for the Offenders

India has also increased penalties and has "tough-on-crime" sentencing against wildlife crime by this amended act. It includes increased fines and longer imprisonment terms for various offences against wildlife and protected areas.

Harsh punishment policies cannot serve as effective instruments to combat wildlife crimes for, generally, criminological research indicates that such policies do not manage to contribute to the decrease of crime rates and may even have some adverse impacts both on individuals and in the community (Akella & Cannon, 2017). The authors argue that punitive policies based upon the principle of deterrence are typically ineffective and that, instead, there is a need for evidence-based legal strategies developed based on empirical data from criminology. It further, therefore, entails an interdisciplinary approach that puts the criminological perspective into the legislation and enforcement of wildlife crime to develop more effective, evidence-based solutions (Akella & Cannon, 2017).

4.4 Importance of Scientific Rigor in Conservation Efforts

The effective conservation of wildlife in the modern era necessitates a robust framework that integrates international standards with robust scientific research and the integration of modern technological tools to inform conservation policies, understand ecological dynamics, mitigate human impact, and ensure the long-term survival of species. Conservation policies that are grounded in evidence-based practices help policymakers, scientists, and conservationists to work together to address the gaps in existing conservation frameworks. This is particularly true for megadiverse countries like India, where biodiversity faces immense pressures due to habitat loss, poaching, and climate change. The Wild Life (Protection) Amendment Act, 2022 is a crucial step toward safeguarding India's unique biodiversity, but its success hinges on aligning with international treaties and incorporating data-driven, scientific approaches. Scientific research serves as the backbone of wildlife management and conservation efforts, providing critical data to guide informed decision-making. The complexity of ecosystems, species interactions, and the impact of human activities requires an evidence-based approach to manage biodiversity effectively.

Data collected through field studies, geographic information systems (GIS), camera traps, DNA barcoding, satellite imagery, remote sensing, genetic analysis, ecological modelling, and population monitoring allows researchers to track species population trends, species behaviour, habitat quality, habitat needs, environmental interactions and estimate the consequences of conservation efforts combating illegal activities such as poaching and wildlife trafficking (Lawler et al., 2009; Pimm et al., 2014). Many countries have successfully integrated international standards and scientific research into their conservation policies, offering valuable lessons for

India. Costa Rica, for example, has implemented policies that blend international commitments with scientific rigour, leading to significant reductions in deforestation and the recovery of biodiversity in key areas (Chazdon, 2014). Similarly, the European Union's Natura 2000 network, which is based on both international conservation obligations and rigorous scientific assessments, has provided an effective model for protecting endangered species and their habitats across multiple countries (Pullin et al., 2009). The reintroduction of California condor (*Gymnogyps californianus*) into the wild via a captive breeding program coupled with intensive monitoring using satellite telemetry has increased the population of 27 individuals to over 400 individuals, is an important testament of combining genetic research, captive breeding, and advanced tracking technologies to recover critically endangered species (Carrilo, 2025). In Africa, technology such as drones equipped with thermal imaging cameras has been used in combating poaching and protecting endangered species such as elephants and rhinos in real time to provide a cost-effective way to patrol large areas of land and respond quickly to illegal activities (Mulero-Pázmány et al., 2014). DNA analysis of ivory, for instance, has been used to identify the geographic origin of poached elephants, enabling law enforcement agencies to target poaching hotspots and disrupt wildlife trafficking networks (Wasser et al., 2007).

Population viability analysis (PVA) has been successfully applied to species like critically endangered black rhinoceros (*Diceros bicornis*), Amur tiger (*Panthera tigris altaica*) and the Florida panther (*Puma concolor coryi*) to inform recovery plans and prioritize conservation actions (Thuo et al., 2015; Carroll & Miquelle, 2006; Ballou et al., 1989). Habitat restoration using scientific methods can increase gene flow and decrease the risk of local extinctions, such as for American bison (*Bison bison*) in prairies in North America (Ranglack et al., 2023). Conservation genetics can provide insights into genetic diversity, population structure, inbreeding depression, assessing the genetic health of endangered species, and guiding breeding programs in captivity such as for the cheetah (*Acinonyx jubatus*) and the African wild dog (*Lycon pictus*) (Dobrynin et al., 2015; Davies-Mostert et al., 2009). Remote sensing has been used to map the distribution of critical habitats for species such as the orangutan (*Pongo pygmaeus*) in Borneo, enabling conservationists to prioritize areas for protection and restoration (Wich et al., 2008). Camera traps have been particularly effective in monitoring elusive and nocturnal species, such as the jaguar (*Panthera onca*) and snow leopard (Panthera uncia), which are difficult to observe through traditional field surveys (Silver et al., 2004; Ugyan, 2020). Long-term camera trap monitoring in the Amazon has revealed shifts in mammal populations and rare behaviors of birds (Arana et al., 2021). DNA barcoding has been used to detect the presence of cryptic or elusive species, monitor biodiversity in aquatic ecosystems, and identify illegal wildlife products in the trade, such as for the European eel (*Anguilla anguilla*) and the Chinese giant salamander (*Andrias davidianus*) (Valentini et al., 2009) (Table 4.1).

Conservation is not restricted to only present times but should also extend to nurturing and protecting the species into the future. This could be achieved by saving genetically diverse populations. Cryobiology can help in preserving genetically

Table 4.1 Highlighting techniques and their effectiveness in conservation of concerned species

S. no.	Species concerned	Techniques used and their effectiveness	References
1.	*Gymnogyps californianus*	A captive breeding program coupled with intensive monitoring using satellite telemetry improved the declining population	Carrilo (2025)
2.	Elephants and Rhinos	Drones equipped with thermal imaging cameras helped in combating poaching and response to illegal activities against wildlife	Mulero-Pázmány et al. (2014)
3.	Elephants	DNA analysis of ivory aided in tracing the geographic origin of poached elephants to disrupt trafficking networks	Wasser et al. (2007)
4.	Black rhinoceros (*Diceros bicornis*), Amur tiger (*Panthera tigris altaica*) and the Florida panther (*Puma concolor coryi*)	Population viability analysis	Thuo et al. (2015), Carroll and Miquelle (2006), Ballou et al. (1989)
5	*Bison bison* (American bison)	Habitat restoration increases gene flow and decreases the risk of local extinctions	Ranglack et al. (2023)
6	*Cheetah* (*Acinonyx jubatus*), African wild dog (*Lycon pictus*)	Conservation genetics provides insights into genetic diversity, population structure, inbreeding depression, assessing the genetic health of endangered species, and guiding breeding programs in captivity	Dobrynin et al. (2015), Davies-Mostert et al. (2009)
7.	*Pongo pygmaeus*	Remote sensing to map the distribution of critical habitats for species helped in the prioritization of areas for protection and restoration	Wich et al. (2008)
8.	Jaguar (*Panthera onca*) and Snow Leopard (*Panthera uncia*)	Camera traps for monitoring elusive and nocturnal species that are difficult to observe using traditional field survey methods	Silver et al. (2004), Ugyan (2020)
9	Birds and mammals	Long-term camera trap monitoring revealed shifts in bird and mammal populations	Arana et al. (2021)

(continued)

Table 4.1 (continued)

S. no.	Species concerned	Techniques used and their effectiveness	References
10	European eel (*Anguilla anguilla*), Chinese giant salamander (*Andrias davidianus)*	DNA barcoding has been used to detect the presence of cryptic or elusive species, monitor biodiversity in aquatic ecosystems, and identify illegal wildlife products in the trade	Valentini et al. (2009)
11	Panthera leo, Amur leopard, Black footed cat, oncilla	Slow freezing and ovarian tissue culture in vitro	Engdawork et al. (2024), Zahmel et al. (2021), Jeong et al. (2018), Herrick et al. (2010)
12	Asian elephant (*Elephas maximus indicus*, Indian rhinoceros (*Rhinoceros unicornis*)	Semen preservation for male animals	Imrat et al. (2013), Stoops et al. (2010)

crucial germplasm and help in reviving animal populations in the future (Engdawork et al., 2024). Various methods can be adopted for preserving female gametes, like slow freezing (SF) and vitrification. SF method has been successfully combined by culturing the ovarian tissue in vitro. This method has been successfully used in animals such as African lion (*Panthera leo*), Amur leopard (*Panthera pardus orientalis*), black-footed cat (*Felis nigripes*), oncilla (*Leopardus tigrinus*) etc. (Zahmel et al., 2021; Jeong et al., 2018; Herrick et al., 2010). Similarly, semen preservation for male animals has also been done in many animals few examples are Asian elephant (*Elephas maximus indicus*) (Imrat et al., 2013) and Indian rhinoceros (*Rhinoceros unicornis*) (Stoops et al., 2010). India can draw on these examples to further strengthen its conservation framework. By fully integrating international standards, and by leveraging scientific tools and methodologies, India can ensure that the Wild Life (Protection) Amendment Act, 2022 is not only a legal instrument but also a scientifically robust tool for protecting its biodiversity, particularly in the management of targeted protected species, invasive species, and vermin. This is particularly important in India, where invasive species like *Lantana camara* have caused significant ecological damage (Sharma et al., 2007). The scientific community has emphasized the need for rigorous environmental impact assessments and long-term monitoring to develop effective management strategies for invasive species. The incorporation of these scientific principles into the Wild Life (Protection) Amendment Act would enable policymakers to make data-driven decisions that mitigate ecological risks.

The captive breeding programmes for the conservation of wildlife have also shown success. A programme was launched by the government of India to preserve the declining populations of vultures. The use of Diclofenac in animals for treating pain has been implicated in mortality and the eventual decline in vulture populations in India. Through the collaborative efforts of the Vulture Conservation Breeding

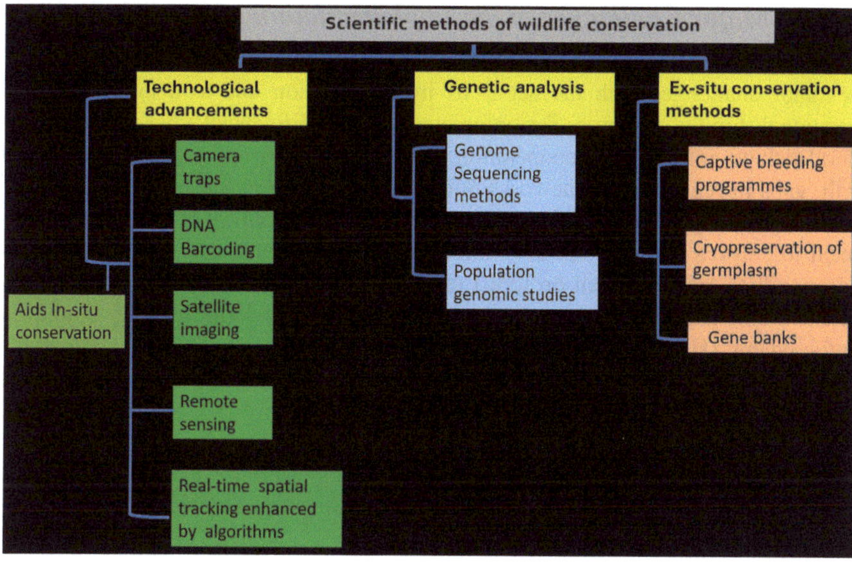

Fig. 4.1 Scientific methods for conservation of wildlife

Centre (VCBC), Haryana, Haryana Forest Department and the Bombay Natural History Society (BNHS), Maharashtra, the dwindling population of three crucial vulture species, namely the White-backed vulture, Long-billed vulture and Slender-billed vulture, have been slowly recovering. As a result of the efforts of the Centre, the use of Diclofenac was banned in India. Studies have shown that the use of drugs like meloxicam, which is not harmful and is readily metabolised in vultures, is a good substitute for diclofenac (Adawaren et al., 2019). However, the vials meant for human use were still being diverted for veterinary use (Galligan et al., 2021). To prevent the use of diclofenac in animals, vials larger than 3 ml were banned but illegal vial manufacturing continues. Hence, strict monitoring and litigation against the miscreants is necessary (Fig. 4.1).

4.5 Challenges and Opportunities

The Wildlife (Protection) Act, 2022 is a pivotal effort to address some of the legislative gaps identified in earlier versions of the Act in alignment with international agreements such as CITES and the CBD, enhanced provisions for wildlife protection, combating wildlife crime and strengthening conservation efforts, and its success will depend on effective implementation and continuous adaptation to evolving scientific and socio-economic contexts. The Act requires further refinement due to discrepancies in the integration of various global standards which need to be addressed to ensure effective compliance and international cooperation.

4.5.1 Major Challenges in the Implementation of Wildlife Act

A major challenges with the act is its implementation due to a lack of trained personnel, limited resources for enforcement, limited financial investment, procedural delays, inter-agency cooperation, coordination, corruption, lack of political will, gaps in regulatory frameworks & enforcement mechanisms, insufficient institutional capacity in terms of manpower, skills, training, political interference, equipment to enforce conservation laws effectively, jurisdictional ambiguities, the local communities are dependent on forest resources, resources, rapid urbanization, different monitoring mechanisms across different states of India leading to uneven enforcement (Haukos, 1997; Karanth et al., 2012; Warchol, 2004; Ferraro & Pattanayak, 2006; Karanth et al., 2013; Ariffin, 2015; Norouzzadeh et al., 2018; Puri et al., 2022).

4.5.1.1 Contrasting Conservation Policies of Developed Nations and Developing Nations

One major issue is the discrepancy between the broad, globally applicable guidelines set by these treaties and the distinct socio-economic, political, and environmental realities of individual developing countries, like India, which is facing particular difficulties in adapting these standards to their legislative frameworks in its expansive and diverse ecological landscapes, while addressing national priorities, such as poverty alleviation and economic development, making consistent monitoring and enforcement logistically difficult, requiring a large and well-equipped workforce that is often unavailable especially in remote conservation hotspots (Ferraro & Pattanayak, 2006; Karanth et al., 2013; Rodrigues et al., 2004).

4.5.1.2 Dilemma of Vermin and Invasive Species

Moreover, classifying invasive species and vermin lacks sufficient scientific rigour which may lead to ineffective management strategies, conflicts with local communities who may view certain species as pests or threats to agriculture and unintended consequences for biodiversity (Puri et al., 2022) and there is an urgent need to adopting standardized international criteria and establishing independent scientific review for this. While the definition of invasive alien species in act aligns with the definitions provided by the IUCN and CBD but Vermin does not. The designation of species as vermin can have significant implications for their protection (often exempt from legal protections and conservation, neglecting their ecological roles and conservation needs, leading to increased persecution, habitat destruction and decline in population while disrupting ecosystem balance and negatively affecting other interacting/ relaying species) (Raj & Victor, 2022). So, due care is needed to be taken while giving this classification. Hence, invasive species and vermin designation can be given as per international standards outlined by the International Union for Conservation of Nature (IUCN) and the Convention on Biological Diversity (CBD) using standardized risk assessment protocols by considering ecological, economic, and social impacts (IUCN, 2020; CBD, 2021) while balancing species protection with practical considerations of human-wildlife interactions and agricultural impacts (IUCN, 2018). Another important aspect is to establish independent scientific review panels, including experts in ecology, taxonomy, and conservation science, to assess and recommend species for inclusion in invasive or vermin categories based on comprehensive and up-to-date scientific knowledge.

4.5.1.3 Incompetent Enforcement Due to Lack of Accountability, Workforce and Resources

To enhance the legitimacy of the designation, a transparent public consultation process may be followed for proposed designations to gather diverse perspectives (Raj & Victor, 2022). The implementation of international standards often requires specialized knowledge and resources, both of which are limited in many parts of India (Karanth & DeFries, 2010). Furthermore, the scientific community faces challenges in translating complex ecological data into actionable policies that can be easily understood and implemented by policymakers. The Act's provision for increased penalties and stricter controls is a positive step. Still, without adequate infrastructure and capacity-building efforts, these legal changes may not translate into tangible conservation outcomes. The Act also falls short of fully aligning with key international frameworks.

4.5.1.4 Inadequately Trained Forest Staff

On the ground level, this act has few lacunae; the inadequacy in the forest officer's training and coordination with the forest department leads to poor implementation. This has further increased exploitation activities and made law implementation difficult (MOEF, 2009).

4.5.1.5 Lack of Coordination Among the Forest and Police Departments

Although the main reason behind the ineffective implementation of wildlife law in India is improper training of forest department officials and lack of resources (Ballabh et al., 2002). There is also a lack of coordination between the forest and police departments. A considerable gap in the availability of wildlife crime data in India, results in laches during the detection of offenders. Two agencies involved in enforcing WPA have different methods of working, which causes impediments while enforcing laws smoothly, and non-coordination between the agencies regarding the data further hinders the prosecution of the criminals. The data on wildlife crime registered by the police is compiled in Crime in India statistics, whereas data from the Forest Departments is unavailable in the public domain. There is an obvious need to set up effective, prompt and integrated wildlife crime data from the police as well as the Forest Departments (Sharm A).

4.5.1.6 Negligence of Scientific Opinions and Infrastructure

One of the most important areas where The Wildlife (Protection) Act, 2022 is lacking is not giving due importance to robust scientific approaches including risk assessment models and protocols (in line with the Risk Analysis Framework of IUCN (Puri et al., 2022; IUCN, 2000) effective monitoring mechanisms and surveillance programs (for early detection and rapid response including online monitoring for wildlife trade) (Roy & Kumar, 2024).

4.5.1.7 Lack of Coordination Between State and Central Authorities

Moreover, different states and agencies involved with wildlife crime use different data collection and reporting mechanisms, leading to unrealizable data. Hence, consistent scientific protocol inclusion can increase evidence-based decision-making and effective management of wildlife resources (Puri et al., 2022) by improving coordination between central and state authorities (Puri et al., 2022). Regular transparent reporting with detailed information on progress, challenges, and outcomes can improve accountability and stakeholder engagement (MoEFCC, 2022).

4.5.1.8 Exclusion of Marine Species

The act focuses more on inland species, and due recognition has not been given to marine biodiversity. A more comprehensive inclusion of marine invertebrates and reef-building corals in Schedule I to restrict unsustainable trade is urgently required. There is also need of better enforcement mechanisms to curb illegal trade at ports and coastal areas. There remain other critical gaps in various addressing issues such as biosafety, plant genetic resources, marine conservation, and forest invasive species and various international treaties which could not find any place in the Act including International Whaling Commission (IWC) (IWC, 1946), International Network for Bamboo and Rattan (INBAR) (INBAR, 1997), Asia-Pacific Forestry Commission (FAO) (APFC, 1949) and Asia-Pacific Forest Invasive Species Network (APFISN) (APFISN, 2004). Future amendments should prioritize these areas to ensure comprehensive alignment with India's international environmental commitments.

4.5.1.9 Unproficient Jurisprudence for Transnational Wildlife Crime

The International Consortium on Combating Wildlife Crime (ICCWC) is the largest global partnership in combating transnational wildlife crime, which is often linked to organized crime, money laundering, and corruption. The Present Act has no provisions targeting the financial aspects of wildlife crime such as following the ICCWC Financial Investigation Guide. ICCWC offers a comprehensive framework for assessing national responses to wildlife and forest crimes through the Wildlife and Forest Crime Analytic Toolkit, to provide a systematic approach to evaluating law enforcement, legal frameworks, prosecution, and cross-border cooperation and ICCWC Indicator Framework, for assessing the effectiveness of a country's wildlife crime enforcement efforts. Despite its proven effectiveness in other jurisdictions, the 2022 Amendment does not incorporate this toolkit and indicator framework into India's wildlife crime prevention efforts. Including this toolkit would improve the ability of Indian law enforcement agencies to detect, investigate, and prosecute wildlife crimes, particularly those with international dimensions. ICCWC Indicator Framework covers areas such as law enforcement capacity, prosecution effectiveness, and judicial outcomes. Not including this shows India lacks a standardized mechanism to evaluate the success of its wildlife crime policies. Moreover, the Act lacks formal cooperation mechanisms with key ICCWC partners for global intelligence-sharing and law enforcement coordination, especially in cases of transnational organized crime. In particular, cooperation with the World Customs Organization (WCO) is crucial for addressing wildlife trafficking through international borders. The act also lacks explicit provisions for addressing the impacts of global warming and climate change on wildlife and ecosystems especially for species which are particularly vulnerable to climate change such as Coral reefs, mangroves and Himalayan ecosystems.

4.5.2 Opportunities in Wildlife Law Enforcement

While significant challenges exist in implementing international conservation standards, numerous opportunities also present themselves, particularly with advancements in scientific research, which require a multifaceted approach. Technological advancements, such as AI-driven data analytics, machine learning and real-time monitoring systems, and ecological and population modelling to inform conservation strategies offer new ways to enhance conservation efforts (IUCN, 2024). Some of these have been successfully applied to species such as the Bengal tiger (*Panthera tigris*), the Asiatic lion (*Panthera leo persica*), Indian Rhino (*Rhinoceros unicornis*) and to track deforestation in the Western Ghats in India (Karanth & DeFries, 2010; Mondol et al., 2013; Singh et al., 2022; Nabeshima et al., 2022). These tools can help bridge the gap between scientific research and policy implementation by providing clear, actionable insights that can inform decision-making processes (Worm et al.,

2006). Furthermore, the creation of conservation "evidence banks," where scientists and policymakers can access a repository of data on past conservation interventions, is another promising development that could help streamline decision-making processes (Pearson, 2024). India needs to invest in building the expertise of their conservation professionals to strengthen institutional capacity through training and capacity-building initiatives, including training in the use of advanced technologies (Biggs et al., 2017). International collaboration can play a pivotal role in this regard, with international cooperation to enhance national conservation capacities (King et al., 2023). Improving governance, reducing corruption and increasing the participation of local communities is another critical strategy (Roe & Booker, 2019). Transparent and accountable governance structures are essential for the effective enforcement of wildlife laws and the implementation of international agreements, while integrating traditional ecological knowledge (TEK) with modern scientific approaches can provide a more holistic understanding of ecosystems and species. The harmonization of laws and clearer jurisdictional frameworks consistent across states with the role of state and federal agencies is delineated to avoid the legal ambiguities that often delay conservation actions and would streamline the enforcement process.

Anti-corruption measures, including the use of technology for real-time monitoring and reporting of illegal activities, can help reduce the influence of criminal networks on conservation efforts (Warchol, 2004). In India, the introduction of digital platforms for reporting wildlife crimes, the recognition of community forest rights under the Forest Rights Act (2006), and the use of mobile applications for real-time data collection have shown promise in reducing illegal activities and improving enforcement efficiency (Lee & Wolf, 2018). Initiatives such as the South Asia Wildlife Enforcement Network (SAWEN) have been instrumental in facilitating cooperation among countries in the region, leading to the successful interception of illegal wildlife consignments (Singh et al., 2022).

The Wildlife (Protection) Act, 2022, has very significant implications for the domestic and international trade of wildlife due to stricter regulations on wildlife trade affecting various lucrative businesses, including pet trade, luxury goods, trophies and traditional medicine. This will not only negatively affect communities involved traditionally in such trade (King et al., 2023) but also questions their livelihoods (Roy & Kumar, 2024). It may lead to a shift in illegal trade routes and practices to avoid penalties and punishments, making use of global geopolitical tensions, conflicting national interests, and differing conservation priorities among neighboring countries, highlighting the need for greater regional cooperation in enforcement (Shepherd & Nijman, 2008). For instance, the lack of cooperation among neighboring countries in Southeast Asia has resulted in the porous border of India, Nepal, and Bangladesh, which facilitates illegal wildlife trade despite CITES regulations, making it difficult for individual nations to combat the issue unilaterally (Shepherd & Nijman, 2008). Hence, alternative livelihoods, adaptive management strategies, latest technology use such as digital tracking systems, remote sensing, geographic information systems (GIS) and proactive international cooperation by strengthening collaborative networks such as the South Asia Wildlife

Enforcement Network (SAWEN) is the need of the hour to prevent this shift (Puri et al., 2022; Singh et al., 2022). Furthermore, the challenge of integrating traditional practices and local communities into the implementation of international standards is often overlooked while they offer promising avenues for enhancing conservation not only in data collection but also in empowering local populations, creating a sense of ownership over conservation efforts (Esmail et al., 2023). Many indigenous and local communities rely on wildlife resources for their livelihoods. International conservation policies often fail to consider these socio-economic dependencies, leading to conflict between local communities and wildlife conservation authorities (Esmail et al., 2023). This issue is prominent in India, where indigenous populations have deep cultural connections with wildlife but are often marginalized in conservation discussions (Agrawal & Redford, 2009). In India, community-led conservation initiatives have been successful in conserving species such as the snow leopard (*Panthera uncia*) through the involvement of local herders and pastoralists in monitoring and anti-poaching activities (Mishra et al., 2003).

India can also take advantage of best practices followed internationally for comprehensive assessment of ecological, social, and economic impacts (Yadav & Pallissery, 2023) by developing an Integrated Conservation Framework for enhancing the effectiveness of the Wildlife (Protection) Act, 2022. One among them is European Council Directives, particularly the Habitats Directive (92/43/EEC) and the Birds Directive (2009/147/EC) which uses a network of protected areas (Natura 2000 network) and conservation measures by European Commission, 2020 by creating a similar network of protected areas and adaptive management to enhance conservation outcomes in India. India can also develop an integrated conservation framework that combines habitat protection, species management, and sustainable use which can enhance the effectiveness of the Wildlife (Protection) Act, 2022 (Zisenis, 2017). WCCB shall also take steps to increase public access to monitoring and reporting data by providing data on wildlife crimes available to researchers, conservationists, and the general public to enhance transparency and facilitate collaborative conservation efforts (IUCN, 2024) (Fig. 4.2).

4.6 Recommendations for Enhanced Conservation

The WLPA, 2022, is a progressive piece of legislation, yet several steps can be taken to ensure it aligns more effectively with international standards with scientific rigor for modern wildlife conservation, providing the data, evidence, and frameworks needed to design effective management plans. To strengthen these, the following policy recommendations should be considered:

Fig. 4.2 Challenges in wildlife conservation

4.6.1 Harmonization of Legal Definitions, Terminology and Adaptive Management Strategies

One of the primary steps towards aligning the WLPA with international standards is harmonizing the definitions and terminologies used within the Act with those accepted globally. For instance, the definitions of terms like "invasive species," "endangered species," and "wildlife crime" should reflect those adopted by CITES and the IUCN to ensure consistent enforcement and interpretation. This step would also support cross-border enforcement and collaborations more effectively. Greater emphasis is needed on enforcing the existing CITES Appendix listings for species in trade, which can be accomplished by automated protection status of CITES listed species without further change in Indian legislation, creating clear legal pathways for managing confiscated wildlife and tightening border control measures.

4.6.2 Introducing Stricter Accountability for Agencies and Involving Indigenous and Local Communities

To align with the Nagoya Protocol on Access and Benefit Sharing, stricter penalties should be introduced for violations involving genetic resources or biopiracy while ensuring that benefits are equitably shared with the communities that serve as custodians of biodiversity. The government agencies working in the field of wildlife crime should also be held accountable for wildlife trade in the area to stop procedural delays & corruption, increase inter-agency cooperation, for strengthening enforcement mechanisms. The positive outcome of the conservation efforts of The Greater Adjutant Stork (*Leptoptilos dubius*), locally known as Hargila (the bone swallower), is a thriving example of wildlife conservation through community endeavour, especially women. This project was initiated in 2007, and a link was established between storks and Assamese traditions and culture. This project not only helped in conserving the Hargila population but also helped in the capacity building of local communities. Under this project, various schemes to enhance the livelihood of nest tree owners and villagers were introduced and specifically did not have any provision for cash incentives. The increase in nests from 28 in 2007–08 nests to 208 nests in the 2019–2020 breeding season in rural areas of Dadara, Pachariya, and Singimari in Kamrup District in Assam makes the success of the involvement of the community in wildlife conservation evident (Barman et al., 2020).

4.6.3 Establishing a Transparent National Wildlife Crime Database

As part of compliance with international law enforcement, India should make data on wildlife crimes public through WCCB to enhance coordination between federal and state enforcement agencies, aid international collaboration to inform the general public, policy makers and researchers.

4.6.4 Enhanced Research Monitoring Mechanisms and Data-Driven Decision Making

The WLPA can support the establishment of biodiversity monitoring programs that employ the latest technology and big data analytics to collect real-time data on wildlife populations, species migration and habitats with regular assessments of endangered species, ecosystem health, and human-wildlife conflict patterns, allowing adaptive management practices based on new findings. One such example is the real-time monitoring of African elephants in Samburu, Laikipia, Mt. Kenya, Chyulu Hills and Mara ecosystems of Kenya and the Kruger-Limpopo ecosystem in

South Africa. The animal-attached tracking units help in the telemetry of positional data of animals. The amalgamation of algorithm-based analytical capability presents opportunities for ecological monitoring and the conservation of wildlife. Proximity, geofencing, movement rate, and immobility are four algorithmic approaches that can be used to monitor various aspects of wildlife spatial activity and behaviour that are not possible with conventional tracking systems. Measuring physiological parameters like the heart rate of animals can reveal spatially distinct metabolic and energy expenditures of animals (Hetem et al., 2025; Cooke et al., 2004). In real-time, this simple physiological information can be of great use in designing conservation frameworks and monitoring animal mortality due to poaching. For example, the absence of a pulse in wild animals can help to nab the poachers in time. Hence, telemetry of physiological parameters like heart rate, body temperature and mobility data can assist in research and management of health, the spread of disease and animal energetics (Cooke et al., 2004).

4.6.5 Genetic Research and Conservation

Genetic research plays a critical role in conserving endangered species by identifying genetic diversity, population structure, and levels of inbreeding. Programs like ex-situ conservation and captive breeding could benefit from stronger legal frameworks supporting genetic studies. For example, DNA fingerprinting can aid in identifying individuals involved in illegal trade, while genomic studies can inform species recovery programs. Population genomics studies can help expedite conservation efforts by identifying particular genetic variants and loci implicated in inbreeding depression and adaptation to changing environmental conditions. Such studies can help in estimating the evolving and adaptive ability of populations in response to environmental change and management of adaptive variations (Hohenlohe et al., 2021).

4.6.6 Strengthening International and Regional Conservation Networks

India is already a part of global and regional conservation networks, but these relationships can be further deepened by fostering partnerships with international and regional conservation organizations, with a focus on sharing research findings, enforcement strategies, and best practices. India could also negotiate bilateral and multilateral agreements with neighboring countries for the establishment of transboundary protected areas along borders shared with Nepal, Bhutan, and Bangladesh to help safeguard migratory species such as elephants and tigers. The Greater Mekong Subregion, for example, offers a successful model of such transboundary cooperation for conservation.

4.6.7 Capacity Building, Knowledge Transfer, Funding and Resource Sharing

Collaborative programs can focus on capacity building and knowledge transfer between countries with advanced wildlife management systems and those facing greater challenges. Training Indian wildlife rangers in international techniques for anti-poaching operations, biodiversity monitoring, and habitat restoration could significantly enhance conservation efforts. International collaboration can also help bridge the financial gap often seen in conservation efforts.

4.7 Conclusion

India is a respected country for wildlife Laws, and the Wildlife (Protection) Amendment Act of 2022 is another great step in this direction. While the enforcement of WLPA faces challenges, depending on the extent to which it incorporates global best practices and scientific insights, particularly in balancing local needs with international standards, the opportunities for leveraging scientific research, technology, and international collaborations are abundant, coupled with strong governance and community involvement. Through focused capacity-building, legal harmonization, and the integration of scientific rigor, India can ensure that its wildlife protection laws not only meet but also exceed global expectations, solidifying its role as a leader in global biodiversity conservation. The study suggests the Integration of Advanced Scientific Techniques, strengthening International Cooperation, Public Engagement and Transparency, and continuous refinement of policies and legislative frameworks to address emerging conservation challenges for balancing conservation with developmental needs. Hence, the act shall act as a bridge to ensure that developmental activities with conservation goals to fulfil sustainable development objectives for the benefit of both wildlife and human populations using adaptive management for India to achieve a balanced approach that promotes both conservation and sustainable development. The study also suggests incorporating the ICCWC Wildlife and Forest Crime Toolkit, establishing formal cooperation with ICCWC partners and integrating the ICCWC Indicator Framework for assessing enforcement effectiveness along with making financial investigations targeting wildlife crime networks a priority for further improvement. More structured approaches are needed to meet CBD's Post-2020 Global Biodiversity Framework and Aichi Targets, especially in areas like marine biodiversity, coral reef health, and ecosystem services. The integration of climate change adaptation strategies in line with the Kyoto Protocol, Doha Amendment, Copenhagen Accord, and Paris Agreement are essential for achieving long-term biodiversity conservation in the face of global environmental challenges. Further improvements are needed in areas such as cross-border enforcement and enhanced collaboration with organizations like UNODC, INTERPOL, and the World Customs Organization (WCO). The Act should focus on strengthening intelligence-sharing, joint operations and tackling the financial underpinnings of wildlife crime to align with international best practices. By aligning the Wild Life (Protection) Amendment Act, 2022, with global treaties and harnessing the power of scientific research, India can set a benchmark for other biodiversity-rich nations. It has been observed that the punishment imposed for illicit acts under the Act is inadequate to curb wildlife abuse. Often, the culprits easily get away by just paying fines, and those who are litigating remain unbothered due to sluggish judicial procedures in India. Wildlife is crucial for a healthy ecosystem, and the imposition of complete restrictions on poaching and killing of animals with strict law implementation with the use of scientific rigor and community-incorporated conservation efforts is a dire necessity.

References

Adawaren, E. O., Mukandiwa, L., Chipangura, J., Wolter, K., & Naidoo, V. (2019). Percentage of faecal excretion of meloxicam in the Cape vultures (Gyps corprotheres). *Comparative Biochemistry and Physiology Part C: Toxicology & Pharmacology, 215*, 41–46.

Agrawal, A., & Redford, K. (2009). Conservation and displacement: An overview. *Conservation and Society, 7*(1), 1.

Akella, A. S., & Cannon, J. B. (2017). Strengthening the weakest links: Strategies for improving the enforcement of environmental laws globally. In *Transnational environmental crime* (pp. 459–492). Routledge.

Arana, A., Esteves, J. F., Arana, C., Carrilo-Percastegui, S., Powell, G. V. N., Tobler, M. W., Zuniga, F., Bowler, M., & Salinas, L. (2021). *Long-term camera trap monitoring of understory birds in the Peruvian Amazon reveals rarely registered behaviors*. Available from https://www.researchgate.net/publication/360939952_Monitoring_report_The_Status_and_Distribution_of_Snow_leopard_Panthera_uncia_through_Camera_Trapping_in_the_Northern_part_of_Reserve_Securing_a_Future_for_JKSNR_Mountain_Ghost?channel=doi&linkId=6294b86a6886635d5caf0778&showFulltext=true. Accessed on 8.4.2025.

Ariffin, M. (2015). Enforcement against wildlife crimes in West Malaysia: The challenges. *Journal of Sustainability Science and Management, 10*(1), 19–26.

Ballabh, V., Balooni, K., & Dave, S. (2002). Why local resources management institutions decline: A comparative analysis of van (forest) panchayats and forest protection committees in India. *World Development, 30*(12), 2153–2167.

Ballou, J. D., Foose, T. J., Lacy, R. C., & Seal, U. S. (1989). *Florida panther (Felis concolor coryi) population viability analysis and recommendations*. Captive Breeding Specialist Group, Species Survival Commission, IUCN.

Barman, P. D., Sharma, D. K., Cockrem, J. F., Malakar, M., Kakati, B., & Melvin, T. (2020). Saving the greater adjutant stork by changing perceptions and linking to Assamese traditions in India. *Ethnobiology Letters, 11*(2), 20–29.

Biggs, D., Cooney, R., Roe, D., Dublin, H. T., Allan, J. R., Challender, D. W., & Skinner, D. (2017). Developing a theory of change for a community-based response to illegal wildlife trade. *Conservation Biology, 31*(1), 5–12.

Carrilo, I. (2025). *Return of the California Condor*. Available from https://earthjournalism.net/stories/return-of-the-california-condor. Accessed on 8.4.2025.

Carroll, C., & Miquelle, D. G. (2006). Spatial viability analysis of Amur tiger Panthera tigris altaica in the Russian Far East: The role of protected areas and landscape matrix in population persistence. *Journal of Applied Ecology, 43*(6), 1056–1068.

CBD (2021) https://www.cbd.int/doc/decisions/cp-mop-10/booklets/cp-mop-10-decision-booklet-en.pdf

Chazdon, R. L. (2014). Second growth: The promise of tropical forest regeneration in an age of deforestation. In *Second growth*. University of Chicago Press.

Cooke, S. J., Hinch, S. G., Wikelski, M., Andrews, R. D., Kuchel, L. J., Wolcott, T. G., & Butler, P. J. (2004). Biotelemetry: A mechanistic approach to ecology. *Trends in Ecology & Evolution, 19*(6), 334–343.

Davies-Mostert, H. T., Mills, M. G., & Macdonald, D. W. (2009). A critical assessment of South Africa's managed metapopulation recovery strategy for African wild dogs. In *Reintroduction of top-order predators* (pp. 10–42). Blackwell Publishing.

Dobrynin, P., Liu, S., Tamazian, G., Xiong, Z., Yurchenko, A. A., Krasheninnikova, K., Kliver, S., Schmidt-Küntzel, A., Koepfli, K. P., Johnson, W., & Kuderna, L. F. (2015). Genomic legacy of the African cheetah, Acinonyx jubatus. *Genome Biology, 16*, 1–20.

Engdawork, A., Belayhun, T., & Aseged, T. (2024). The role of reproductive technologies and cryopreservation of genetic materials in the conservation of animal genetic resources, a review. *Ecological Genetics and Genomics, 31*, 100250.

Esmail, N., McPherson, J. M., Abulu, L., Amend, T., Amit, R., Bhatia, S., Bikaba, D., Brichieri-Colombi, T. A., Brown, J., Buschman, V., & Fabinyi, M. (2023). What's on the horizon for community-based conservation? Emerging threats and opportunities. *Trends in Ecology & Evolution, 38*(7), 666–680.

Ferraro, P. J., & Pattanayak, S. K. (2006). Money for nothing? A call for empirical evaluation of biodiversity conservation investments. *PLoS Biology, 4*(4), e105.

Galligan, T. H., Mallord, J. W., Prakash, V. M., Bhusal, K. P., Alam, A. S., Anthony, F. M., Dave, R., Dube, A., Shastri, K., Kumar, Y., & Prakash, N. (2021). Trends in the availability of the vulture-toxic drug, diclofenac, and other NSAIDs in South Asia, as revealed by covert pharmacy surveys. *Bird Conservation International, 31*(3), 337–353.

Gayen, D., Bhattacharya, T., Banerjee, S., Chatterjee, A., & Chaudhuri, S. (2024). A review on the ecology, conservation and current research priorities of Indian pangolins Manis crassicaudata in South Asia with special emphasis to India. In *Proceedings of the Zoological Society* (Vol. 77, No. 1, pp. 21–34). Springer India.

Gerard, J. D., Veerappan, D., Christopher, P., & Fritz, U. (2023). New records of Leith's softshell turtle (Nilssonia leithii, Gray 1872) after 30 years from Kaveri River delta in Tamil Nadu, India: Nilssonia leithii resurfaces in Tamil Nadu. *Hamadryad, 40*(1 and 2), 24–28.

Haukos, D. A. (1997). Expanding partnerships in conservation (book review). *Journal of Wildlife Management, 61*, 976–978.

Herrick, J. R., Campbell, M., Levens, G., Moore, T., Benson, K., D'Agostino, J., West, G., Okeson, D. M., Coke, R., Portacio, S. C., & Leiske, K. (2010). In vitro fertilization and sperm cryopreservation in the black-footed cat (Felis nigripes) and sand cat (Felis margarita). *Biology of Reproduction, 82*(3), 552–562.

Hetem, R. S., Haylock, K. A., Boyers, M., Parrini, F., Owen-Smith, N., Beytell, P., & Strauss, W. M. (2025). Integrating physiology into movement ecology of large terrestrial mammals. *Journal of Experimental Biology, 228*(Suppl_1), JEB248112.

Hohenlohe, P. A., Funk, W. C., & Rajora, O. P. (2021). Population genomics for wildlife conservation and management. *Molecular Ecology, 30*(1), 62–82.

Hughes, A. C. (2021). Wildlife trade. *Current Biology, 31*(19), R1218–R1224.

Imrat, P., Suthanmapinanth, P., Saikhun, K., Mahasawangkul, S., Sostaric, E., Sombutputorn, P., Jansittiwate, S., Thongtip, N., Pinyopummin, A., Colenbrander, B., & Holt, W. V. (2013). Effect of pre-freeze semen quality, extender and cryoprotectant on the post-thaw quality of Asian elephant (Elephas maximus indicus) semen. *Cryobiology, 66*(1), 52–59.

IUCN. (2000). *IUCN guidelines for the prevention of biodiversity loss caused by alien invasive species*. Prepared by the SSC Invasive Species Specialist Group. Approved by the 51st Meeting of the IUCN Council, Gland Switzerland, February 2000.

IUCN. (2018). *Guidelines for invasive species planning and management on islands*. IUCN. Available from https://portals.iucn.org/library/sites/library/files/documents/2018-030-En.pdf. Accessed on 11.4.2025.

IUCN. (2020). https://portals.iucn.org/library/sites/library/files/documents/2021-042-En.pdf

IUCN. (2024). *Embracing technology for improved wildlife conservation.* Available from https://iucnsos.org/embracing-technology-for-improved-wildlife-conservation/. Accessed on 11.4.2025.

Jain, A. K., Mercy, T. V., & Jain, A. (2022). Issues on the inclusion of Puntius denisonii (Day), a freshwater ornamental fish of global value, as schedule-I species under the Wild Life (Protection) Amendment Act, 2021 of India. *Frontiers in Marine Science, 8*, 944680.

Jeong, D. H., Kim, J. H., & Na, K. J. (2018). Characterization and cryopreservation of Amur leopard cats (Prionailurus bengalensis euptilurus) semen collected by urethral catheterization. *Theriogenology, 119*, 91–95.

Joshi, B. D. (2022). Ambient ecological conditions of the Gangetic dolphin-Platanista gangetica of river ganga, between Varanasi to Farakka. In *Advances in animal experimentation and modeling* (pp. 253–266). Academic Press.

Karanth, K. K., & DeFries, R. (2010). Conservation and management in human-dominated landscapes: Case studies from India. *Biological Conservation, 143*(12), 2865–2869.

Karanth, K. K., Gopalaswamy, A. M., Defries, R., & Ballal, N. (2012). Assessing patterns of human-wildlife conflicts and compensation around a central Indian protected area. *PLoS One, 7*(12), e50433.

Karanth, K. K., Gopalaswamy, A. M., Prasad, P. K., & Dasgupta, S. (2013). Patterns of human—Wildlife conflicts and compensation: Insights from Western Ghats protected areas. *Biological Conservation, 166*, 175–185.

King, J. S., Manning, J., & Woodward, A. (2023). In this together: International collaborations for environmental and human health. *Journal of Law, Medicine & Ethics, 51*(2), 271–286.

Kothari, A. (2006). Community conserved areas: Towards ecological and livelihood security. *Parks, 16*(1), 3–13.

Krishnasamy, K., & Zavagli, M. (2020). *Southeast Asia: At the heart of wildlife trade* (TRAFFIC ed.). Southeast Asia Regional Office.

Kulkarni, J., Mehta, P., & Sajan, S. (2022). *Ecological factors influencing site occupancy of the Forest Owlet (Athene blewitti) in East Kalibhit Forests, Madhya Pradesh, India.* Listed in UGC-CARE, 1141.

Kumar, K., Singh, N. M., & Rao, Y. G. (2017). Promise and performance of the Forest Rights Act. *Economic & Political Weekly, 52*, 25–26.

Lawler, J. J., Shafer, S. L., White, D., Kareiva, P., Maurer, E. P., Blaustein, A. R., & Bartlein, P. J. (2009). Projected climate-induced faunal change in the Western Hemisphere. *Ecology, 90*(3), 588–597.

Lee, J. I., & Wolf, S. A. (2018). Critical assessment of implementation of the Forest Rights Act of India. *Land Use Policy, 79*, 834–844. https://doi.org/10.1016/j.landusepol.2018.08.024

Mishra, C., Allen, P., McCarthy, T. O., Madhusudan, M. D., Bayarjargal, A., & Prins, H. H. (2003). The role of incentive programs in conserving the snow leopard. *Conservation Biology, 17*(6), 1512–1520.

MOEF. (2009). *Meeting on Institution building for managing our environment & forests August 18, 2009.* New Delhi Ministry of Environment & Forests, Government of India. Available from https://moef.gov.in/uploads/2019/06/procd_FP_meeting.pdf. Accessed on 8.4.2025.

MoEFCC. (2022). *Annual report on wildlife protection.* Ministry of Environment, Forest and Climate Change.

Mondol, S., Bruford, M. W., & Uma, R. (2013). Demographic loss, genetic structure and the conservation implications for Indian tigers. *Proceedings of the Royal Society B: Biological Sciences.* https://doi.org/10.1098/rspb.2013.0496

Mulero-Pázmány, M., Stolper, R., Van Essen, L. D., Negro, J. J., & Sassen, T. (2014). Remotely piloted aircraft systems as a rhinoceros anti-poaching tool in Africa. *PLoS One, 9*(1), e83873.

Nabeshima, K., Nakajima, N., Ogata, M., & Onuma, M. (2022). Draft genome sequence data of Indian rhinoceros, *Rhinoceros unicornis. Data in Brief, 41*, 107857. https://doi.org/10.1016/j.dib.2022.107857. PMID: 35141371; PMCID: PMC8814301.

Nakamura, J. N., & Kuemlangan, B. (2023). *Implementing the Convention on International Trade in Endangered Species of Wild Fauna and Flora (CITES) through national fisheries legal frameworks: A study and a guide* (2nd ed.). Food & Agriculture Organization.

Nijman, V. (2010). An overview of international wildlife trade from Southeast Asia. *Biodiversity and Conservation, 19*(4), 1101–1114.

Norouzzadeh, M. S., Nguyen, A., Kosmala, M., Swanson, A., Palmer, M. S., Packer, C., & Clune, J. (2018). Automatically identifying, counting, and describing wild animals in camera-trap images with deep learning. *National Academy of Sciences of the United States of America, 115*(25), E5716–E5725.

Pascal, A., Banerjee, D., Poojitha, D., Kalita, H., Rao, K. V., Ahmad, M., Priya, G., Chouhan, S. K., & Zeal, J. (2025). A walk-through the Wildlife (Protection) Act, 1972: Journey till now and the way ahead. *Zoo's Print, 40*(1), 69–83.

Pearson, H. (2024). Scientists are building giant 'evidence banks' to create policies that actually work. *Nature, 634*(8032), 16–17.

Pimm, S. L., et al. (2014). The biodiversity of species and their rates of extinction, distribution, and protection. *Science, 344*(6187), 1246752.

Poonia, S., et al. (2022). Policy reform and wildlife conservation in India: An analysis of the Wild Life (Protection) Amendment Act, 2022. *Conservation and Society, 20*(3), 112–118.

Pullin, A. S., Baldi, A., Can, O. E., Dieterich, M., Kati, V., et al. (2009). Conservation focus on Europe: Major conservation policy issues that need to be informed by conservation science. *Conservation Biology, 23*, 818–824. http://conbio.onlinelibrary.wiley.com/doi/epdf/10.1111/j.1523-1739.2009.01283.x

Puri, M., Srivathsa, A., Karanth, K. K., Patel, I., & Kumar, N. S. (2022). Links in a sink: Interplay between habitat structure, ecological constraints and interactions with humans can influence connectivity conservation for tigers in forest corridors. *Science of the Total Environment, 809*, 151106. https://doi.org/10.1016/j.scitotenv.2021.151106

Raj, A., & Victor, A. (2022). *Vermin' politics: Human-wildlife conflict mitigation needs to be based on science, not historical biases*. Down to Earth.

Rana, A. K., & Kumar, N. (2023). Current wildlife crime (Indian scenario): Major challenges and prevention approaches. *Biodiversity and Conservation, 32*(5), 1473–1491.

Ranglack, D. H., Plumb, G. E., & Rogers, L. R. (2023). American bison (Bison bison): A rangeland wildlife continuum. In *Rangeland wildlife ecology and conservation* (pp. 791–827). Springer International Publishing.

Rodrigues, A. S., Akcakaya, H. R., Andelman, S. J., Bakarr, M. I., Boitani, L., Brooks, T. M., Chanson, J. S., Fishpool, L. D., Da Fonseca, G. A., Gaston, K. J., & Hoffmann, M. (2004). Global gap analysis: Priority regions for expanding the global protected-area network. *Bioscience, 54*(12), 1092–1100.

Roe, D., & Booker, F. (2019). Engaging local communities in tackling illegal wildlife trade: A synthesis of approaches and lessons for best practice. *Conservation Science and Practice, 1*(5), e26.

Roy, D., & Kumar, V. (2024). An analysis of illegal wildlife trade with the aid of social media and prevention strategies. *Journal of Wildlife and Biodiversity, 8*(1), 386–401.

Sharma, O. P., Sharma, S., Pattabhi, V., Mahato, S. B., & Sharma, P. D. (2007). A review of the hepatotoxic plant Lantana camara. *Critical Reviews in Toxicology, 37*(4), 313–352.

Shepherd, C. R., & Nijman, V. I. (2008). *Elephant and ivory trade in Myanmar*. TRAFFIC Southeast Asia.

Silver, S. C., Ostro, L. E., Marsh, L. K., Maffei, L., Noss, A. J., Kelly, M. J., Wallace, R. B., Gómez, H., & Ayala, G. (2004). The use of camera traps for estimating jaguar Panthera onca abundance and density using capture/recapture analysis. *Oryx, 38*(2), 148–154.

Singh, A., Thakur, M., Kumar Sharma, L., & Banerjee, D. (2022). African or Asiatic origin? Genome analyses solved the mystery of the hybrid origin of the rescued lion cub. *Gene, 817*, 146228. https://doi.org/10.1016/j.gene.2022.146228. Epub 2022 Jan 19. PMID: 35063578.

Sinha, S. (2010). *Handbook on wildlife law enforcement in India*. Natraj Publishers.

Srikanth, V., Das, A., Elangbam, S., Singh, B., & Rafiq, M. M. (2023). Biodiversity and wildlife conservation efforts in India. *National Parks, 106*(44,402.95), 1–35.

Srinivas, Y., Yumnam, B., Dutta, S., & Jhala, Y. (2022). Assessing genetic diversity and population structure for prioritizing conservation of the critically endangered Great Indian Bustard (Aredotis nigriceps). *Global Ecology and Conservation, 40*, e02332.

Stoops, M. A., Atkinson, M. W., Blumer, E. S., Campbell, M. K., & Roth, T. L. (2010). Semen cryopreservation in the Indian rhinoceros (Rhinoceros unicornis). *Theriogenology, 73*(8), 1104–1115.

Talukdar, A., Panda, A., Srivastav, A., Hussain, S. A., Malik, P. K., & Nigam, P. (2021). Growth patterns of critically endangered, head-started three-striped roofed turtle, Batagur dhongoka (Gray, 1834). *Biologia, 76*(12), 3705–3710.

Thakur, S., Pal, R., Kahera, N. S., & Sathyakumar, S. (2023). Forced sympatry? Spatiotemporal interactions of ursids, the Himalayan brown bear and the Asiatic black bear, along a gradient of anthropic disturbances in Western Himalaya. *Journal of Zoology., 321*(1), 59–74.

Thuo, D. N., Junga, J. O., Kamau, J. M., Amimo, J. O., Kibegwa, F. M., & Githui, K. E. (2015). Population viability analysis of black rhinoceros (Diceros bicornis michaeli) in Lake Nakuru National Park, Kenya. *Journal of Biodiversity & Endangered Species, 3*(1). https://doi.org/10.4172/2332-2543.1000150

Tian, M., Potter, G. R., & Phelps, J. (2023). What is "wildlife"? Legal definitions that matter to conservation. *Biological Conservation, 287*, 110339.

Ugyan, U. (2020). *Monitoring report: The status and distribution of snow leopard (Panthera uncia) through camera trapping in the Northern part of reserve: "Securing a future for JKSNR mountain ghost"* (Technical report 1–20). Available from: https://www.researchgate.net/publication/360939952_Monitoring_report_The_Status_and_Distribution_of_Snow_leopard_Panthera_uncia_through_Camera_Trapping_in_the_Northern_part_of_Reserve_Securing_a_Future_for_JKSNR_Mountain_Ghost?channel=doi&linkId=6294b86a6886635d5caf0778&showFulltext=true. Accessed April 8, 2025.

Valentini, A., Pompanon, F., & Taberlet, P. (2009). DNA barcoding for ecologists. *Trends in Ecology & Evolution, 24*(2), 110–117.

Wadhwa, S. (2022). *Navigating the wildlife (protection) amendment bill, 2021 a comprehensive analysis of the need for scientific methods in wildlife protection*. Available from https://ceerapub.nls.ac.in/navigating-the-wildlife-protection-amendment-bill-2021-a-comprehensive-analysis-of-the-need-for-scientific-methods-in-wildlife-protection/. Accessed on 11.4.2025.

Warchol, G. L. (2004). The transnational illegal wildlife trade. *Criminal Justice Studies, 17*(1), 57–73.

Wasser, S. K., Mailand, C., Booth, R., Mutayoba, B., Kisamo, E., Clark, B., & Stephens, M. (2007). Using DNA to track the origin of the largest ivory seizure since the 1989 trade ban. *National Academy of Sciences of the United States of America, 104*(10), 4228–4233.

Wich, S. A., Meijaard, E., Marshall, A. J., Husson, S., Ancrenaz, M., Lacy, R. C., Van Schaik, C. P., Sugardjito, J., Simorangkir, T., Traylor-Holzer, K., & Doughty, M. (2008). Distribution and conservation status of the orang-utan (Pongo spp.) on Borneo and Sumatra: How many remain? *Oryx, 42*(3), 329–339.

Worm, B., Barbier, E. B., Beaumont, N., Duffy, J. E., Folke, C., Halpern, B. S., Jackson, J. B., Lotze, H. K., Micheli, F., Palumbi, S. R., & Sala, E. (2006). Impacts of biodiversity loss on ocean ecosystem services. *Science, 314*(5800), 787–790.

Yadav, N., & Pallissery, F. (2023). International investments and environmental protection in India – Policy and implementation gaps in mitigating the carbon footprints. *Global Transitions, 5*, 217–224.

Zahmel, J., Jänsch, S., Jewgenow, K., Sandgreen, D. M., Simonsen, K. S., & Colombo, M. (2021). Maturation and fertilization of African lion (Panthera leo) oocytes after vitrification. *Cryobiology, 98*, 146–151.

Zisenis, M. (2017). Is the Natura 2000 network of the European Union the key land use policy tool for preserving Europe's biodiversity heritage? *Land Use Policy, 69*, 408–416.

Chapter 5
Bridging Borders for Wildlife Protection: India's Legal Efforts and International Cooperation

Kapish Samota, Sharda Kalra, Devinder Singh, Amit Kumar, and Asha Poonia

Contents

Abstract India is a mega diversity region having varied ecosystems from western Ghats to Himalayas, from Sundarbans to Thar desert. The Human-wildlife conflict

---

K. Samota · S. Kalra · A. Kumar
Department of Zoology, Chaudhary Bansi Lal University, Bhiwani, Haryana, India

D. Singh
Department of Laws, Panjab University, Chandigarh, India

A. Poonia (✉)
Faculty of Life Sciences, Chaudhary Bansi Lal University, Bhiwani, Haryana, India
e-mail: asha.poonia@cblu.ac.in

A. Poonia et al. (eds.), *Wildlife Trade in India*,
https://doi.org/10.1007/978-981-95-3694-8_5

and ecosystem degradation have heightened the need for comprehensive and actionable conservation strategies. However, presently its diversity is being threatened due to various reasons including poaching, habitat loss, climate change, habitat fragmentation, illegal wildlife trade amongst others. As India alone cannot fight this issue largely due to its transboundary nature, global alliances and harmonised legal frameworks with neighbouring countries are essential. Moreover, wildlife conservation has transitioned from being primarily a local or national issue to a global priority, addressing biodiversity challenges through international collaboration. This chapter examines various harmonized transboundary laws and treaties designed to safeguard wildlife in India. India actively participates in numerous international and multinational conservation efforts to protect biodiversity. Case studies like, the Cheetah Reintroduction Program in India, The Great Limpopo Transfrontier Park, The Green Belt, and the W-Arly-Pendjari (WAP) Complex, illustrate the success of coordinated legal frameworks in transboundary conservation. By coordinating legal standards, promoting collaboration and strengthening enforcement mechanisms these examples focus attention on the critical role of legal harmonization in achieving effective transboundary conservation outcomes.

Keywords Wildlife conservation · International harmonization · Treaties · Transboundary conservation · CITES · Cheetah Reintroduction · The Wildlife Protection Act 1972 · Convention on biological diversity · Ramsar convention · Biological Diversity Act 2002

5.1 Introduction

India is recognized among 17 most biodiverse countries. With only 2.4% of worlds land area, India boasts 7–8% of world's recorded flora and fauna. India is home to 4 of the 36 global biodiversity hotspots, which further strengthen its position as a biodiverse country (Hoffman et al., 2016). "To qualify as a biodiversity hotspot, an area must meet two strict criteria: (1) Contain at least 1500 species of vascular plants found nowhere else on Earth (known as "endemic" species). (2) Have lost at least 70% of its primary native vegetation" (Hoffman et al., 2016). It shows the true concern India is facing towards its endemic diversity. To cope up, India requires not only internal efforts but also international cooperation. Ecosystems and wildlife extend beyond political borders so the need for international cooperation has become a key tenet of successful conservation. Wildlife conservation transcends the local and even national boundaries so now it demands global coordination to address the complex issues affecting biodiversity (Thomsen & Thomsen, 2021). Ecosystems cross national borders so the necessity for partnerships among governments, organizations, and communities dedicated to protecting the natural resources we all share is indisputable (Loescher et al., 2022). The need for joint effort is exceptional in the field of wild animal protection as the actions of a particular country can shape the biodiversity and ecological welfare of neighboring countries (Bawa et al., 2021).

Hence, it is a need of the hour that countries engage in meaningful communication and joint initiatives to unfold effective strategies that manage the challenges caused by illegal poaching, habitat destruction, and climate change (Gale & Báez, 2024).

As global teamwork is an absolute essential in preserving of wildlife across the nations. The international treaties, bilateral agreements and the regional efforts have the potential to merge the assets of a countries and share their expertise while implementing the coordinated action plan (Jiao et al., 2021). Integral challenges such as the devastation of habitats, the risk of species extinction and the unrestrained illegal wildlife trade are addressed by these cumulative plans (Marcet, 2021). A commanding example of this is seen in international agreement like the Convention on Biological Diversity (CBD) and the Convention on International Trade in Endangered Species (CITES) which have proven an excellent instance of how international agreements achieve global integration and establish responsible biodiversity management frameworks (Bibi et al., 2024).

Transboundary conservation projects encounters a variety of obstacles which can be resolved through international efforts between countries (Laako et al., 2022). Political belief systems and governance structures varies among the different countries. Inequalities in Legal and law imposition between countries cause uncooperative outcomes that impair conservation project (Bodin, 2017). The limited financial capacity of nations leads to poor execution of treaties across different countries (Brummer, 2010). Different species migratory patterns make it tougher to proceed with transboundary conservation efforts (Bhargava, 2022). Survival of these migrant animals requires considerable international cooperation between different countries because they cross through various countries (Sakapaji et al., 2024). The differing national agendas coupled with political tensions hinder the development and enactment of strong and comprehensive conservation strategies (Adanma & Ogunbiyi, 2024).

While the conservation efforts face many obstacles the success stories of many protective measures demonstrate the extraordinary influence that international collaboration can achieve. Overcoming important environmental concerns affecting the entire globe demands robust legal frameworks along with collaborative governance and joint projects. Among them The Cheetah Reintroduction Program (2022) highlights the power of international cooperation between India and Namibia which breaks geographical borders to restore species to their native habitats. These efforts illustrate the essential role of cohesive legal systems, and shared initiatives in tackling the conservation challenges that impact the planet as a whole.

The significant position of international treaties together with bilateral agreements in wildlife protection of India will be discussed in this chapter. The analysis includes example of specific venture including Cheetah Reintroduction Program. The chapter inquire into international coordination through both achievements and blockages to demonstrate why united actions remain essential for biodiversity preservation for India.

5.2 Need for Harmonization of Legal Frameworks for Transboundary Conservation

Transboundary conservation implies the systematic management of natural resources that traverse political lines to protect biodiversity, uphold ecosystem services and encourage sustainable consumption (Kotzé & Marauhn, 2014). The ecosystem and wildlife exist independently of human-imposed boundaries which necessitates the need of combined efforts among neighbouring nations to fulfil conservation objectives (Erg et al., 2015). The legal harmonization process is indispensable for the effective implementation and administration of transnational preservation initiatives because it organize legislations, regulations and administration treaty across different legal territories (Lim, 2016).

The conservation of biodiversity, the assistance of international alliance, the improvement of administration compliance and the assistance of sustainable resource management is because of the legal harmonization (Adanma & Ogunbiyi, 2024). Multiple national boundaries are cut across by flowing rivers, mountain ranges, forests and vital wetlands specially for India as it shares its land boundaries with seven countries i.e. Pakistan, Afghanistan, China, Nepal, Bhutan, Myanmar, and Bangladesh and marine boundaries also with seven countries i.e. Pakistan, Bangladesh, Myanmar, Thailand, Indonesia, Sri Lanka, and Maldives. Legal harmonization enhances the conservation of biodiversity by ensuring uniform protection measures across these diverse regions, as it helps to mitigate habitat fragmentation and afford protection to migratory species (Fischman & Hyman, 2010). Additionally harmonized legal frameworks play a vital role in minimizing misunderstandings and argument among nations thereby cultivating an atmosphere of trust and structured procedure for conservation initiatives (Weiss & Jacobson, 2000). Consistent legal regulation process enable superior monitoring, surveillance and sustainable resource management by establishing clear parameters for access, quotas, and resource management practices essential for sustaining ecosystems (Norfolk, 2004).

Legal harmonization encounters abundance of challenging obstacles. Countries frequently place a premium on their sovereignty and economic agendas, which stimulate reluctance toward the embrace of common regulatory frameworks (Boubakri et al., 2023). The existence of cultural and legal discrepancy among nations further complicates the pursuit of uniform regulations, rendering their enactment an difficult task (Berman, 2006). Many nations face constraints in financial and technical resources, which obstruct their ability to participate effectively in harmonization initiatives. The political instability and tense relationships among nations jeopardize the conservation projects. The varying conservation priorities driven by an array of social, economic, and political factors further complicate joint efforts rendering agreement even more difficult to achieve (Nie, 2003).

Transboundary conservation includes bilateral and multilateral treaties, global conventions, regional cooperation mechanisms, collective management entities, capacity enhancement programs, and the assistance of legal templates (Mitrotta,

2019; Mumme, 2015). Bilateral and multilateral agreements help establish common conservation objectives and procedures, as seen in initiatives like the Convention on Migratory Species (CMS) (Caddell, 2011). Global agreements such as the Convention on Biological Diversity (CBD) provide essential principles and requirements for countries to align their conservation strategies (Chandra & Idrisova, 2011). Regional initiatives, including the Southern African Development Community (SADC), facilitate synchronized legal frameworks at broader scale (Joseph, 2016). Transboundary management entities oversee coordinated actions across regions to ensure efficient administration and regulatory alignment. Moreover, capacity-building programs provide critical training and technical expertise to support effective legal integration, while legal toolkits serve as valuable references for formulating compatible legislation (World Health Organization, 2001).

5.3 International Treaties and Bilateral Agreements in Wildlife Protection

Important International treaties and bilateral agreements in wildlife protection are given in the Table 5.1. This Table consist of the list of the agreements and treaties in which there is involvement of the many countries and these all are working together for the wildlife protection.

There are some treaties and agreement which are specific to the regions and followed by these regions only. In Table 5.2 the brief overview of such treaties is discussed.

India is also part of multiple international agreements which establish a common framework for nations to cooperate on issues of biodiversity and conservation (Fig. 5.1).

The integration of CITES regulations into the *Wild Life (Protection) Amendment Act, 2022* demonstrates India's commitment of aligning national legislation with international obligations. The amendment enhances penalties for wildlife crimes and aims to curb illegal trade, reflecting India's compliance with CITES provisions. India's National Biodiversity Action Plan (NBAP), formulated under the CBD framework, is designed to address biodiversity loss through conservation, sustainable use, and equitable sharing of benefits derived from biological resources (Sodhi et al., 2010). By aligning the *Wild Life (Protection) Amendment Act* with international standards, India can ensure that its conservation policies are not only comprehensive but also globally relevant.

The Wild Life (Protection) Amendment Act, 2022 represents a significant improvement over the Wild Life (Protection) Act, 1972 by being a stepping stone when it comes to fulfil the international treaties and obligations India is a member of by modernizing India's wildlife protection framework and by incorporating provisions for sustainable use, species recovery, and ecosystem restoration. It

Table 5.1 International agreement and convention for the wildlife protection

International agreement/ convention	Countries involved	Functions of the agreement/ convention
Convention on biological Diversity (Convention on Biological Diversity, 2024)	196 parties (nearly all UN Member states)	1. The preservation of biological diversity 2. The sustainable utilisation of biological diversity's fundamental components 3. The equitable and impartial distribution of advantages resulting from the use of genetic resources
Convention on migratory species (Convention on Migratory Species, 2020a)	133 parties, including Germany, South Africa, Brazil, Australia, and India	1. Promotes international collaboration for the preservation of migratory animals and their ecosystems. 2. Maintains migratory species throughout their migratory ranges.
Ramsar Convention on Wetlands (The Convention on Wetlands, 1971)	172 parties, including Nigeria, Japan, Argentina, Canada	1. Emphasises the preservation and sustainable utilisation of wetlands. 2. Fosters international cooperation for the governance of transboundary wetland ecosystems.
Convention on International Trade in Endangered Species of Wild Fauna and Flora (Convention on International Trade in Endangered Species of Wild Fauna and Flora, 2023)	184 parties, including the USA, Russia, India and China	1. Regulates the international trade of endangered animals to safeguard their survival. 2. Addresses the issue of illegal cross-border wildlife trading.
World Heritage Convention (UNESCO, 1975)	196 parties, including 195 UN member states, 2 UN Observer states, and 2 States in Free association with New Zealand.	1. Safeguards cultural and natural treasures of exceptional worldwide significance. 2. Promotes the preservation of transboundary natural areas such as the Great Limpopo Transfrontier Park.
UNESCO Man and the Biosphere Programme (UNESCO, 1971)	120 countries having the biosphere reserve which counts 669, including 16 transboundary sites.	1. Advocates for the preserving of biodiversity and cultural diversity. 2. Facilitates sustainable development and scientific inquiry in transboundary biosphere reserves.
African-Eurasian Migratory Waterbird Agreement (Convention on Migratory Species, 2020b)	80 countries and the European Union.	1. Safeguards migratory waterbirds and their habitats throughout Africa and Eurasia.

(continued)

Table 5.1 (continued)

International agreement/ convention	Countries involved	Functions of the agreement/ convention
		2. Promotes collaborative conservation efforts among range states.
Alpine Convention (Alpine Convention, 1995)	European Union and 8 countries including, Austria Switzerland, Italy and France.	1. Safeguards the natural ecosystem of the Alps. 2. Promotes sustainable development in the region, encompassing transboundary conservation of species.

significantly improves international cooperation on wildlife crime by aligning with global and integrating provisions for stronger penalties and enforcement mechanisms. Some of the major international treaties India is partnered with are listed in Table 5.3. The Act has introduced a New Chapter VB (Sections 49A–49F) in Indian legislation for regulating international trade in species covered under CITES which has placed India amongst selected countries to follow CITES regulations in toto. The act has introduced specific provisions and comprehensive mechanism with rigorous enforcement and prosecution against wildlife crimes for fulfilling CITES obligations including Section 49A to establish procedures for dealing with the import, export, and re-export of wildlife in such a way that it does not threaten their survival (Ministry of Environment, Forest and Climate Change, 2022). The act has also for the first time Introduced a CITES Management Authority and Scientific Authority for implementing CITES and CMS and expanded the protected species list as per CITES and IUCN. However, India need to work closely with CITES for ensuring the execution of the act (Gautam & Kaur, 2023).

The Wildlife (Protection) Act, 2022 marks the first formal recognition Ramsar convention, IPBES, CMS of The Wildlife (Protection) Act, 2022 has tried to accommodate Convention on Biological Diversity (CBD) especially Post-2020 Global Biodiversity Framework established at COP15 for conservation and sustainable use of biological diversity while equitable benefit sharing by including CITES listed species and important habitats by including them among protection schemes (Singh, 2023) while integrating scientific knowledge (Gautam & Kaur, 2023) and engaging local communities (Noe & Kangalawe, 2015). While the Wild Life Protection Act, 1972 had no direct reference to the Ramsar Convention on Wetlands or specific wetland protection provisions, present act has made new Provisions for Wetlands in Section 26A and 26B by designating Ramsar sites as protected areas and introducing habitat restoration, contributing to the conservation of critical wetland ecosystems and their biodiversity to reflects best practices and strengthens India's role in global conservation (Gautam & Kaur, 2023; Singh, 2022).

Table 5.2 Region specific convention and treaty for the conservation

Region	Multinational agreement/ convention	Countries involved	Functions of the agreement/convention
Africa	Nairobi Convention (Nairobi Convention, 2022)	Comoros, Kenya, France, Madagascar, Mauritius, Mozambique, Seychelles, Somalia, South Africa, Tanzania	1. Safeguards marine and coastal habitats 2. Tackles pollution and biodiversity preservation.
	Zambezi Watercourse Commission (Zambezi Warercourse Commission (ZAMCOM), 2014)	Angola, Botswana, Malawi, Mozambique, Namibia, Tanzania, Zambia, Zimbabwe	1. Fair utilisation of communal water resources. 2. Preserves habitats throughout the Zambezi Basin.
	Bangweulu Swamps Conservation Initiative (Africa Geographic Stories, 2024)	Zambia (with neighboring countries)	1. Conserves wetland ecosystems 2. Preserves critical species such as the shoebill stork and black lechwe.
	Great Limpopo Transfrontier Conservation Area Treaty (Transfrontier Conservation Areas Southern African Development Community, 2024)	South Africa, Mozambique, Zimbabwe.	1. Preserving iconic species, such as rhinos and elephants. 2. Anti-poaching and ecotourism programs.
Europe	European Green Belt Initiative (European Green Belt, 2018)	Germany, Poland, Romania, Finland, and others.	1. Creating a transnational ecological network that encourages ecotourism and habitat preservation.
	Carpathian Convention (Carpathian Convention, 2006)	Czech Republic, Hungary, Poland, Romania, Serbia, Slovakia, Ukraine.	1. Mountain ecosystem conservation 2. Control of large animals, such as wolves and bears.
America	Mesoamerican Biological Corridor (The Mesoamerican Biological Corridor (English), 2011)	Belize, Costa Rica, El Salvador, Guatemala, Honduras, Mexico, Nicaragua, Panama	1. Creates corridors between protected areas and encourages sustainable development and biodiversity conservation.
	Cartagena Convention (Cartgena Convention, 2022)	28 contracting parties, including Cuba, Jamaica, Trinidad and Tobago, USA	1. Coordinates responses to marine pollution and safeguards marine and coastal habitats.
	Yellowstone to Yukon Conservation Initiative (Yellowstone to Yukon Conservation Initiative, 1877)	USA and Canada	1. Encourages ecological connection and protects vital areas for animals like grizzly bears.
	Amazon Cooperation Treaty Organization (Amazon Cooperation Treaty Organization, 1978)	Brazil, Colombia, Peru, Venezuela, Ecuador, Bolivia, Guyana, Suriname.	1. Preserving the biodiversity of the Amazon 2. Combating illegal wildlife trade and destruction.

(continued)

Table 5.2 (continued)

Region	Multinational agreement/ convention	Countries involved	Functions of the agreement/convention
	Migratory Bird Treaty Act (Migratory Bird Treaty Act, 1918)	USA and Canada.	1. Protection of cross-border migratory bird species. 2. The preservation of areas that are essential for migration and nesting.
Asia	Heart of Borneo Initiative (WWF, 2007)	Brunei, Indonesia, Malaysia	1. Reduces habitat loss and preserves Borneo's forests and biodiversity.
	Mekong River Commission (Mekong River Commission for Sustainable Development, 1995)	Cambodia, Laos, Thailand, Vietnam	1. Handles fisheries and sediment control while promoting sustainable management of water resources.
	Altai-Sayan Ecoregion Conservation Initiative (WWF, 2021)	Russia, Kazakhstan, Mongolia, China	1. Preserves mountain ecosystems' biodiversity and safeguards argali sheep and snow leopards.
	ASEAN Agreement on Conservation of Nature and Natural Resources (Centre for International Law, 1985)	Brunei Darussalam, Cambodia, Lao PDR, Indonesia, Malaysia, Philippines, Singapore, Myanmar, Thailand, Vietnam.	1. Preserving biodiversity in the ecosystems of Southeast Asia. 2. Organising conservation activities in common areas.
	Central Asian Mammals Initiative (Central Asian Mammals & Climate Adaptation Project, 2021)	Kazakhstan, Mongolia, Kyrgyzstan, Tajikistan, Uzbekistan, China, Turkmenistan, India, Islamic Republic of Iran, Pakistan, Russian federation	1. Harmonisation of habitat management techniques. 2. Preservation of migratory mammals, such as snow leopards and saiga antelope.
	South Asia Wildlife Enforcement Network (South Asia Wildlife Enforcement Network, 2024)	Afghanistan, Bangladesh, Bhutan, Maldives, India, Nepal, Pakistan, Sri Lanka	1. Enhances regional collaboration in combating against South Asia's illegal wildlife trade.
	Global Tiger Initiative (The Global Tiger Initiative, 2008)	Bangladesh, Bhutan, Cambodia, China, India, Indonesia, Lao PDR, Malaysia, Myanmar, Nepal, Russia, Thailand, Vietnam	1. International collaboration to protect tiger populations across their range.
Polar Regions	Agreement on the Conservation of Albatrosses and Petrels (Agreement on the Conservation of Albatrosses and Petrels, 2001)	Australia, Argentina, Brazil, Chile, France, Ecuador, New Zealand, South Africa, Spain, Norway, Peru, UK and Uruguay	1. Organises conservation efforts and defends petrel and albatross species against hazards including habitat deterioration and bycatch.

(continued)

Table 5.2 (continued)

Region	Multinational agreement/ convention	Countries involved	Functions of the agreement/convention
	Arctic Council (Arctic Council, 1991)	Canada, Denmark, Finland, Iceland, Norway, Russia, Sweden, USA.	1. Arctic environment conservation 2. Coordination for migratory species, such as whales and polar bears.
	Antarctic Treaty System (The Antartic Treaty, 1961)	58 parties, including the USA, UK, China, and Russia.	1. Prohibits mining and military presence. 2. Maintains Antarctica as a natural reserve for peace and science.

Fig. 5.1 Convention and treaties in which India is a part

5.4 Case Studies

Successful case studies illustrate the potential of harmonized legal frameworks for transboundary conservation. Example of some successful case studies is:

5.4.1 Cheetah Reintroduction Program in India

A notable instance of wildlife restoration is the Cheetah Reintroduction Program in India, which aims to revive a species that has been absent from the country for over

Table 5.3 India and its neighbouring countries wildlife laws and policies

Parameter	India	Pakistan	Afghanistan	China	Nepal	Bhutan	Myanmar	Bangladesh	Thailand	Indonesia	Sri Lanka	Maldives
Primary Law	Wildlife Protection Act, 1972	The Pakistan Trade Control of Wild Fauna and Flora Rules, 2012*	Environmental Law, 2005	Wildlife Protection Law, 1989	**National Parks and Wildlife Conservation Act, 2029 (1973), Aquatic Animals Protection Act (1961), National Parks and Wildlife Conservation Rules, 2030 (1974)**	Forest and Nature Conservation Act, 1995	Protection of Wildlife and Conservation of Natural Areas Law, 1994	Wildlife (Preservation and Security) Act, 2012	Wildlife Conservation and Protection Act, B.E. 2562 (2019)	Law No. 5/1990 on Conservation of Natural Resources and Ecosystems	Fauna and Flora Protection Ordinance, 1937	Environmental Protection and Preservation Act of Maldives (Law No 4/93) along with Fisheries Law of Maldives (5/87)
Amendments/ Updates	Amended in 2002, 2006, 2022	Amended in 2018	No recent amendment	Revised in 2016	Revised in 1993, 2017	Amended in 2006, 2023	No major amendment	Amended in 2017	–	Revised in 1999	Amended in 2009	No recent amendment
Protection of Flagship Species	Strong protection for tigers, elephants, and rhinos	Focus on snow leopards and Indus dolphins	Snow leopards, Marco Polo sheep	Panda, tigers, and pangolins heavily protected	Tigers, rhinos, and snow leopards prioritized	Protection for takin and snow leopards	Elephants, pangolins, and tigers protected	Bengal tigers, dolphins, and migratory birds	Elephants, gibbons, and big cats	Orangutans, tigers, and sea turtles	Elephants, leopards, and blue whales	Whale sharks, dolphins, and coral reefs
Wildlife Crime Control Units	Wildlife Crime Control Bureau (WCCB)	Provincial Wildlife Control Units	Under Environmental Directorate	National Forestry and Grassland Administration (NFGA)	Wildlife Crime Control Bureau (WCCB)	Forest Protection Units	Wildlife Crime Task Force	Wildlife Crime Control Unit	National Park Rangers & Anti-Poaching Units	Wildlife Enforcement Network	Wildlife Protection Force	Marine Conservation and Surveillance Units

(continued)

Table 5.3 (continued)

Parameter	India	Pakistan	Afghanistan	China	Nepal	Bhutan	Myanmar	Bangladesh	Thailand	Indonesia	Sri Lanka	Maldives
Judicial Mechanisms	Specialized courts for wildlife crimes	General courts handle wildlife cases	General judiciary involvement	Environmental courts established	Specialized environmental courts	Environmental Tribunal established	No specialized courts for wildlife	Environmental Courts and Tribunals	Environmental Courts & Wildlife Task Forces	Specialized Environmental Courts	Environmental Tribunal established	No specialized courts
Transboundary Conservation Programs	Active with SAWEN, CITES, CMS	Minimal transboundary efforts	Weak transboundary programs	Participates in cross-border conservation efforts	Involved in Terai Arc Landscape and Himalaya Conservation Program	Bhutan Tiger Conservation Program with India	Weak transboundary collaboration	Participates in Sundarbans protection programs	Active in ASEAN Wildlife Enforcement Network	ASEAN Wildlife Law Enforcement Network	Bilateral conservation with India	CITES, Limited transboundary collaboration
Reference	Poonia et al. (2022)	Usman et al. (2022)	Maheshwari (2022)	Fang et al. (2022), Huang et al. (2021)	Government of Nepal (2025)	FAO (2025a)	FAO (2025b)	Islam and Zheng (2023), Suvra and Ontar (2024)	FAO (2025c)	Febrian et al. (2021)	FAO (2025d)	Techera and Cannell-Lunn (2019)

Along with this Pakistan has various regional acts including Sindh Wildlife (Protection, Preservation, Conservation and Management) Act, 2020, Khyber Pakhtunkhwa Wildlife and Biodiversity (Protection, Preservation, Conservation and Management) Act, 2014, Azad Jammu and Kashmir Wildlife (Protection, Preservation, Conservation and Management) Act, 2014, Balochistan Wildlife (Protection, Preservation, Conservation and Management) Act, 2014, Punjab Wildlife (Protection, Preservation, Conservation and Management) (Amendment) Act, 2007, Islamabad Wildlife (Protection, Preservation, Conservation and Management) Ordinance, 1979 and Gilgit-Baltistan Wildlife Preservation Act, 1975 seven decades (Singh, 2022). The Cheetah Reintroduction program was made possible through the combined effort between India and African nations like Namibia and South Africa (Biswas, 2024). To clear the way for the safe transfer of these animals the legal agreements and protocols were carefully implemented (Soni et al., 2023). The translocation of Cheetah into the India is facing many challenges including the differences in prey species and habitat. This program highlights the power of international cooperation in conservation efforts demonstrating how legal frameworks can actively contribute to species recovery while promoting broader goals of biodiversity conservation (Thomsen & Thomsen, 2021).

5.4.2 The Great Limpopo Transfrontier Park

The Great Limpopo Transfrontier Park is a remarkable example of the successful legal harmonization for conservation (Wolmer, 2003). It spans the borders of South Africa, Mozambique, and Zimbabwe. This extensive park apprenticed cooperative venture to unify wildlife management regulations and anti-poaching initiatives (Spenceley, 2006). A transfrontier park is a viable alternative in achieving sustainable development along with the protection of the environment (Bhatasara et al., 2013). A large protected zone was established, through the collaborative agreements which help in the preservation of biodiversity and also promote eco-tourism prospects for surrounding communities (Bhatasara et al., 2013).

5.4.3 The Green Belt

The European Green Belt is a central ecological network which stretch from the Barents Sea to the Black Sea. It is serving as a global hallmark of trans-national cooperation for nature preservation and sustainable development (Terry et al., 2006). The Green Belt initiative entail a complex geopolitical realignment bringing together 24 European nations and stakeholders at several levels including governmental and non-governmental organizations (Geidezis & Kreutz, 2012). It has the potential to serve as a core feature of Europe's Green Infrastructure and as a backbone for a Pan-European ecosystem (Yang & Yang, 2019). Along with its environmental significance, the Green Belt is also a remarkable historical and cultural landscape. It symbolizes European unity while fostering transboundary cooperation, sustainable development, and international understanding (Vasilijević & Pezold, 2011).

5.4.4 W-Arly-Pendjari (WAP) Complex

The WAP Complex is shared by Benin, Burkina Faso and Niger, is another example of a successful project (Konrad, 2015). The anti-poaching and resource management has been improved in this conservation area by putting the effort in harmonizing laws and institutions (Sogbohossou, 2023). By the collaborative framework and effort, the ecological management and region's biodiversity conservation have made the most of benefit (Amahowé et al., 2013).

These case studies highlight the significance of legal harmonisation in attaining successful transboundary conservation objectives by encouraging collaboration, harmonising rules and strengthening enforcement mechanisms.

5.5 Challenges Forced by India for Cross-Border Collaboration

The global demand for wildlife and wildlife products per year is approximately between 10 to 20 billion dollars each year (WWF, 2016). Illegal wildlife trade (green crime) (Wyatt, 2009), is a major challenge and an urgent global conservation priority as it is among largest sources of criminal income globally (Yang & Lye, 2018) and especially in Asia (Esmail et al., 2020). Although legislation exists in all Asian countries to regulate illegal wildlife trade, its transboundary nature makes this issue complicated, especially for India (Biggs et al., 2017). As India is home to many significant species such as tiger, leopard, snow leopard, elephant, great Indian bustard, rhino and many other migratory birds and animals, cross-border collaboration is necessary for their protection and conservation. As India shares various land and marine ecosystems with neighbouring countries which are very important for transboundary migration/movement of important fauna, legal harmony with these countries is essential.

As already depicted in Fig. 5.1, India is part of various international and regional treaties and collaborations for cooperation on issues of biodiversity and conservation, it is still facing numerous challenges when it comes to cross-border collaboration.

5.5.1 Inconsistent Laws

As India and its neighbouring countries (Pakistan, Afghanistan, China, Nepal, Bhutan, Myanmar, Bangladesh, Thailand, Indonesia, Sri Lanka, and Maldives) all are member of CITES, still their wildlife laws and policies remain incoherent (Table 5.3). These differences decrease the effectiveness of coordinated regional responses to all threats related to biodiversity, especially for species which have transboundary habitats or migrate to these areas. It is true that CITES provides a basic framework for regulating international wildlife trade, documents the trade between nations on CITES listed species, but the implementation of CITES by these countries is left to the discretion of national governments, which leads to different legal interpretations, compliance, and enforcement.

One strong example of this is India's The Wildlife (Protection) Act, 1972, and the Biological Diversity Act, 2002, which is a relatively robust legal structure, while many neighbouring countries lack such comprehensive legislation such as Afghanistan and Pakistan and also face institutional weaknesses in enforcement. In addition to this, there is significant difference in different countries for priories for different species, penalties for wildlife crimes, judicial mechanism, wildlife control units along with procedural delays in the repatriation of seized fauna, and the no uniform cross-border protocols. Such differences give opportunity to wildlife traffickers to exploit legal loopholes by operating across borders where monitoring is weak or penalties are less stringent. Without regional harmonisation of legal frameworks, aligned enforcement capacities, and regular intergovernmental coordination, the shared commitment to CITES remains only partially effective. A more unified approach is imperative to protect species that transcend political boundaries and to disrupt sophisticated networks engaged in illegal wildlife trade across South and Southeast Asia.

5.6 Conclusion

Wildlife conservation has transcended national boundaries and emerged as a global imperative, necessitating robust legal frameworks, diplomatic coordination, and multilateral cooperation. India, as a mega-diverse country with vast ecological landscapes and a complex history of human-wildlife interactions, continues to strengthen its domestic legislation, including the Wildlife (Protection) Act, 1972 and the Biological Diversity Act, 2002. It has also actively engaged in international agreements such as CITES, the Convention on Biological Diversity (CBD), and the Ramsar Convention, signalling its commitment to harmonised biodiversity governance. However, the persistent challenge lies in the lack of congruence among the wildlife protection laws of India and its neighboring countries. Despite being members of CITES, countries in South and Southeast Asia often operate under disjointed legal regimes, resulting in inconsistencies in species protection, penalties, and enforcement procedures. These discrepancies create exploitable gaps that facilitate illegal wildlife trade, hinder species repatriation, and compromise conservation efforts in shared ecosystems.

The success of initiatives like the Cheetah Reintroduction Programme in India and international models such as the Great Limpopo Transfrontier Park and the W-Arly-Pendjari Complex demonstrates that legal harmonisation, cooperative enforcement mechanisms, and shared conservation objectives can yield positive outcomes. Strengthening cross-border collaboration requires not only aligning legal frameworks but also establishing bilateral and multilateral protocols for intelligence-sharing, joint patrols, scientific research, and capacity-building. Furthermore, regional platforms such as the South Asian Wildlife Enforcement Network (SAWEN) and ASEAN-WEN must be empowered and adequately resourced to act as catalysts for coordinated policy implementation.

In essence, legal instruments must evolve beyond static national statutes to reflect the dynamic, interconnected nature of wildlife and ecosystems. India's leadership in wildlife diplomacy, supported by science-based policymaking and community engagement, can serve as a model for fostering transboundary conservation governance. The focus on reconciling domestic legal systems with international commitments, enabling a seamless framework for protecting biodiversity across borders is needed. Only through integrated, cooperative, and legally coherent action can the region hope to reverse biodiversity loss and ensure a sustainable future for both wildlife and human communities.

5.7 The Way Forward

To ensure effective transboundary conservation and protection of wildlife across India and its neighbouring countries, there is an urgent need for a shift from fragmented national approaches to an integrated regional conservation strategy. The following pathways are proposed as part of the way forward to strengthen legal, institutional, and collaborative frameworks:

5.7.1 Legal Harmonisation and Policy Alignment

While all countries in the region are parties to CITES and other biodiversity-related international agreements, the disparity in their national wildlife laws creates loopholes for traffickers and undermines coordinated enforcement. It is imperative that regional governments initiate formal dialogues aimed at harmonising their legal definitions of wildlife crimes, standardising the categorisation of protected species, and aligning sentencing provisions. An agreed regional protocol on wildlife offences, similar to the SAARC Convention on Mutual Legal Assistance, could be developed to guide cross-border enforcement actions.

5.7.2 Strengthening Regional Cooperation Mechanisms

Regional institutions such as the South Asia Wildlife Enforcement Network (SAWEN), ASEAN Wildlife Enforcement Network (ASEAN-WEN), and the BIMSTEC platform must be revitalised through adequate funding, political support, and technical capacity. These platforms can act as convergence points for intelligence-sharing, joint training of enforcement personnel, development of common databases on offenders and seizures, and conducting coordinated anti-trafficking operations. India, with its institutional experience and geopolitical position, can take a leadership role in fostering such cooperative efforts.

5.7.3 Capacity Building and Technology Integration

Capacity constraints among enforcement and forest personnel continue to hamper surveillance and timely prosecution. There is a need for continuous training programmes focused on transboundary law enforcement, digital forensics, wildlife crime investigation, and the use of technologies such as drones, camera traps, GIS mapping, and DNA barcoding. Cross-border training modules involving enforcement agencies from multiple countries should be institutionalised under a regional conservation academy or collaborative university network.

5.7.4 Bilateral and Multilateral Agreements on Wildlife Corridors and Migratory Routes

Given the migratory nature of several endangered species such as the snow leopard, elephant, tiger, and various avian species, bilateral and multilateral Memorandums of Understanding (MoUs) must be strengthened to formally designate and protect cross-border wildlife corridors. Such agreements should outline joint responsibilities for habitat management, research, and community-based conservation, especially in ecologically sensitive areas like the Eastern Himalayas, Indo-Myanmar border, and Sundarbans delta.

5.7.5 Public Awareness and Community Engagement

A long-term solution to transboundary wildlife conservation cannot rely solely on legal enforcement. Borderland communities must be integrated into conservation strategies through awareness campaigns, incentives for conservation stewardship, and alternative livelihood programmes to reduce dependence on wildlife trade. Transnational awareness campaigns and community-level conservation pacts involving civil society organisations from adjoining countries can enhance trust and grassroots participation.

5.7.6 Creating a Regional Wildlife Crime Tracking System

A real-time, interoperable wildlife crime tracking and monitoring system accessible to all South and Southeast Asian nations would allow for early warning, seizure tracking, and coordinated responses to illicit activities. This system could be modelled on INTERPOL's I-24/7 platform but adapted to the regional context, with national focal points for timely data reporting and analysis.

5.7.7 Promoting Wildlife Diplomacy

Wildlife conservation must become a key theme in diplomatic discussions among the nations of South and Southeast Asia. Just as river water sharing and border security dominate bilateral agendas, wildlife protection should be integrated into foreign policy dialogues. Establishing a dedicated "Wildlife Diplomacy Forum" or including biodiversity cooperation in regional summits could elevate the political importance of this issue and ensure sustained commitment.

In conclusion, a secure future for South and Southeast Asia's wildlife hinges on a concerted regional effort that bridges legal gaps, builds institutional trust, and empowers communities across borders. India's evolving legal framework and active participation in international conventions places it in a strong position to champion a model of collaborative conservation that others in the region can emulate. The time is ripe to convert political will and legal instruments into meaningful action for biodiversity preservation on a transboundary scale.

References

Adanma, U. M., & Ogunbiyi, E. O. (2024). Evaluating the effectiveness of global governance mechanisms in promoting environmental sustainability and international relations. *Finance Accounting Research Journal, 6*(5), 763–791.

Africa Geographic Stories. (2024). https://africageographic.com/stories/bangweulu-wetlands-shoebills-swamps-and-shhhh/. Accessed 22 Jan 2025.

Agreement on the Conservation of Albatrosses and Petrels. (2001). https://www.acap.aq/about-acap/parties-to-acap. Accessed 23 Jan 2025.

Alpine Convention. (1995). https://www.alpconv.org/en/home/. Accessed 22 Jan 2025.

Amahowé, I. O., Houessou, L. G., Ashanti, S., & Tehou, A. C. (2013). Transboundary protected areas management: Experiences from W-Arly-Pendjari Parks in West Africa. *Parks, 19*(2), 95–105.

Amazon Cooperation Treaty Organization. (1978). https://otca.org/en/about-us/. Accessed 22 Jan 2025.

Arctic Council. (1991). https://arctic-council.org. Accessed 23 Jan 2025.

Bawa, K. S., Sengupta, A., Chavan, V., Chellam, R., Ganesan, R., Krishnaswamy, J., Mathur, V. B., Nawn, N., Olsson, S. B., Pandit, N., & Quader, S. (2021). Securing biodiversity, securing our future: A national mission on biodiversity and human well-being for India. *Biological Conservation, 253*, 108867.

Berman, P. S. (2006). Global legal pluralism. *Southern California Law Review, 80*, 1155.

Bhargava, R. (2022). The opportunities and challenges of transboundary conservation–solutions in adaptive management. In *Regional political ecologies and environmental conflicts in India* (pp. 29–48). Routledge.

Bhatasara, S., Nyamwanza, A. M., & Kujinga, K. (2013). Transfrontier parks and development in southern Africa: The case of the Great Limpopo Transfrontier Park. *Development Southern Africa, 30*(4–5), 629–639.

Bibi, C., Tabassum, S., & Roussel, Y. (2024). International environmental law: Challenges and opportunities for biodiversity conservation. *Journal of Energy and Environmental Policy Options, 7*(2), 28–35.

Biggs, D., Cooney, R., Roe, D., Dublin, H. T., Allan, J. R., Challender, D. W., & Skinner, D. (2017). Developing a theory of change for a community-based response to illegal wildlife trade. *Conservation Biology, 31*(1), 5–12.
Biswas, A. (2024). Re-interrogating the India-Africa relations: Knitting the pieces of India's emerging strategic partnerships and pathways. In *75 years of India's foreign policy: Bilateral, conventional and emerging trends* (Vol. 17, pp. 201–229). Palgrave Macmillan.
Bodin, Ö. (2017). Collaborative environmental governance: Achieving collective action in social-ecological systems. *Science, 357*(6352), eaan1114.
Boubakri, N., Fotak, V., Guedhami, O., & Yasuda, Y. (2023). The heterogeneous and evolving roles of sovereign wealth funds: Issues, challenges, and research agenda. *Journal of International Business Policy, 6*(3), 241–252.
Brummer, C. (2010). Why soft law dominates international finance—and not trade. *Journal of International Economic Law, 13*(3), 623–643.
Caddell, R. (2011). The integration of multilateral environmental agreements: Lessons from the biodiversity-related conventions. *Yearbook of International Environmental Law, 22*(1), 37–75.
Carpathian Convention. (2006). http://www.carpathianconvention.org/convention/history/. Accessed 22 Jan 2025.
Cartgena Convention. (2022). https://wesr-cartagena.unepgrid.ch. Accessed 22 Jan 2025.
Central Asian Mammals & Climate Adaptation Project. (2021). https://camcaproject.org/resources/cms-central-asian-mammals-initiative-cami-english/. Accessed 23 Jan 2025.
Centre for International Law. (1985). https://cil.nus.edu.sg/databasecil/1985-agreement-on-the-conservation-of-nature-and-natural-resources/. Accessed 23 Jan 2025.
Chandra, A., & Idrisova, A. (2011). Convention on biological diversity: A review of national challenges and opportunities for implementation. *Biodiversity and Conservation, 20*, 3295–3316.
Convention on Biological Diversity. (2024). https://www.cbd.int. Accessed 20 Jan 2025.
Convention on International Trade in Endangered Species of Wild Fauna and Flora. (2023). https://cites.org/eng/disc/parties/chronolo.php. Accessed 21 Jan 2025.
Convention on Migratory Species. (2020a). https://www.cms.int/en/parties-range-states. Accessed 20 Jan 2025.
Convention on Migratory Species. (2020b). https://www.cms.int/en/legalinstrument/aewa. Accessed 22 Jan 2025.
Erg, B., Groves, C., McKinney, M., Michel, T. R., Phillips, A., Schoon, M. L., Vasilijevic, M., & Zunckel, K. (2015). *Transboundary conservation: A systematic and integrated approach*. IUCN.
Esmail, N., Wintle, B. C., t Sas-Rolfes, M., Athanas, A., Beale, C. M., Bending, Z., Dai, R., Fabinyi, M., Gluszek, S., Haenlein, C., & Harrington, L. A. (2020). Emerging illegal wildlife trade issues: A global horizon scan. *Conservation Letters, 13*(4), e12715.
European Green Belt. (2018). https://www.europeangreenbelt.org. Accessed 22 Jan 2025.
Fang, G., Liu, H., Wu, K., Wei, T., & Wang, Q. (2022). Changing legislative thinking in China to better protect wild animals and human health. *Conservation Biology, 36*(1), 13837.
FAO. (2025a). As accessed from https://www.fao.org/faolex/results/details/en/c/LEX-FAOC224230/ on 5 Apr 2025.
FAO. (2025b). As accessed from https://www.fao.org/faolex/results/details/en/c/LEX-FAOC139132/ on 5 Apr 2025.
FAO. (2025c). As accessed from https://www.fao.org/faolex/results/details/en/c/LEX-FAOC201932/ on 5 Apr 2025.
FAO. (2025d). https://www.fao.org/faolex/results/details/en/c/LEX-FAOC022041/. Accessed 5 Apr 2025.
Febrian, F., Apriyani, L., & Novianti, V. (2021). Rethinking Indonesian legislation on wildlife protection: A comparison between Indonesia and the United States. *Sriwijaya Law Review, 5*(1), 143–160.

Fischman, R. L., & Hyman, J. B. (2010). The legal challenge of protecting animal migrations as phenomena of abundance. *Virginia Environmental Law Journal, 28*(2), 173–239.
Gale, T., & Báez Montenegro, A. (2024). Toward understanding research evolution on indirect drivers of ecosystem change along the interface of protected and non-protected lands. *Sustainability, 16*(17), 7572.
Gautam, A., & Kaur, S. (2023). Management and conservation of wildlife in India: With special reference to Wildlife Protection (Amendment) Act, 2022. *Indian Journal of Integrated Research in Law, 3*, 1.
Geidezis, L., & Kreutz, M. (2012). Green Belt Europe–Structure of the initiative and significance for a pan European ecological network. In *The Green Belt as a European Ecological Network: Strengths and gaps* (Vol. 10, p. 12). University of Applied Sciences Erfurt.
Government of Nepal. (2025). As accessed from https://dnpwc.gov.np/en/acts/ on 5 Apr 2025.
Hoffman, M., Koenig, K., Bunting, G., & Costanza, J. (2016, April 25). *Biodiversity hotspots* (version 2016.1).
Huang, G., Ping, X., Xu, W., Hu, Y., Chang, J., Swaisgood, R. R., Zhou, J., Zhan, X., Zhang, Z., Nie, Y., & Cui, J. (2021). Wildlife conservation and management in China: Achievements, challenges and perspectives. *National Science Review, 8*(7), 042.
Islam, M. Z., & Zheng, X. (2023). Combating illegal wildlife trade and conserving biodiversity in Bangladesh through legal provisions. *Israel Journal of Ecology & Evolution, 69*(3–4), 127–138.
Jiao, Y., Yeophantong, P., & Lee, T. M. (2021). Strengthening international legal cooperation to combat the illegal wildlife trade between Southeast Asia and China. *Frontiers in Ecology and Evolution, 9*, 645427.
Joseph, J. E. (2016). *The Southern African Development Community (SADC) as a regional security community: An assessment of common defence and security institutions* [Master's thesis, University of Johannesburg].
Konrad, T. (2015). *Governance of protected areas in West Africa: The case of the W-Arly-Pendjari (WAP) Complex in Benin and Burkina Faso* [Doctoral dissertation, Universität Würzburg].
Kotzé, L. J., & Marauhn, T. (Eds.). (2014). *Transboundary governance of biodiversity* (Vol. 19). Martinus Nijhoff Publishers.
Laako, H., Pliego Alvarado, E., Ramos Muñoz, D., & Marquez, B. (2022). Transboundary conservation and nature states in the Maya Forest: International relations, challenged. *Globalizations, 19*(8), 1288–1310.
Lim, M. (2016). Governance criteria for effective transboundary biodiversity conservation. *International Environmental Agreements: Politics, Law and Economics, 16*, 797–813.
Loescher, H. W., Vargas, R., Mirtl, M., Morris, B., Pauw, J., Yu, X., Kutsch, W., Mabee, P., Tang, J., Ruddell, B. L., & Pulsifer, P. (2022). Building a global ecosystem research infrastructure to address global grand challenges for macrosystem ecology. *Earth's Futures, 10*(5), e2020EF001696.
Maheshwari, A. (2022). Biodiversity conservation in Afghanistan under the returned Taliban. *Nature Ecology & Evolution, 6*(4), 342–343.
Marcet, E. E. N. (2021). *The current and future state of the international wildlife trade* [Thesis].
Mekong River Commission for Sustainable Development. (1995). https://www.mrcmekong.org. Accessed 22 Jan 2025.
Migratory Bird Treaty Act. (1918). https://www.fws.gov/law/migratory-bird-treaty-act-1918#:~:text=The%20Migratory%20Bird%20Treaty%20Act%20(MBTA)%20prohibits%20the%20take%20(,U.S.%20Fish%20and%20Wildlife%20Service. Accessed 22 Jan 2025.
Ministry of Environment, Forest and Climate Change. (2022). *Annual report on wildlife protection.* https://moef.gov.in/uploads/2022/03/Annual-report-2021-22-Final.pdf. Accessed 23 Jan 2025.
Mitrotta, E. (2019). *Decentralised international cooperation: Enhancing conservation and sustainable management of transboundary natural resources* [Doctoral dissertation, University of Trento].

Mumme, S. P. (2015). The evolution of natural resource conservation capacity on the US–Mexico border: Bilateral and trilateral environmental agreements since La Paz. *Review of Policy Research, 32*(1), 19–39.

Nairobi Convention. (2022). https://www.nairobiconvention.org/nairobi-convention/who-we-are/. Accessed 22 Jan 2025.

Nie, M. (2003). Drivers of natural resource-based political conflict. *Policy Sciences, 36*(3), 307–341.

Noe, C., & Kangalawe, R. Y. (2015). Wildlife protection, community participation in conservation, and (dis)empowerment in southern Tanzania. *Conservation and Society, 13*(3), 244–253.

Norfolk, S. (2004). *Examining access to natural resources and linkages to sustainable livelihoods: A case study of Mozambique* (Livelihood Support Programme Working Paper 17) (p. 69). FAO.

Poonia, A., Ghanghas, A., Sharma, S., Jangra, M., & Sindhu, V. (2022). Trends in CITES listed bird's trade in South Asian countries in view of evolution of Indian laws during last four decades. *International Journal of Ecological Economics and Statistics™, 43*(9), 115.

Sakapaji, S. C., García Molinos, J., Parilova, V., Gavrilyeva, T., & Yakovleva, N. (2024). Navigating legal and regulatory frameworks to achieve the resilience and sustainability of indigenous socioecological systems. *Resources, 13*(4), 56.

Singh, U. S. (2022). Kuno National Park is not yet ready for cheetahs. *The Applied Biology & Chemistry Journal, 3*(3), 56–61.

Singh, S. (2023). *A comparative analysis of wildlife conservation in the US and India* [Master's thesis, Tufts University].

Sodhi, N. S., Koh, L. P., Clements, R., Wanger, T. C., Hill, J. K., Hamer, K. C., Clough, Y., Tscharntke, T., Posa, M. R. C., & Lee, T. M. (2010). Conserving Southeast Asian forest biodiversity in human-modified landscapes. *Biological Conservation, 143*(10), 2375–2384.

Sogbohossou, E. A. (2023). Transboundary conservation of large carnivores in West Africa: The case of the WAP complex. In *Managing transnational UNESCO world heritage sites in Africa* (pp. 167–175). Springer International Publishing.

Soni, V. K., Tomar, A. S., & Tomar, S. S. (2023). A case study for introducing African cheetah in Kuno National Park, Sheopur, Madhya Pradesh, India. *Uttar Pradesh Journal of Zoology, 44*(9), 23–30.

South Asia Wildlife Enforcement Network. (2024). https://www.sawen.org. Accessed 23 Jan 2025.

Spenceley, A. (2006). Tourism in the Great Limpopo Transfrontier Park. *Development Southern Africa, 23*(5), 649–667.

Suvra, A. N., & Ontar, M. H. (2024). Illegal wildlife trade in Bangladesh and India: A study of comparative laws. *Uti Possidetis Journal of International Law, 5*(3), 488–547.

Techera, E. J., & Cannell-Lunn, M. (2019). A review of environmental law in Maldives with respect to conservation, biodiversity, fisheries and tourism. *Asia Pacific Journal of Environmental Law, 22*(2), 228–256.

Terry, A., Ullrich, K., & Riecken, U. (2006). *The green belt of Europe: From vision to reality*. IUCN.

The Antartic Treaty. (1961). https://www.ats.aq/e/antarctictreaty.html. Accessed 23 Jan 2025.

The Convention on Wetlands. (1971). https://www.ramsar.org. Accessed 21 Jan 2025.

The Global Tiger Initiative. (2008). *Feature brief*. World Bank Group. http://documents.worldbank.org/curated/en/714651637819049529/The-Global-Tiger-Initiative. Accessed 23 Jan 2025.

The Mesoamerican Biological Corridor (English). (2011). *Regional Program Review (RPR)* (Vol. 5, No. 2). World Bank. https://documents.worldbank.org/en/publication/documents-reports/documentdetail/378391468330902777/the-mesoamerican-biological-corridor. Accessed 22 Jan 2025.

Thomsen, B., & Thomsen, J. (2021). Multispecies livelihoods: Partnering for sustainable development and biodiversity conservation. In *Partnerships for the goals* (pp. 758–768). Springer.

Transfrontier Conservation Areas Southern African Development Community. (2024). https://tfcaportal.org/node/23. Accessed 22 Jan 2025.

UNESCO. (1971). *Man and the Biosphere Programme (MAB).* https://www.unccd.int/resources/knowledge-sharing-system/unesco-man-and-biosphere-programme-mab#:~:text=Launched%20in%201971%2C%20UNESCO's%20Man,between%20people%20and%20their%20environments. Accessed on 21 Jan 2025.

UNESCO. (1975). *World heritage convention.* https://whc.unesco.org/en/statesparties. Accessed 21 Jan 2025.

Usman, A., Hassan, M., Rehman, Z. U., & Sial, A. Q. (2022). Legal framework in aid of biological diversity and statutory deficiencies in Pakistan. *Brazilian Journal of Biology, 84*, e262695.

Vasilijević, M., & Pezold, T. (Eds.). (2011). *Crossing borders for nature: European examples of transboundary conservation.* IUCN.

Weiss, E. B., & Jacobson, H. K. (Eds.). (2000). *Engaging countries: Strengthening compliance with international environmental accords.* MIT Press.

Wolmer, W. (2003). Transboundary conservation: The politics of ecological integrity in the Great Limpopo Transfrontier Park. *Journal of Southern African Studies, 29*(1), 261–278.

World Health Organization. (2001). *What do we know about capacity building? An overview of existing knowledge and good practice.*

WWF. (2007). https://wwf.panda.org/discover/knowledge_hub/where_we_work/borneo_forests/. Accessed 22 Jan 2025.

WWF. (2016). https://wwf.panda.org/discover/knowledge_hub/all_publications/living_planet_report_timeline/lpr_2016/. Accessed 5 Apr 2025.

WWF. (2021). https://asiapacific.panda.org/priority_places/altai_sayan_ecoregion/#:~:text=In%20order%20to%20protect%20this,snow%20leopards%2C%20saiga%20and%20argali. Accessed 23 Jan 2025.

Wyatt, T. (2009). Exploring the organization of Russia Far East's illegal wildlife trade: Two case studies of the illegal fur and illegal falcon trades. *Global Crime, 10*(1–2), 144–154.

Yang, S. C. W., & Lye, L. H. (2018). Wildlife trade in China and Vietnam: A comparative study of the implementation of the Convention on International Trade in Endangered Species of Wild Fauna and Flora (CITES). In *Sustainability matters: Environmental management in the anthropocene* (pp. 25–57). National University of Singapore: World Scientific Publishing Co. Pte. Ltd.

Yang, F., & Yang, M. (2019). Greening the one belt and one road initiative. *Mitigation and Adaptation Strategies for Global Change, 24*, 735–748.

Yellowstone to Yukon Conservation Initiative. (1877). https://y2y.net. Accessed 22 Jan 2025.

Zambezi Warercourse Commission (ZAMCOM). (2014). https://www.zambezicommission.org/zsp/zambezi-watercourse-commission-zamcom. Accessed 22 Jan 2025.

Part III
Ecological Impacts of Wildlife Trade in India

Chapter 6
Indian Species Under Threat: Focus on Endangered Fauna

Ramneek Kaur, Radhika Sharma, and Daamini Arora

Contents

R. Kaur (✉) · D. Arora
PG Department of Zoology, Dev Samaj College for Women, Ferozepur, Punjab, India

R. Sharma
Department of Biology and Environmental Sciences, College of Basic Sciences, Chaudhary Sarwan Kumar Himachal Pradesh Krishi Vishvavidyalaya, Palampur, Himachal Pradesh, India

A. Poonia et al. (eds.), *Wildlife Trade in India*,
https://doi.org/10.1007/978-981-95-3694-8_6

Abstract India is home to a wide variety of rare animals, many of which are seriously threatened by illicit wildlife trade and poaching. With an emphasis on the animals that are the targets of wildlife trafficking, this chapter examines the complex problem of endangered animal species in India. The first part of the chapter looks at the main factors that influence the wildlife trade, such as the desire for exotic pets, traditional remedies, and fashion. It draws attention to the predicament of a number of seriously endangered animals that are at the center of the illegal trade, including rhinoceroses, pangolins, and tigers. The chapter goes into greater detail into the ecological effects of overexploitation, examining how species extinction affects community structures, interferes with ecosystem processes, and hastens habitat degradation. Among the main ecological effects of poaching are the loss of keystone species, species co-extinction, and the spread of invasive species. The chapter describes key conservation tactics to address these issues, ranging from community-based conservation initiatives to the creation of national parks and protected regions. It also examines how crucial laws, captive breeding initiatives, restoration ecology, and public awareness efforts are to reducing the risks to India's threatened wildlife. Through a thorough examination of these problems and solutions, the chapter highlights how urgently coordinated efforts are needed to protect India's biodiversity.

Keywords Wildlife trade · Endangered species · Conservation strategies · Ecological impact

Abbreviations

CITES	Convention on International Trade in Endangered Species of Wild Fauna and Flora
COVID-19	Coronavirus Disease 2019
H1N1	A type-1 hemagglutinin (H) protein and a type-1 neuraminidase (N) protein
H1N5	A type-5 hemagglutinin (H) protein and a type-1 neuraminidase (N) protein
HIV	Human Immunodeficiency viruses
IUCN	International Union for Conservation of Nature
IUWT	Illegal or unsustainable wildlife trade
IWT	Illegal Wildlife Trade
MoEF	Ministry of Environment and Forests
NGOs	Non-Governmental Organizations
NTCA	National Tiger Conservation Authority
UNODC	United Nations Office on Drugs and Crime
USA	United States of America

6.1 Introduction

Numerous species that have adapted to India's diverse environments make up the country's astoundingly diverse wildlife. Wildlife plays a crucial role in maintenance of ecological balance and biodiversity (Zhou et al., 2016). However, this rich biodiversity is increasingly under threat, with many species facing the dire prospects of extinction. According to the International Union for Conservation of Nature (IUCN), a species is considered endangered if it faces the possibility of going extinct in the near future. This classification also includes critically endangered and vulnerable species, other than endangered each facing varying levels of threat. Major reasons for extinction are poaching and illegal trade of wild animals (Carter & Linnell, 2016; Anagnostou & Doberstein, 2022). There is a demand of various goods made from wild species their body parts which increase the rate of wildlife crime (Toland et al., 2020). Currently, most threatened species include Tigers, Rhinoceros, Elephants, Pangolins, and many more (Gaubert et al., 2017). The increase in demand of skin of tiger, horns of rhinoceros, ivory of elephant, scales of pangolin responsible for increase in poaching of these species. Out of total of 31,500 about 7500 species of amphibians, reptiles, aves and mammals are traded illegally and pushing these species at the risk of extinction (Scheffers et al., 2019). The Wildlife Protection Act 1972, prohibited the trade of around 2000 species of flora and fauna. India has made significant efforts to protect and increase the population of tigers up to 3000 tigers which mainly found in Madhya Pradesh, Uttarakhand and Karnataka. Despite of all the laws and efforts, the trade of wildlife species and their derivatives are still going on (Nittu et al., 2023).

According to the United Nations Office on Drugs and Crime (UNODC) classification, the primary classes of animals included illegal trade are:

- **Mammalia**: This includes Indian pangolins, Asian elephants, Bengal tigers, one-horned rhinoceros, leopards, snow leopards, red panda, Indian giant squirrel rhesus macaque, Himalayan black bear etc. These species traded for their body parts such as ivory, horns, bones etc.
- **Aves**: Mostly birds traded as exotic pets. The reasons of trade also include their eggs and feathers. The birds involve in illegal wildlife trade includes alexandrine parakeet, owls, hill mynah, Indian grey hornbill etc.
- **Reptilia**: The reptiles such as Indian rock python, king cobra, Indian star tortoise, Indian flapshell turtle, black pond turtle, gharial, Bengal monitor lizard etc. are traded for medicinal purposes, skin etc.
- **Amphibia**: Amphibians are not commonly traded but Asian common toad, Indian bullfrog, fire salamander and Himalayan newt are traded for perceived medicinal values.
- **Pisces**: Fishes are mainly traded for aquariums but also used food, decorative and medicinal purposes. Indian glass fish, tiger barb, rainbow snakeheads, gold fish, red line torpedo barb and others are a part of fish trade (UNODC, World Wildlife Crime Report, 2020).

6.2 Reasons for Wildlife Trade

According to UNODC, main reasons of illegal wildlife trade (IWT) are (Fig. 6.1):

- **Luxury, Goods and Fashion**: This involves the accessories, clothes and footwears made from animal products such as fur, leather, skin, shells, even bones. Belts, shoes, fur coats etc. (Sosnowski & Petrossian, 2020) are mainly made from the animal products. About 90% trade for fur is legal but illegal trade includes the skin of pythons and crocodiles (Andersson et al., 2021).
- **Food**: The animal and their body parts traded as a source of wild food. This includes indian gaur, deer, elephants, turtle, tortoise, pangolins, grey francolin and ducks (Nasi & Fa, 2015). In rural and tribal areas, and also in lower incomes, the consumption of wild meat to get essential nutrients for body is very common (Shairp et al., 2016). The elite or upper class consume wild meat as a status symbol (Koppert & Hladik, 1990).

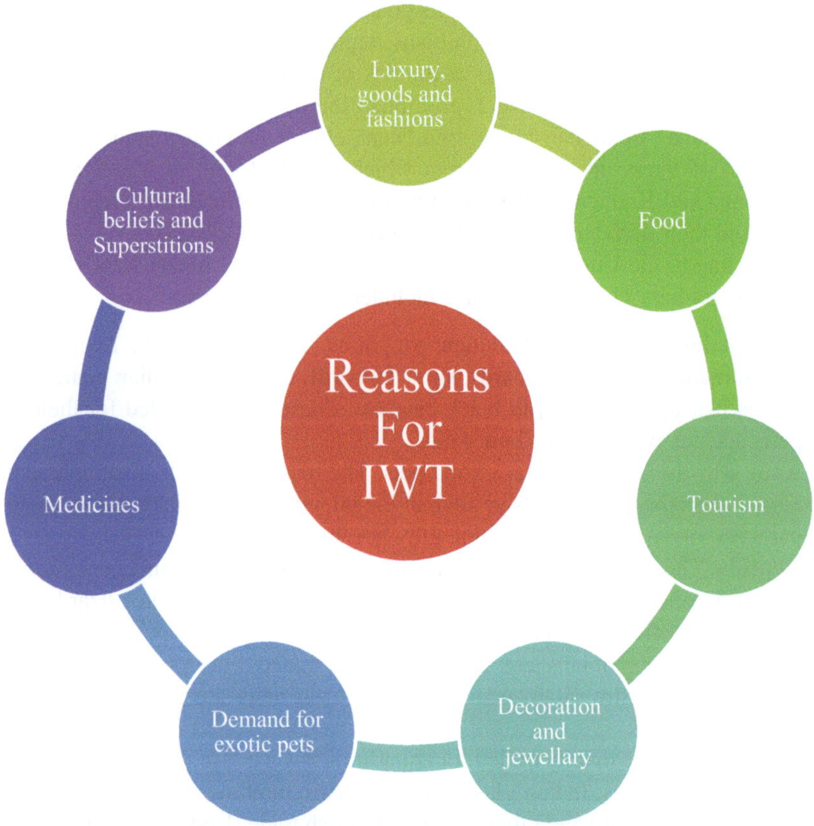

Fig. 6.1 Reasons for Illegal Wildlife Trade (UNODC, World Wildlife Crime Report: Trafficking in protected species, 2016)

- **Decoration and Jewellery**: The parts of animal body such as ivory, shells, corals etc. are used to make decorative things and jewelleries. Tiger skin, otter fur, antlers, pinned butterflies, corals, ivory, owl and eagle feathers, tortoise shell are used as decorative purposes. Turtle shell is also used in making jewellery (Proulx & Rodtka, 2019).
- **Exotic Pets**: Many species of Aves, Reptilia and Mammalia traded as pets (Bush et al., 2014). People kept exotic pets' status symbol (Tricorache et al., 2018), some used these pets for competition. Indian alexandrine parakeet, indian star tortoise, green tea python, fennec fox, slow loris are organisms which traded as exotic pets (Ribeiro et al., 2019). Sometimes exotic pets are kept in unlicensed zoos (WWF, More Tigers in American Backyards than in the Wild, 2014).
- **Medicines**: The body parts of wild animals such as horn of rhinoceros, bones of tiger and scales of pangolin are used as traditional medicines (UNODC, World Wildlife Crime Report, 2020). About 500 animals' species used for medicinal purposes (da Nóbrega Alves et al., 2008). About 53% species of reptiles which are used in medicines are endangered now. Pangolin scales, bear bile, bones of tiger, musk gland secretion of musk deer and brain of rhesus macaque are used in making of traditional medicines (UNODC, World Wildlife Crime Report: Trafficking in protected species, 2016).
- **Tourism**: The tourism industry also fueled a IWT (Rosen & Smith, 2010). Every coin has two sides as with eco-tourism, on the one hand government and other agencies trying to promote eco-tourism for enhance the efforts of conservation but on other hand it also increases exploitation of animal species (Buckley et al., 2016). Profit making management agencies also involved in illegal wildlife trade to gain benefit from eco-tourism. For example, Tigers are a centre of attraction for tourists. They tried to capture photographs of tigers, also tried to touch and hold them. For this, agencies cut claws and teeth of tigers also drugged the tigers (Hayman & Brack, 2002).
- **Cultural beliefs and Superstitions**: Various countries, especially India, is a hub of illegal wildlife trade for numerous rituals, black magic and supersitipn beliefs (Vyas, 2012). Many wild species of animals used in black magic practices (Desai, 2011). Animals like owls, snakes, hyenas, dogs, cats, bats and vultures. Beliefs related with black magic is curing of diseases like HIV, asthma, to destroy enemies, to gain good luck (Vyas, 2012). The demand of these animals also increases in international market which eventually leads to the smuggling of preferred species at very high prices (Marshall et al., 2020).

6.3 Endangered Animal Species Targeted by Trade

There are numerous numbers of animals belonging from different phylum and classes traded illegally worldwide (Table 6.1). Some of them are:

Table 6.1 List of some critically endangered, endangered and vulnerable species of India (Birdlife International, 2000)

Common name	Scientific name	Class	Category
Asian Brown Tortoise	*Manouria emys*	Reptilia	Endangered
Chitra Turtle	*Chitra indica*		Endangered
Gharial	*Gavialis gangeticus*		Endangered
Mugger or Marsh Crocodile	*Crocodylus palustris*		Vulnerable
Three-striped Roofed Turtle	*Kachuga dhongoka*		Endangered
Black-necked Crane	*Grus nigricollis*	Aves	Vulnerable
Forest Owlets	*Athene blewitti*		Critically endangered
Green Peafowl	*Pavo muticus*		Vulnerable
Indian Skimmer	*Rynchops albicollis*		Vulnerable
Manipur Bush-quail	*Perdicula manipurensis*		Vulnerable
Purple Wood-Pigeon	*Columba punicea*		Vulnerable
Rufous-necked Hornbill	*Aceros nipalensis*		Vulnerable
Spotted Greenshank	*Tringa guttifer*		Endangered
White-bellied Heron	*Ardea insignis*		Endangered
White-rumped Vulture	*Gyps bengalensis*		Critically endangered
Wood Snipe	*Gallinago nemoricola*		Vulnerable
Asian Elephant	*Elephas maximus*	Mammalia	Endangered
Clouded Leopard	*Neofelis nebulosa*		Vulnerable
Gangetic Dolphin	*Platanista gangetica*		Endangered
Golden Langur	*Trachypithecus geei*		Endangered
Great Indian One-horned Rhinoceros	*Rhinoceros unicornis*		Endangered
Himalayan Black Bear	*Ursus thibetanus*		Vulnerable
Indian Pangolin	*Manis crassicaudata*		Endangered
Kashmir Cave Bat	*Myotis longipes*		Vulnerable
Red Panda	*Ailurus fulgens*		Endangered
Sloth Bear	*Melursus ursinus*		Vulnerable
Stump-tailed Macaque	*Macaca arctoides*		Vulnerable
Tiger	*Panthera tigris*		Endangered
Wild Yak	*Bos grunniens*		Vulnerable

*6.3.1 Tigers (*Panthera tigris*)*

Tigers are one of the strongest and most popular mammalian species in the world belongs to the order carnivora and family felidae. They acquire a high trophic level in ecosystem. Few centuries ago, there were about 58,000 tigers present in India (Mondol et al., 2009; Sharma et al., 2014). In 1972, the population of tigers decreased to 1800 (Panwar, 1987). In 2008, it only remained about 1400 (Jhala et al., 2008). The main reasons of decline of tiger population is poaching, habitat loss and decrease in prey population (Chapron et al., 2008). In 1960, the U.S. and Europe became a fashion hub for coats made from tiger skin. In 1968, the government issued licenses for the export of 3000 tiger skins and only permitted the hunting of 500 tigers which indicates that 2500 skins obtained illegally. In 1967, the International Union for Conservation of Nature (IUCN) recommended to put restrictions on import and export of tiger skin. In 1968, India banned the export of tiger and leopard skins. Later they relaxed ban to fulfil the commitments made prior to the ban which resulted in trade of 5000 skins of tigers and leopard only by traders of Delhi (Wright, 1969). In 1970, a total ban on shooting tigers and significant act was passed in 1972, named, Wildlife Protection Act. In 1973, the most important project was launched to conserve tigers, named, Project Tiger. Despite of all the efforts, the population of tigers decreased continuously. The poaching and illegal trade of tigers and their skin still continued due to its high demand in China and South East Asia (Wikramanayake et al., 2010). This put the species at the risk of extinction (Dinerstein et al., 2007). To protect tigers and their habitat, many conservation areas have been created where human interference was totally prohibited. After all the efforts, project tiger proved very fruitful in increasing the population of tigers from about 1200 to 5000. The project tiger is still in continuation and will continue until tigers do not come out from the category of endangered species (Yashaswini, 2021). Currently, tigers across their range are primarily threatened by direct poaching and secondly by prey depletion (Chapron et al., 2008). As long as there is a demand for tiger parts and products, there will be poachers who will be willing to take high risks. The global community needs to build pressure to eradicate the demand for wildlife parts and products.

*6.3.2 Pangolins (***Manis crassicaudata***)*

Pangolin is a mammal which belongs to order Pholidota and family Manidae. About eight species of pangolins widely distributed in Asia and Africa. Earlier Indian pangolins were distributed in Myanmar and southwest China. The hunting, poaching and habitat destruction leads to the extinction of pangolins in some regions. Other reasons for loss of pangolins are forest fires, construction of dams, agriculture expansion and urbanization (Molur et al., 2005; Mahmood et al., 2012). The pangolins are mainly hunted for their scales as well as for consumption as its meat contains a rich amount of protein and also for traditional medicines. Due to its nutritional value and medicinal value, pangolins a traded illegally at extreme level. Pangolins also have ecological importance as they help in aerating soil, control pests and rotate mineral cycle (Hansell, 1993). Indian pangolins along with other species of pangolins (Sunda Pangolin and Chinese Pangolin) included in endangered and critically endangered respectively in IUCN red list (Baillie et al., 2014). Tracking and trapping are the common methods for pangolin poaching (Simo et al., 2023). Currently, pangolins are considered as most trafficked animal in the world (Sutter, 2016). Uttarakhand in India becomes a hotspot of illegal trade of pangolins

(Mohapatra et al., 2015). In India, due to declining population of pangolins, they are included under a Wildlife Protection Act, 1972. Efforts to protect pangolins from extinction and to remove this from the list of endangered species are required.

*6.3.3 Rhinoceros (***Rhinoceros unicornis***)*

Rhinoceros is one of the most iconic mammalian species belongs to the order perissodactyla and family rhinocerotidae. By the early of twentieth century, only 200 rhinoceros left in India and were at the risk of extinction (Rookmaaker et al., 2016). The main reason of extinction of rhinoceros is poaching, habitat loss, climate change and illegal trade of their horns (Talukdar et al., 2008). Some people also consider possession of rhinoceros horn to be a symbol of social status. Also, the inadequacy of suitable habitat for the survival of rhinoceros hinders the conservation efforts as they need tall grasslands (Kafley et al., 2015). They also require a waterhole to cool down their body in hot weather conditions (Talukdar et al., 2008). In 1908, less than 20 rhinoceros was remained in Kaziranga National Park and then conservation efforts were started (Laurie, 1982). These efforts have proved successful, resulting in an increase in the population of rhinoceros to 1800 by 2006 (Martin et al., 1987).

*6.3.4 White-Rumped Vulture (***Gyps bengalensis***)*

White-rumped vulture of class aves falls under accipitriformes order and accipitridae family. Before 1990, vultures are widely distributed throughout the northern states of India. It was also found in Afghanistan, Pakistan, Bhutan, Bangladesh, Thailand, Myanmar and others. During last 20 years, maximum vulture species disappeared. About 50% of decline in population of vulture was seen due to the excessive use diclofenac. Various studies also showed there is a poaching, hunting and illegal trade of vultures and their parts for bushmeat and also for traditional medicines. Over time, vultures have developed into highly skilled scavengers. The stomach of vultures is extremely acidic, with a pH ranging from little over zero to 3.9. The majority of germs cannot survive in vultures' digestive systems as a result of this adaptation, which is also one of the main factors that permits them to safely eat carrion. Vultures are therefore crucial keystone species (Roggenbuck et al., 2014). Birdlife International identified them as severely endangered in 2011. The creation of a Vulture Conservation Center in Pinjore, Nagaon, and Buxa, the implementation of a vulture advocacy program, and the nationwide prohibition of diclofenac were among the steps done to preserve white-rumped vultures. The Wildlife Protection Act of 1972 also covered white-rumped vultures. Programs for captive breeding and vulture reintroduction are further conservation efforts (Jayakumar, 2018).

6.3.5 *Forest Owlet (***Athene blewitti***)*

This critically endangered animal falls under a class aves, order strigiformes and family strigidae. These are only found in dry deciduous forests in central Indian forests. The threats to forest owlets are hunting for their body parts, and collection of eggs to use in local customs. Other threats include deforestation, forest fires and construction of irrigation dams. Usage of pesticides may also impose an additional threat to the declination of animal. This species also included in Wildlife Protection Act, 1972. Government also imposed a ban on cutting of trees in forest and hunting of these species. Various awareness programmes were also started for local communities (Jayakumar, 2018).

6.3.6 *Gharial (***Gavialis gangeticus***)*

The gharial falls under a class reptilia, family gavialidae and order crocodilia is gaint among crocodiles. The gharial (*Gavialis gangeticus*), a critically endangered freshwater reptile endemic to the Indian subcontinent, faces severe threats from various anthropogenic stressors (Dwivedi & De, 2024). There was quick decline in the population of gharials from 5000 to 10,000 in 1940 to less than 200 in 1976 (Whitaker et al., 1974; Whitaker, 1975). The main reasons for decline in their population was poaching for skin, construction of dams, irrigation channels, grazing and agricultural practices. The body parts of gharial included skin, penis, fat and meat used in making of traditional medicines. Some efforts to increase the population of gharials included introduction of species since 1977 but no positive result was seen due to lack of proper planning and monitoring (Sharma & Basu, 2004; Andrews, 2006). Gharials added in a list of critically endangered species in red data book of IUCN (IUCN, 1975). This species was also included under a Wildlife Protection Act, 1972. Conservation efforts also include a total ban on hunting of animal and export of their skin and other derivative products.

6.3.7 *Asian Elephant (***Elephas maximus***)*

Asian elephants comes under class mammalia, elephantidae family and proboscidea order. There was time when asian elephants prominently found in western asia but now these are not even present western India. Asian elephants were widely distributed in nine million square kilometer which now limited only upto 5 lakh square kilometer (Sukumar, 2003). A main reason of declination of elephant population is habitat loss and poaching. Habitat loss is due to increase in human population, development projects, deforestation and urbanization. Illegal poaching of elephants mainly because of the trade of their ivory. Human elephant conflict also had one of the significant impact on declining population of elephants. As human population is growing rapidly increases their competition with elephant population which led to a killing of elephants (AsERSM, 2006). Also capturing, training and poor management of elephants led to their deaths (Bandara & Tisdell, 2003). In 2006, IUCN listed asian elephant as an endangered animal in the red list of threatened species (Fernando et al., 2000). Currently, there are numerous efforts at various levels to conserve asian elephants include reduction of deforestation and complete prohibition on poaching of elephants. Goverments also framed various policies to protect elephants but still there is a need of good monitoring, more strict penalties and punishments, and fully trained and equipped persons (Stephenson & Ntiamoa-Baidu, 2010).

*6.3.8 White-Bellied Heron (*Ardea insignis*)*

White-bellied Heron falls under class aves, ardeidae family and pelecaniformes order. This rare species is now considered as a critically endangered species in the world. About 60 white-bellied heron species are existing today. The major reason of declination of white-bellied heron is loss of their natural habitat and construction of various hydroelectric projects (Stanley Price & Goodman, 2015). Many incidences highlighted that forest fires also threatens their nests (Dendup, 2016). The construction of dams change the habitat of river which affects the piscivorous birds (fish eating birds). These birds are also under the threat of hunting (Aiyadurai et al., 2010). In 2007, they are listed as critically endangered bird (BirdLife International, 2017). Numerous efforts are still proceeding to protect white-bellied herons and increase their number.

*6.3.9 Golden Langur (*Trachypithecus geei*)*

This endangered species comes under class mammalia, order primates and family ceracopithecidae. The main reason of reducing number of golden langur is a loss of their natural habitat. Other reasons include their illegal hunting and environmental stress. In 2006, this included in the red list of threatened species of IUCN as critically endangered species. This species included in a Wildlife Protect Act, 1972 (Srivastava, 1999). This prohibits the illegal capturing and poaching of golden langurs. Many Non-Governmental Organizations (NGOs) are participating in conserving golden langurs and their habitat mainly by restoring a forests (reforestation) (Community Conservation, 2006). NGOs and Government also doing efforts to provide awareness to local communities about conservation of golden langurs.

6.4 Influence of Loss of Community Structure on Ecosystem Function

The intricate relationships between community structure and ecosystem functions are vital for maintaining ecological balance and resilience. Community structure refers to the composition and arrangement of species within an ecosystem, while ecosystem function pertains to the processes that occur within an ecosystem, such as nutrient cycling, energy flow, and interactions among organisms. The loss of community structure, particularly the decline or extinction of key animal species, has profound implications for ecosystem functioning. There are various components of biodiversity which are important for the functioning of ecosystem. These components include richness, evenness, composition of species their interactions and their temporal and spatial variations (Symstad et al., 2003). Species richness is essential for differentiating ecological roles for ecosystem functioning such as decomposition of organic matter, retention of minerals and primary productivity. This is also important of composition of species (Tilman et al., 1996). Change in species is very common and possess significant consequences for functioning of ecosystem (Estes et al., 1989). Karp et al. (2012) discuss how invasive species can outcompete native organisms, leading to reduced biodiversity and altered food webs. This disruption can diminish ecosystem services, such as water quality and habitat stability, emphasizing the importance of managing invasive species to protect native biodiversity.

Species composition also influence a functioning of ecosystem. Particular species has specific effect on ecosystem which able to alter the functioning of ecosystem (Hooper & Vitousek, 1997). Burkle and Alarcón (2011) illustrate how the loss of large herbivores can alter tree population dynamics and reduce species diversity within forest ecosystems. This disruption not only affects tree regeneration but also impairs nutrient cycling, leading to a decline in overall forest health and ecosystem resilience.

Species interaction mutualism, parasitism, predation and others may affect functioning of ecosystem by changing the efficiency of energy, flow of materials and energy pathways (Power et al., 1996). Reau and Suen (2018) note that alterations in species composition and community dynamics due to climate stressors can lead to significant shifts in ecosystem services, affecting human well-being. The interconnectedness of community structure and ecosystem function is evident, as the loss of biodiversity can compromise ecosystems' ability to provide essential services.

6.5 Ecological Consequences of Over-Exploitation of Species

Over exploitation of various species has direct impact on other species, includes, animals, plants, fungi, humans and many others. It also alters the abiotic components of ecosystem. Over exploitation and extinction of species disturbs the functioning of

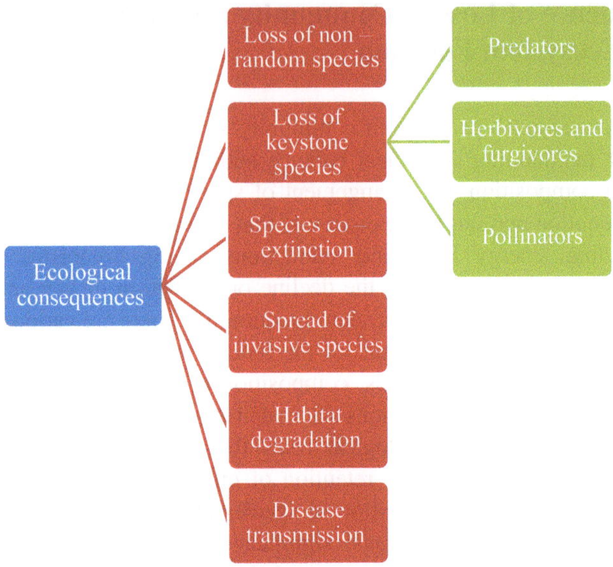

Fig. 6.2 Ecological Consequences of over exploitation and extinction of Species

ecosystem (Fig. 6.2). It threatens their existence and sometimes eventually leads to their extinction.

6.5.1 Loss of Non-random Species

Non-random species are those species that can alter in composition in a non-random way. This includes a rare species, dominant species etc. If these species is over-exploited or become extinct it has an unavoidable consequence for functioning of ecosystem. For example, large vertebrate consumers are more at the risk of extinction. These species have significant impact on structure and function of ecosystem (Larsen et al., 2005). These targets the processes like predation, selective grazing, regeneration of nutrients and others. Declination or extinction of specific species also determine the Extinction of these species also responsible for the alteration of behaviour of remaining species which tries to compensate the loss of extinct species. (Walker, 1992).

6.5.2 Loss of Keystone Species

Keystone species are living species which plays an important role in the maintenance of structure and function of an ecosystem. This includes predators, prey and pollinators etc. These are present in every ecosystem, may or may not inhibit higher trophic level and influence various communities through ecological interactions, pollination and altering abiotic factors and habitat loss (Bond, 1993). Loss of keystone species impact the structure and function of an ecosystem to a great extent (Mills et al., 1993). Removal of predators of higher trophic level results in increased herbivory which affects primary productivity of ecosystem. For example, studies suggested that sea otters play an important role in maintaining densities of herbivorous population of sea urchin. Sea urchins mainly feed on kelp beds (seaweeds). Removal of sea otters due to poaching for their fur resulted in high densities of herbivores i.e. sea urchin which increases herbivory and leads to the depletion of kelp beds (Estes et al., 1998).

Densities of small predators i.e. mesopredators also released after the loss of predators which leads to the declination and extinction of prey species (Soule et al., 1988). For example, the declination of one of the most large predators coyote (American Jackal) resulted in release of mesopredators including racoon, grey fox and others (native species) and opposum, cats and others (exotic species) resulted in increased mortality and extinction of scrub-breeding birds such as Scrubfowl, Scrub Jay and others. These studies revealed that the loss of these birds was much more in the presence of mesopredators than in the presence of coyotes (Crooks & Soule, 1999).

Loss of herbivores and frugivores disrupts herbivory and seed dispersal (Redford, 1992). For example, there is a comparison between two tropical forests, one with composed of large mammals and other in which species are eradicated by hunters. Studies in Central Panama revealed that the hunters reduced the density of herbivorous mammals which altered the seed dispersal and seedling recruitment for *Attalea butyraceae* (Wright et al., 2000). In the Philippine islands, the removal of frugivore species represented an impact on regeneration of forests in wet tropical rainforests. Studies revealed that the dispersal of seeds of climax species was done by endangered species like fruit pigeons and hornbill. A dispersal species was at the risk of extinction due to their over-hunting. These climax species also suspected to extinction due to the loss of their dispersers (Hamann & Curio, 1999). A study in a rainforest of Peru revealed an inter-relation among seed predators, seed dispersal and seed dispersers which decide the fate of seeds. The seed dispersal was done by primates including howler monkeys and spider monkeys which dispersed the seeds of various plant species (Andresen, 1993). Studies in deciduous forest of Madagascar indicated the importance of primates in dispersal of seeds. Loss of pollinators affects the mating system of plants and produce more vulnerable plants. Absence of pollinators increased self-pollination and decreased heterozygosity and increased the trait of inbreeding. Their loss reduced production of seeds and fruits which lead to the extinction of plant species (Kearns & Inouye, 1997). For example, flying foxes (mammal) have great ecological importance but are at the risk of extinction due to over-exploitation. Some species of flying foxes were already extinct. About 290 species of plants rely on this mammalian species (Fujita & Tuttle, 1991). These plants also have a great economic importance i.e. they produce about 450 valuable products. There are various evidences are available which showed the importance of flying foxes as pollinators (Cox et al., 1991).

6.5.3 Species Co-extinction

The species co-extinction means the loss of parasite species along with their host (Stork & Lyal, 1993). This concept also describes in case of loss of predators with their prey and herbivorous animal with their host plant. For example, specificity of larval host plant was proved an important ecological factor of butterfly disturbance in Singapore (Koh et al., 2004). The speed of extinction of butterfly rapidly increased with their host plant. Butterflies may extinct earlier than their host plant. The declination of host plant decreased the number of butterflies. The loss of butterflies also decreases the number of various flowering plants which depends on butterflies for their pollination. The loss of host plants derives a extinction of butterfly as their caterpillars (larval form of butterfly) are dependent more on plants for their feeding than the dependence of plants on butterfly for pollination (Corlett, 2004). Studies revealed that about 6300 species are co-endangered and at the risk of co-extinction with their host species. Other examples of species co-extinction is *Ficus* wasp and *Ficus*, butterflies and their larval host plants, parasite and their hosts and many others (Purvis et al., 2000).

6.5.4 Spread of Invasive Species

Impacts seen on areas where traded species are introduced. One of the most significant impacts of introducing invasive species are on ecosystem services which can suffer because of trophic cascade (Pyšek et al., 2020). Invasive or alien species also a threat to native system. In many cases we have seen the extinction of native species due to introduction of invasive species. These species compete with native species for various resources. For example, the introduction of trout *Salvelinus namaycush* in 1937 in Alpine lake of North America led to the extinction of endemic fish *Orestias cuvieri* (Villwock, 1972). *Salvelinus namaycush* also been depleted in its natural habitat due to the introduction of sea lampreys (Schneider et al., 1996).

6.5.5 Habitat Degradation

Over-exploitation often leads to habitat degradation, which occurs when natural habitats are altered or destroyed, making them less suitable for wildlife. Logging for timber, agricultural expansion, and urban development can result in significant habitat loss. The conversion of forests into agricultural land not only destroys the habitats of countless species but also disrupts the soil structure and leads to erosion.

For example, the over-fishing of shark species in marine ecosystem disturbs the food web, mass increase in number of few other species, particularly herbivores which resulted in the destruction of seagrass and coral reef habitats and changes the structure of marine habitat (Ferretti et al., 2010).

6.5.6 *Disease Transmission*

Encroachment of natural habitats due to hunting and IUWT increases a direct contact between humans and wild animals. This eases the transmission of various zoonotic diseases (Allen et al., 2017). Most of emerging infectious disease outbreak affecting humans include a type-1 hemagglutinin (H) protein and a type-1 neuraminidase (N) protein (H1N1) or swine flu, a type-5 hemagglutinin (H) protein and a type-1 neuraminidase (N) protein (H5N1) or bird flu, Ebola, Human Immunodeficiency viruses (HIV), possibly Coronavirus Disease 2019 (COVID-19) have zoonotic origin. The global trade of wildlife species breaks the barriers ecologically and geographically (Jones et al., 2008). For example, Beak and Feather Disease Virus which causes diseases to the birds globally and negatively impacting wild population of threatened wild parrots. These parrots have now spread world-wide due to their global trade (Fogell et al., 2018).

6.6 Conservation Priorities and Strategies

In order to mitigate the effects of overexploitation and safeguard endangered species and their habitats, conservation initiatives are essential. Many nations, including India, have achieved notable progress in animal conservation through a combination of legislative action, community engagement, habitat restoration, and creative tactics. The majority of the world's biodiversity is found in developing nations that need assistance from donors to increase their capacity for conservation (Smith et al., 2003). Proper scientific quantification and areas where focus must be maintained on an as-needed basis are necessary for donor assistance.

India faces different conservation challenges than other major nations like China, Brazil, and the United States. These nations' sparsely populated areas, which also have relatively low biodiversity values in the USA and China, are home to sizable protected areas (Pimm et al., 2018). Rather, millions of people in India live within a few kilometers of protected areas, and an estimated four million of them dwell there (Narain et al., 2005), while the exact number is unknown and hasn't been updated this century. For those in charge of managing India's biodiversity, this poses significant difficulties (Ghosh-Harihar et al., 2019).

6.6.1 Protected Areas and National Parks

Up to 15.5% of the earth's land surface is made up of protected areas, which are one of the most crucial tools for preserving species variety and habitat integrity (Geldmann et al., 2013). India formally preserves 5% of its land (Dinerstein et al., 2017). A key component of conservation initiatives is the creation of protected areas, such as national parks and wildlife sanctuaries. These places aid in the preservation of vital habitats and offer safe havens for endangered animals. According to the Geldmann et al. review, protected areas are a crucial component of conservation plans aimed at maintaining tropical forests. There are currently 551 Wildlife Sanctuaries and 104 National Parks in India that are officially protected areas (www.wiienvis.nic.in); however, 5 National Parks and 79 Sanctuaries are less than 1 km^2 (many of which are islands).

According to the International Union for the Conservation of Nature (IUCN), Category II (www.iucn.org), National Parks are legally exempt from human interference and enjoy the greatest level of protection. IUCN category IV permits restricted usage of wildlife sanctuaries. In 2002, India added two more legal classifications to its protected areas (Saigal et al., 2018). Approved management plans serve as the primary foundation for the establishment of Conservation Reserves on public land and Community Reserves on private land (IUCN Categories V and VI, respectively). More than 70% of the 214 reserves (a total area of 4811 km^2) created by 2019 are located in just three states: 34 Conservation Reserves in Jammu & Kashmir, and 122 Community Reserves in Meghalaya and Nagaland.(Anil et al., 2014) and Ghosh-Harihar et al. (2019).

In India, well-known parks like Jim Corbett National Park, which is home to the Bengal tiger, and Kaziranga National Park, which is renowned for its achievements in rhino conservation, serve as examples of how protected places can aid in population recovery.

6.6.2 Community-Based Conservation

Engaging local communities in conservation efforts has proven effective in achieving sustainable outcomes. By involving communities in decision-making processes and providing alternative livelihoods, conservation initiatives can reduce reliance on resource exploitation. Traditional human relationships like beliefs, faith, taboos, customs and preferences played an important role in conservation of habitats and individual species. The cultural ethos of the Indian people is amply demonstrated by such conservation efforts (Gadgil, 1991). Frequently, species selected by the local people for social significance turn out to be also of ecological significance (Ramakrishnan, 1996; Anil et al., 2014). Programs such as the "Eco-Development Committees" in India empower local populations to manage and benefit from nearby protected areas. Success stories include the involvement of indigenous communities in the conservation of the Great Indian Bustard, where local knowledge and practices have been integrated into conservation strategies.

6.6.3 Legislation and Policy Frameworks

Effective legal frameworks are essential to the preservation of wildlife. In India, the conservation of endangered animals and their habitats is supported by regulations like the Wildlife Protection Act of 1972. Sustainable practices are encouraged and the trade in endangered species is regulated by international agreements such as the Convention on International Trade in Endangered Species of Wild Fauna and Flora (CITES). Numerous species, such as tigers, elephants, and different bird species, now have better protection as a result of the effective enforcement of these laws (Naniwadekar et al., 2015; Velho et al., 2016).

The National Council for Environmental Policy and Planning, an environmental protection body inside the Department of Science and Technology, was established in 1972 to carry out the decisions made at the 1972 Conference on the Human Environment in Stockholm. In 1985, this Council changed its name to the Ministry of Environment and Forests (MoEF), which is currently in charge of overseeing and protecting India's environment (Sharma et al., 2014). India has formally committed to the Nagoya Protocol by ratifying it. The idea that all three of the Convention on Biological Diversity's goals-conservation, sustainable use, and benefit sharing from the use of genetic resources should be given sufficient and equal attention has served as the foundation for the strategy for preserving and advancing biodiversity. The Biological Diversity Act of 2002 in India is based on this methodology. Various government entities have identified specific action points in the 2008 National Biodiversity Action Plan. Although all of these rules and regulations have been helpful, wildlife protection still requires more stringent law enforcement.

6.6.4 Restoration Ecology

The science of restoring the biodiversity and ecological functioning of degraded, damaged, or destroyed ecosystems to civilization is known as restoration ecology. It has offered a variety of economic and ecological theories and instruments for hastening the restoration of ecosystems that have been harmed and turned out to be extremely important to people. For instance, meta-analyses of the published literature indicate that ecological restoration can enhance the provisioning of biodiversity and ecosystem services (Casazza et al., 2016), and ecosystem service valuations indicate that the economic benefits of restoration may exceed the costs (Bullock et al., 2011).

For endangered species to recover, deteriorated habitats must be restored. To restore ecosystems, governments and conservation groups have started a number of habitat restoration initiatives. For instance, initiatives to re-establish mangrove forests along India's coastlines provide coastal protection against erosion and the effects of climate change while simultaneously promoting local biodiversity. Preservation of wetlands in India is also a step in same direction. The Olive Ridley turtle, which depends on healthy beaches for nesting, benefits from successful restoration efforts that enhance habitat quality. Many endangered animals, including the Bengal tiger, depend on mangrove forests, especially the Sundarbans, for vital habitat. The necessity of conservation strategies that give priority to the preservation and restoration of these vital ecosystems is emphasized by Ghosh et al. (2015).

6.6.5 Captive Breeding and Reintroduction Programs

One facet of species conservation initiatives is captive breeding, which involves raising animals of endangered species in captivity with the intention of eventually reintroducing them to the wild. For the conservation of endangered wildlife species, captive breeding is frequently required. It keeps a species, subspecies, or population from going extinct. It's likely that small confined populations will go extinct. Their prospects of long-term survival can be significantly raised by using thorough breeding programs that are founded on the concepts of population genetics and demography. Preserving as much genetic diversity as possible within a species is a primary objective of captive breeding programs. The methods of genomics and genetics can be helpful in stopping the extinction process.

It is now feasible to directly evaluate a species genetic diversity and determine its population structure thanks to modern sequencing and genotyping techniques. For example, sequencing a small sample of geographically separated individuals can reveal several genetic variants, and genotyping techniques can effectively ascertain the genetic composition of a large number of individuals in a population (Wakchaure & Ganguly, 2016). Critically endangered animals have seen successful population increases because to captive breeding efforts. For animals like the Asiatic lion and the Pygmy hog, organizations like the Wildlife Institute of India and several zoos run breeding initiatives.

The Asiatic lion population in Gujarat's Gir Forest National Park has successfully returned as a result of reintroduction efforts. 168 bird species are raised in captivity for population augmentation and reintroduction worldwide (Wakchaure & Ganguly, 2016). These initiatives support genetic variety, increase population size, and facilitate the return of animals to their native environments. For these endangered species, either suitable breeding programs must be created or the current programs must be evaluated and improved.

6.6.6 Research and Monitoring Initiatives

Effective conservation requires constant study and observation. Management tactics are informed by scientific research that shed light on species behaviour, habitat usage, and population dynamics. Adaptive management techniques are made possible by organizations like India's National Tiger Conservation Authority (NTCA), which regularly evaluates tiger numbers via camera trapping and monitoring. In order to inform future conservation efforts, research projects also concentrate on comprehending the effects of habitat loss and climate change. Innovations in technology are essential to improving conservation methods. One interesting strategy for enhancing data gathering in difficult areas is the use of drones for monitoring critically endangered species, like the riverine rabbit (Burke et al., 2018). Furthermore, it has been noted that remotely sensed imagery is a useful tool for tracking and evaluating animal population.

Furthermore, although there are still issues with detection capabilities, remotely sensed imaging has been emphasized as a useful tool for tracking ecosystems and assessing animal abundance (Hollings et al., 2018). In order to enhance monitoring and management initiatives, these studies support ongoing conservation technology research and development. A potential approach to the protection of India's endangered species is provided by recent developments in population genetics. According to Hohenlohe et al. (2020), genomic methods can clarify population structure and genetic diversity, both of which are essential for managing threatened species effectively. Conservationists can identify at-risk populations and create focused interventions by evaluating variables like effective population size and inbreeding percentages. Additionally, knowing the genetic loci linked to adaptation may increase a species' resistance to environmental changes, guaranteeing the scientific validity of conservation initiatives.

6.6.7 Public Awareness and Education Campaigns

Gaining support requires educating the public about the value of biodiversity and conservation. The importance of preserving endangered species and their ecosystems is emphasized in education programs aimed at stakeholders, communities, and schools. To engage the public, organizations run media campaigns, outreach initiatives, and workshops. For wildlife conservation initiatives to be successful, conservation laws and regulations must be communicated effectively. Compliance may be hampered by a lack of awareness in many areas on species protection laws (Pearson et al., 2014). Effective programs like "Earth Hour" and "Wildlife Week" have inspired citizens to be more involved in conservation efforts by encouraging a sense of activism and responsibility. In a similar vein, other holidays could raise public awareness of the need to conserve these species. For example, World Rhino Day is celebrated annually on September 22 (Dutta & Das, 2022).

6.7 Conclusion

To conclude, India is recognized as a mega-diverse country with numerous types of species belonging to different phylums and classes. Due to increasing threats of illegal poaching, climate change and habitat destruction have led to extinction of few animal species and many species are at the risk of extinction as they are consider critically endangered or endangered. The loss of wildlife also disrupting an ecosystem balance drastically. The continued conservation efforts to protect to protect wildlife by centre government, state governments, NGOs and other local communities resulted in survival of critically endangered animals like Asian lion, Bengal tiger and one-horned rhinoceros. These conservation efforts include restoration of natural habitat, strict enforcement of wildlife protection laws including long term imprisonment and hefty fines, public awareness and their involvement in conservation efforts. Protection of India's endangered fauna is a necessary step to maintain an ecosystem balance which is important for the survival of life on earth and also a moral responsibility of each individual of a country to ensure the continued survival of these species for future generations.

References

Aiyadurai, A., Singh, N. J., & Milner-Gulland, E. J. (2010). Wildlife hunting by indigenous tribes: A case study from Arunachal Pradesh, North-East India. *Oryx, 44*(4), 564–572.

Allen, T., Murray, K. A., Zambrana-Torrelio, C., Morse, S. S., Rondinini, C., Di Marco, M., et al. (2017). Global hotspots and correlates of emerging zoonotic diseases. *Nature Communications, 8*, 1124. https://doi.org/10.1038/s41467-017-00923-8

Anagnostou, M., & Doberstein, B. (2022). Illegal wildlife trade and other organised crime: A scoping review. *Ambio, 51*(7), 1615–1631. https://doi.org/10.1007/s13280-021-01675-y

Andersson, A. A., Tilley, H. B., Lau, W., Dudgeon, D., Bonebrake, T. C., & Dingle, C. (2021). CITES and beyond: Illuminating 20 years of global, legal wildlife trade. *Global Ecology and Conservation, 26*, e01455. https://doi.org/10.1016/j.gecco.2021.e01455

Andresen, E. (1993). Seed dispersal by monkeys and the fate of dispersed seeds in a Peruvian rainforest. *Biotropica, 31*(1), 145–158. https://doi.org/10.1111/j.1744-7429.1999.tb00127.x

Andrews, H. (2006). *Status of the Indian gharial (Gavialis gangeticus), conservation action and assessment of key locations in North India.* Madras Crocodile Bank Trust Report.

Anil, M. N. V., Kumari, K., & Wate, S. R. (2014). Loss of biodiversity and conservation strategies: An outlook of Indian scenario. *Asian Journal of Conservation Biology, 3*(2), 105–114.

Asian Elephant Range States Meeting (AsERSM). (2006). *Asian Elephant Range States Meeting, January 2006, Kuala Lumpur, Malaysia.* IUCN-SSC Report.

Baillie, J., Challender, D., Kaspal, P., Khatiwada, A., Mohapatra, R., & Nash, H. (2014). Manis crassicaudata. *The IUCN Red List of Threatened Species, 2014*, e.T12761A45221874. [Internet]. [cited 2019 Feb 10]. Available from: https://www.iucnredlist.org

Bandara, R., & Tisdell, C. (2003). Comparison of rural and urban attitudes to the conservation of Asian elephants in Sri Lanka: Empirical evidence. *Biological Conservation, 110*(3), 327–342.

BirdLife International. (2000). *Threatened birds of the world.* BirdLife International and Lynx Edicions.

BirdLife International. (2017). Ardea insignis. *The IUCN Red List of Threatened Species, 2018*, e. T22697021A134201407. [Internet]. [cited 2025 Apr 17]. Available from: https://doi.org/10.2305/IUCN.UK.2018-2.RLTS.T22697021A134201407.en

Bond, W. J. (1993). Keystone species. In D. Schulze & H. A. Mooney (Eds.), *Ecosystem function and biodiversity* (pp. 237–253). Springer.

Buckley, R. C., Morrison, C., & Castley, J. G. (2016). Net effects of ecotourism on threatened species survival. *PLoS One, 11*(12), e0147988. https://doi.org/10.1371/journal.pone.0147988

Bullock, J. M., Aronson, J., Newton, A. C., Pywell, R. F., & Rey-Benayas, J. M. (2011). Restoration of ecosystem services and biodiversity: Conflicts and opportunities. *Trends in Ecology & Evolution, 26*(10), 541–549.

Burke, C., Rashman, M., Wich, S., Symons, A., Theron, C., & Longmore, S. (2018). Optimizing observing strategies for monitoring animals using drone-mounted thermal infrared cameras. *International Journal of Remote Sensing, 40*(2), 439–467. https://doi.org/10.1080/01431161.2018.1558372

Burkle, L., & Alarcón, R. (2011). The future of plant-pollinator diversity: Understanding interaction networks across time, space, and global change. *American Journal of Botany, 98*(3), 528–538. https://doi.org/10.3732/ajb.1000391

Bush, E. R., Baker, S. E., & Macdonald, D. W. (2014). Global trade in exotic pets 2006–2012. *Conservation Biology, 28*(3), 663–676. https://doi.org/10.1111/cobi.12240

Carter, H. N., & Linnell, D. C. J. (2016). Co-adaptation is key to coexisting with large carnivores. *Trends in Ecology & Evolution, 31*(8), 575–578. https://doi.org/10.1016/j.tree.2016.05.006

Casazza, M. L., Overton, C. T., Bui, T. V. D., Hull, J. M., Albertson, J. D., Bloom, V. K., et al. (2016). Endangered species management and ecosystem restoration: Finding the common ground. *Ecology and Society, 21*(1), 19.

Chapron, G., Miquelle, D. G., Lambert, A., Goodrich, J. M., Legendre, S., & Clobert, J. (2008). The impact on tigers of poaching versus prey depletion. *Journal of Applied Ecology, 45*(6), 1667–1674. https://doi.org/10.1111/j.1365-2664.2008.01538.x

Community Conservation, Inc. (2006). *Community conservation assists in golden langur conservation*. Community Conservation Report.

Corlett, R. T. (2004). Flower visitors and pollination in the oriental (Indo-Malayan) region. *Biological Reviews of the Cambridge Philosophical Society, 79*(3), 497–532.

Cox, P. A., Elmqvist, T., Pierson, E. D., & Rainey, W. E. (1991). Flying foxes as strong interactors in South Pacific Island ecosystems: A conservation hypothesis. *Conservation Biology, 5*(4), 1–7.

Crooks, K. R., & Soule, M. E. (1999). Mesopredator release and avifaunal extinctions in a fragmented system. *Nature, 400*, 563–566. https://doi.org/10.1038/22941

Dendup, S. (2016). *Fire impact on White-bellied Heron habitats along the Bertichu, Zhemgang Dzongkhag* [Undergraduate dissertation]. Royal University of Bhutan, College of Natural Resources, Lobesa, Punakha, Bhutan.

Desai, A. (2011). *Sarp Sandarbh: Information about the snakes of Gujarat*. Prakruti Mitra Mandal.

Dinerstein, E., Loucks, C., Wikramanayake, E., Ginsberg, J., Sanderson, E., Seidensticker, J., & Songer, M. (2007). The fate of wild tigers. *Bioscience, 57*(6), 508–514. https://doi.org/10.1641/B570608

Dinerstein, E., Olson, D., Joshi, A., Vynne, C., Burgess, N. D., Wikramanayake, E., et al. (2017). An ecoregion-based approach to protecting half the terrestrial realm. *Bioscience, 67*(6), 534–545. https://doi.org/10.1093/biosci/bix014

Dutta, H., & Das, S. (2022). A historic event marks World Rhino Day 2021 in India. *Oryx, 56*(3), 331.

Dwivedi, A. K., & De, K. (2024). Proposal on species-specific synergistic conservation measures for critically endangered gharial (*Gavialis gangeticus*). *Aquatic Conservation, 34*(11), e70010. https://doi.org/10.1002/aqc.70010

Estes, J. A., Duggins, D. O., & Rathbun, G. B. (1989). The ecology of extinctions in kelp forest communities. *Conservation Biology, 3*(3), 251–264.

Estes, J. A., Tinker, M. T., Williams, T. M., & Doak, D. F. (1998). Killer whale predation on sea otters: Linking oceanic and nearshore ecosystems. *Science, 282*(5388), 473–476. https://doi.org/10.1126/science.282.5388.473

Fernando, P., Pfrender, M. E., Enclada, S. E., & Lande, R. (2000). Mitochondrial DNA variation, phylogeography, and population structure of the Asian elephant. *Heredity, 84*(3), 362–372. https://doi.org/10.1046/j.1365-2540.2000.00688.x

Ferretti, F., Worm, B., Britten, G. L., Heithaus, M. R., & Lotze, H. K. (2010). Patterns and ecosystem consequences of shark declines in the ocean. *Ecology Letters, 13*(8), 1055–1071. https://doi.org/10.1111/j.1461-0248.2010.01489.x

Fogell, D. J., Martin, R. O., Bunbury, N., Lawson, B., Sells, J., McKeand, A. M., et al. (2018). Trade and conservation implications of new beak and feather disease virus detection in native and introduced parrots. *Conservation Biology, 32*(6), 1325–1335. https://doi.org/10.1111/cobi.13214

Fujita, M. S., & Tuttle, M. D. (1991). Flying foxes (Chiroptera: Pteropodidae): Threatened animals of key ecological and economic importance. *Conservation Biology, 5*(4), 455–463.

Gadgil, M. (1991). Conserving India's biodiversity: The societal context. *Evolution Trends Plants, 5*(1), 38.

Gaubert, P., Patel, R. P., Veron, G., Goodman, S. M., Willsch, M., Vasconcelos, R., et al. (2017). Phylogeography of the small Indian civet and origin of introductions to Western Indian Ocean islands. *The Journal of Heredity, 108*(3), 270–279. https://doi.org/10.1093/jhered/esw085

Geldmann, J., Barnes, M., Coad, L., Craigie, I. D., et al. (2013). Effectiveness of terrestrial protected areas in reducing habitat loss and population declines. *Biological Conservation, 161*, 230–238.

Ghosh, A., Schmidt, S., Fickert, T., & Nüsser, M. (2015). The Indian Sundarban mangrove forests: History, utilization, conservation strategies, and local perception. *Diversity, 7*(2), 149–169. https://doi.org/10.3390/D7020149

Ghosh-Harihar, M., An, R., Athreya, R., Borthakur, U., Chanchani, P., Chetry, D., et al. (2019). Protected areas and biodiversity conservation in India. *Biological Conservation, 237*, 114–124. https://doi.org/10.1016/j.biocon.2019.06.003

Hamann, A., & Curio, E. (1999). Interactions among frugivores and fleshy fruit trees in a Philippine submontane rainforest. *Conservation Biology, 13*(4), 766–773.

Hansell, M. H. (1993). The ecological importance of animal nests and burrows. *Functional Ecology, 7*, 5–12.

Hayman, G., & Brack, D. (2002). *International environmental crime: The nature and control of environmental black markets*. Royal Institute of International Affairs.

Hohenlohe, P. A., Funk, W., & Rajora, O. (2020). Population genomics for wildlife conservation and management. *Molecular Ecology, 30*, 62–82. https://doi.org/10.1111/mec.15720

Hollings, T., Burgman, M., van Andel, M., Gilbert, M., Robinson, T., & Robinson, A. (2018). How do you find the green sheep? A critical review of the use of remotely sensed imagery to detect and count animals. *Methods in Ecology and Evolution, 9*, 881–892. https://doi.org/10.1111/2041-210X.12973

Hooper, D. U., & Vitousek, P. M. (1997). The effects of plant composition and diversity on ecosystem processes. *Science, 277*(5325), 1302–1305.

IUCN. (1975). *The red data book of the IUCN, 1975: Lists the gharial and other crocodiles of India as endangered species*. International Union for Conservation of Nature.

Jayakumar, S. (2018). *Endangered birds in India* [Internet]. [cited 2025 Apr 17]. Available from: https://www.researchgate.net/publication/327189623

Jhala, Y. V., Qureshi, Q., & Gopal, R. (2008). *Status of tigers, co-predators, and prey in India*. National Tiger Conservation Authority & Wildlife Institute of India.

Jones, K. E., Patel, N. G., Levy, M. A., Storeygard, A., Balk, D., Gittleman, J. L., et al. (2008). Global trends in emerging infectious diseases. *Nature, 451*, 990–994. https://doi.org/10.1038/nature06536

Kafley, H. A., Shrestha, S. S., & Shrestha, R. K. (2015). The status of greater one-horned rhinoceros (*Rhinoceros unicornis*) in Nepal: Current conservation challenges and opportunities. *International Journal of Ecology, 2015*, 1–11.

Karp, D., Rominger, A., Zook, J., Ranganathan, J., Ehrlich, P., & Daily, G. (2012). Intensive agriculture erodes β-diversity at large scales. *Ecology Letters, 15*(9), 963–970. https://doi.org/10.1111/j.1461-0248.2012.01815.x

Kearns, C. A., & Inouye, D. W. (1997). Pollinators, flowering plants, and conservation biology. *Conservation Biology, 47*, 297–307.

Koh, L. P., Sodhi, N. S., & Brook, B. W. (2004). Co-extinctions of tropical butterflies and their hostplants. *Biotropica, 36*(3), 272–274. https://doi.org/10.1111/j.1744-7429.2004.tb00340.x

Koppert, G., & Hladik, C. M. (1990). Measuring food consumption. In C. M. Hladik, S. Bahuchet, & I. de Garine (Eds.), *Food and nutrition in the African rain forest* (pp. 59–63). UNESCO/CNRS.

Larsen, T. H., Williams, N. M., & Kremen, C. (2005). Extinction order and altered community structure rapidly disrupt ecosystem functioning. *Ecology Letters, 8*(5), 538–547. https://doi.org/10.1111/j.1461-0248.2005.00749.x

Laurie, A. T. (1982). The status of the greater one-horned rhinoceros (*Rhinoceros unicornis*) in Kaziranga National Park, Assam. *Oryx, 16*(4), 422–424.

Mahmood, T., Hussain, R., Irshad, N., Akrim, F., & Nadeem, M. S. (2012). Illegal mass killing of Indian pangolin (*Manis crassicaudata*) in Potohar Region, Pakistan. *Pakistan Journal of Zoology, 44*, 1457–1461.

Marshall, B. M., Strine, C., & Hughes, A. C. (2020). Thousands of reptile species threatened by under-regulated global trade. *Nature Communications, 11*, 4738. https://doi.org/10.1038/s41467-020-18558-7

Martin, R. R., Karanth, K. U., & Sivasubramanian, M. (1987). The status of the greater one-horned rhinoceros (*Rhinoceros unicornis*) in Assam, India. *Biological Conservation, 40*(1), 23–40.

Mills, L. S., Soulé, M. E., & Doak, D. F. (1993). The keystone species concept in ecology and conservation. *Bioscience, 43*, 219–224.

Mohapatra, R. K., Panda, S., Nair, M. V., Acharjyo, L. K., & Challender, D. W. S. (2015). A note on the illegal trade and use of pangolin body parts in India. *TRAFFIC Bulletin, 27*(1), 33–40.

Molur, S., Srinivasulu, C., Srinivasulu, B., Walker, S., Nameer, P. O., & Ravikumar, L. (2005). *Status of South Asian non-volant small mammals: Conservation assessment and management plan (C.A.M.P.) workshop report*. Zoo Outreach Organisation/CBSG-South Asia.

Mondol, S., Karanth, K. U., & Ramakrishnan, U. (2009). Why the Indian subcontinent holds the key to global tiger recovery. *PLoS Genetics, 5*(8), e1000585. https://doi.org/10.1371/journal.pgen.1000585

Naniwadekar, R., Shukla, U., Isvaran, K., & Datta, A. (2015). Reduced hornbill abundance associated with low seed arrival and altered recruitment in a hunted and logged tropical forest. *PLoS One, 10*(3), e0120062. https://doi.org/10.1371/journal.pone.0120062

Narain, S., Panwar, H. S., Gadgil, M., Thapar, V., & Singh, S. (2005). *Joining the dots: The report of the Tiger Task Force*. Project Tiger Directorate, Union Ministry of Environment, Government of India.

Nasi, R., & Fa, J. E. (2015). The role of bushmeat in food security and nutrition. In: *XIV World Forestry Congress*.

Nittu, G., Shameer, T. T., Nishanthini, N. K., & Sanil, R. (2023). The tide of tiger poaching in India is rising! An investigation of the intertwined facts with a focus on conservation. *GeoJournal, 88*(1), 753–766. https://doi.org/10.1007/s10708-022-10633-4

da Nóbrega Alves, R. R., da Silva Vieira, W. L., & Santana, G. G. (2008). Reptiles used in traditional folk medicine: Conservation implications. *Biodiversity and Conservation, 17*(9), 2037–2049. https://doi.org/10.1007/s10531-007-9305-0

Panwar, H. S. (1987). Project Tiger: The reserves, the tigers, and their future. In R. L. Tilson & U. S. Seal (Eds.), *Tigers of the world: The biology, biopolitics, management, and conservation of an endangered species* (pp. 110–117). Noyes Publications.

Pearson, E. L., Lowry, R., Dorrian, J., & Litchfield, C. (2014). Evaluating the conservation impact of an innovative zoo-based educational campaign: 'Don't palm us off' for orangutan conservation. *Zoo Biology, 33*(2), 184–196. https://doi.org/10.1002/zoo.21120

Pimm, S. L., Jenkins, C. N., & Li, B. V. (2018). How to protect half of earth to ensure it protects sufficient biodiversity. *Science Advances, 4*(8), eaat2616. https://doi.org/10.1126/sciadv.aat2616

Power, M. E., Tilman, D., Estes, J. A., Menge, B. A., Bond, W. J., Mills, L. S., et al. (1996). Challenges in the quest for keystones. *Bioscience, 46*(8), 609–620.

Proulx, G., & Rodtka, D. (2019). Killing traps and snares in North America: The need for stricter checking time periods. *Animals (Basel), 9*(8), 570. https://doi.org/10.3390/ani9080570

Purvis, A., Agapow, P. M., Gittleman, J. L., & Mace, G. M. (2000). Nonrandom extinction and the loss of evolutionary history. *Science, 288*(5464), 328–330.

Pyšek, P., Hulme, P. E., Simberloff, D., Bacher, S., Blackburn, T. M., Carlton, J. T., et al. (2020). Scientists' warning on invasive alien species. *Biological Reviews of the Cambridge Philosophical Society, 95*(5), 1511–1534. https://doi.org/10.1111/brv.12627

Ramakrishnan, P. S. (1996). Conserving the sacred: From species to landscapes. *Nature and Resources, 32*(1), 11–19.

Reau, A. L., & Suen, G. (2018). The ruminococci: Key symbionts of the gut ecosystem. *Journal of Microbiology, 56*, 199–208. https://doi.org/10.1007/s12275-018-8024-4

Redford, K. H. (1992). The empty forest. *Bioscience, 42*(6), 412–422.

Ribeiro, J., Reino, L., Schindler, S., Strubbe, D., Vall-llosera, M., Araújo, M. B., et al. (2019). Trends in legal and illegal trade of wild birds: A global assessment based on expert knowledge. *Biodiversity and Conservation, 28*, 1–19.

Roggenbuck, M., Schnell, I. B., Blom, N., Bælum, J., Bertelsen, M. F., Sicheritz-Pontén, T., et al. (2014). The microbiome of New World vultures. *Nature Communications, 5*, 5498. https://doi.org/10.1038/ncomms6498

Rookmaaker, L. C., Suri, S., & Biswas, S. (2016). Revised population and distribution status of the greater one-horned rhinoceros (Rhinoceros unicornis) in South Asia. *Oryx, 50*(2), 275–283.

Rosen, G. E., & Smith, K. F. (2010). Summarizing the evidence on the international trade in illegal wildlife. *EcoHealth, 7*, 24–32.

Saigal, S., Chaliha, S., Nahata, C., & Kaur, S. (2018). Conservation and development: Beyond national parks and sanctuaries. In A. Chhatre & R. Lakerveld (Eds.), *Conservation and development in India* (pp. 142–163). Routledge.

Scheffers, B. R., Oliveira, B. F., Lamb, I., & Edwards, D. P. (2019). Global wildlife trade across the tree of life. *Science, 366*(6461), 71–76. https://doi.org/10.1126/science.aav5327

Schneider, C. P., Owens, R. W., Bergstedt, R. A., & O'Gorman, R. (1996). Predation by sea lamprey (Petromyzon marinus) on lake trout (Salvelinus namaycush) in southern Lake Ontario, 1982–1992. *Canadian Journal of Fisheries and Aquatic Sciences, 53*(9), 1921–1932.

Shairp, R., Veríssimo, D., Fraser, I., Challender, D., & MacMillan, D. (2016). Understanding urban demand for wild meat in Vietnam: Implications for conservation actions. *PLoS One, 11*(8), e0134787. https://doi.org/10.1371/journal.pone.0134787

Sharma, R., & Basu, D. (2004). Recent reversals in the population trends in the population of gharial in the National Chambal Sanctuary in North India; implications and a suggested strategy for the conservation of one of the world's most endangered crocodilians. In: Crocodile Specialist Group. *Crocodiles: Proceedings of the 17th working meeting of the crocodile specialist group* (pp. 180–186). IUCN.

Sharma, K., Wright, B., Joseph, T., & Desai, N. (2014). Tiger poaching and trafficking in India: Estimating rates of occurrence and detection over four decades. *Biological Conservation, 179*, 33–39. https://doi.org/10.1016/j.biocon.2014.08.014

Simo, F. T., Difouo, G. F., Kekeunou, S., Ichu, I. G., Ingram, D. J., & Olson, D. (2023). Pangolin hunting and trafficking in the forest–savannah transition area of Cameroon. *Oryx, 57*(6), 704–713.

Smith, R., Muir, R. D., Walpole, M. J., Balmford, A., & Leader-Williams, N. (2003). Governance and the loss of biodiversity. *Nature, 426*(6962), 67–70. https://doi.org/10.1038/nature02027

Sosnowski, M. C., & Petrossian, G. A. (2020). Luxury fashion wildlife contraband in the USA. *EcoHealth, 17*, 94–110. https://doi.org/10.1007/s10393-020-01467-y

Soule, M. E., Bolger, D. T., Alberts, A. C., Wrights, J., Sorice, M., & Hill, S. (1988). Reconstructed dynamics of rapid extinctions of chaparral-requiring birds in urban habitat islands. *Conservation Biology, 2*(1), 75–92.

Srivastava, A. (1999). *Primates of Northeast India.* Megadiversity Press.

Stanley Price, M. R., & Goodman, G. L. (2015). *White-bellied Heron (Ardea insignis): Conservation strategy.* IUCN Species Survival Commission White-bellied Heron Working Group, part of the IUCN SSC Heron Specialist Group.

Stephenson, P. J., & Ntiamoa-Baidu, Y. (2010). Conservation planning for a widespread, threatened species: WWF and the African elephant Loxodonta africana. *Oryx, 44*(2), 194–204.

Stork, N. E., & Lyal, C. H. C. (1993). Extinction or co-extinction rates. *Nature, 366*(6456), 307.

Sukumar, R. (2003). *The living elephants: Evolutionary ecology, behavior and conservation.* Oxford University Press.

Sutter, J. D. (2016). The most trafficked mammal you've never heard of [Internet]. *CNN* [cited 2025 Apr 17]. Available from: http://edition.cnn.com/interactive/2014/04/opinion/sutter-change-the-list-pangolin-trafficking/index.html

Symstad, A. J., Chapin, F. S., III, Wall, D. H., Gross, K. L., Huenneke, L. F., Mittelbach, G. G., Peters, D. P. C., & Tilman, D. (2003). Long-term and large-scale perspectives on the relationship between biodiversity and ecosystem functioning. *Bioscience, 53*(1), 89–98.

Talukdar, B. K., Khan, M. L., & Suri, S. (2008). Conservation status of the greater one-horned rhinoceros (Rhinoceros unicornis) in Northeast India. *Journal of Threatened Taxa, 6*(1), 31–38.

Tilman, D., Wedin, D., & Knops, J. (1996). Productivity and sustainability influenced by biodiversity in grassland ecosystems. *Nature, 379*, 718–720.

Toland, E., Bando, M., Hamers, M., Cadenas, V., Laidlaw, R., Martínez-Silvestre, A., & van der Wielen, P. (2020). Turning negatives into positives for pet trading and keeping: A review of positive lists. *Animals, 10*(12), 1–45. https://doi.org/10.3390/ani10122371

Tricorache, P., Nowell, K., Wirth, G., Mitchell, N., Boast, L. K., & Marker, L. (2018). Pets and pelts: Understanding and combating poaching and trafficking in cheetahs. In L. Marker, L. K. Boast, & A. Schmidt-Kuentzel (Eds.), *Cheetahs: Biology and conservation* (pp. 191–205). Elsevier.

United Nations Office on Drugs and Crime (UNODC). (2016). *World wildlife crime report: Trafficking in protected species.* United Nations.

United Nations Office on Drugs and Crime (UNODC). (2020). *World wildlife crime report 2020: Trafficking in protected species.* United Nations.

Velho, N., Srinivasan, U., Singh, P., & Laurance, W. F. (2016). Large mammal use of protected and community-managed lands in a biodiversity hotspot. *Animal Conservation, 19*(2), 199–208.

Villwock, W. (1972). Gefahren für die endemische Fischfauna durch Einbürgerungsversuche und Akklimatisation von Fremdfischen am Beispiel des Titicaca-Sees (Peru/Bolivien) und des Lanao-Sees (Mindanao/Philippinen). *Verhandlungen des Internationalen Verein Limnologie, 18*, 1227–1234.

Vyas, R. (2012). Notes on the trade in, and its effects on, the red sand boa (Eryx johnii johnii) in Gujarat, India. *Sauria, 34*(2), 3–8.

Wakchaure, R., & Ganguly, S. (2016). Captive breeding in endangered wildlife: A review. *Journal of Biological and Scientific Opinion, 4*, 186.

Walker, B. (1992). Biodiversity and ecological redundancy. *Conservation Biology, 6*, 18–23.

Whitaker, R. (1975). Status and conservation of the gharial. *Herpetological Review, 6*(1), 1–3.

Whitaker, R., Rajamani, V., Basu, D., & Balakrishnan, V. (1974). *Preliminary survey of the gharial, Gavialis gangeticus.* Madras Snake Park Trust Report, 1–16.

Wikramanayake, E., McKnight, M., Dinerstein, E., Joshi, A., Gurung, B., & Smith, D. (2010). Designing a conservation landscape for tigers in human-dominated environments. *Conservation Biology, 24*(3), 633–642. https://doi.org/10.1111/j.1523-1739.2009.01324.x

World Wildlife Fund (WWF). (2014). *More tigers in American backyards than in the wild* [Internet]. [cited 2022 May 6]. Available from: https://www.worldwildlife.org/stories/more-tigers-in-american-backyards-than-in-the-wild

Wright, B. (1969). *The tiger and its conservation in India*. Government of India.

Wright, S. J., Zeballos, H., Dominguez, I., Gallardo, M. M., Moreno, M. C., & Ibanez, R. (2000). Poachers alter mammal abundance, seed dispersal, and seed predation in a Neotropical forest. *Conservation Biology, 14*(1), 227–239.

Yashaswini, B. S. (2021). A study on Project Tiger in India. *National Journal of Environmental Law, 4*(2). https://doi.org/10.37591/NJEL

Zhou, Z., Newman, C., Buesching, C. D., Meng, X., Macdonald, D. W., & Zhou, Y. (2016). Revised taxonomic binomials jeopardize protective wildlife legislation. *Conservation Letters, 9*(5), 313–315. https://doi.org/10.1111/conl.12223

Chapter 7
Wildlife Trade and Ecosystem Imbalances in India

Ajit Kumar, Pranjal Sharma, and Priyanshi Chaudhary

Contents

Abstract India, recognized for its rich native biodiversity, faces even greater challenges. India represents four major biodiversity hotspots—the Western Ghats, the Himalayas, the Sundarbans, and the Indo-Burma region and has lost almost 90% of its unique habitats in these areas. As per International Union for Conservation of Nature (IUCN) Red List, In India out of the 1212 animal species, 12% are currently listed under endangered category. This emphasizes how urgent it is to combat with the threats like habitat destruction and illegal trade.

Illegal wildlife trade (IWT) is a major concern and growing global issue that threatens both plant and animals, disrupting their habitat, lead change ecosystem's and native biodiversity. Between 2015 and 2021, enforcement agencies were seized around 13 million of wildlife products across 162 countries. These products represented around 4000 species from them around 3000 species are listed under

A. Kumar (✉) · P. Sharma · P. Chaudhary
School of Forensic Science, Uttar Pradesh State Institute of Forensic Science, Aurawan, Lucknow, Uttar Pradesh, India

A. Poonia et al. (eds.), *Wildlife Trade in India*,
https://doi.org/10.1007/978-981-95-3694-8_7

CITES. Such highly and widespread exploitation of wildlife puts immense pressure on native biodiversity, species survival, and disturbs ecological structure.

The illegal harvesting of wildlife for commercial gains—whether for traditional medicines, exotic pets, or luxury items—worsens the problem. Iconic species such as tigers and elephants, as well as precious woods like sandalwood and red sanders, are being targeted. This not only threatens their survival but also destabilizes the ecosystems they support.

Animals such as the Indian Star Tortoise (*Geochelone elegans*), Monitor Lizards (*Varanus*) serve as stark examples of how illegal wildlife trades can wave through entire ecosystems. Losing a single species can disrupt the entire food chains and natural system. Moreover, wildlife trafficking increases the risk of zoonotic diseases, that transfer from animals to humans and posing serious health threats.

Another significant issue in India is the introduction of non-native and exotic species. By accident or design, these species are widely used in the pet and aquarium industries. Invasive species like water hyacinth and Parthenium have crowded out native flora and fauna, throwing ecosystems off balance. The pet trade has made things worse, turtle species like the Red-eared Slider are outcompeting native turtles' species, leading to further biodiversity harm.

Besides the ecological impact, IWT affects societies that heavily depend on local resources. It undermines local economies and sustainable development. Undertaking this issue calls for strong protection laws, international cooperation, and local people involvement. With only such combined efforts can we hope to protect native biodiversity and reduced IWT.

Keywords Illegal wildlife trade · Biodiversity conservation · Ecosystem imbalance · Endangered species in India · Invasive alien species · Habitat loss · CITES-listed fauna · Zoonotic disease emergence · Wildlife trafficking · Community-based conservation

7.1 Introduction

Illegal Wildlife trade (IWT) referred to as green crimes or environmental crime, involves the illegal activities that harming the environment benefitting individuals or groups or companies from the exploitation of natural resource (Nellemann et al., 2016). However, the definition of environmental crime is not universally agreed by many researchers. The IWT estimates value around 7–23 billion dollars annually makes these green crimes the fourth largest crime in the world after drug trafficking (estimated around 344 billion USD), counterfeit crimes (288 billion USD)and human trafficking (157 billion USD) (Nellemann et al., 2016). The illicit trade of animal and plants products poses severe threats to the global biodiversity, disturbing ecological processes, and forcing species towards extinction risk and weakening the resilience of entire ecosystems. Despite 20 years, many national and international efforts have been made to prevent IWT, but wildlife trafficking persists. As per the United Nations Office on Drugs and Crime's (UNODC) World Wildlife Crime Report 2024, during 2015 and 2021, approximately 4000 plant and animal species were illegally traded (UNODC, 2024). It represents the seizures of 13 million items across 162 countries and territories. The commonly encountered species were Pangolins, rhinos, elephants, crocodilians, eels, cedar, rosewoods, and corals.

The scale and complexity of the challenge of IWT have been broadly recognised in various meeting of Convention on International Trade in Endangered Species of Wild Fauna and Flora (CITES), which reflecting the global concern to address this emerging issue. As per the CITES report, more than 38,700 plants and animals are listed in its appendices, highlighting that international trade of all these species is strictly prohibited and under law. To date, IWT is rising by 5–7% annually represents approx. 2–3 times the rate of the global economy (Nellemann et al., 2016). Beyond endangering the species, environmental crime severely disrupts ecosystems, reduce their ability to adapt in changing climate and maintain ecological balance.

7.1.1 Biodiversity and Conservation Measure

India has four biodiversity hotspots—the Western Ghats, Sundarbans, Himalayan ranges, and Indo-Burma region. As per the data of International Union for Conservation of Nature (IUCN) Red List, India is home to 1212 animal species, with over 12% classified as endangered. Given the complexity of Illegal exploitation of natural resources, including ITW, has negative consequence on potential revenues in safeguarding its rich natural heritage. Recent estimates indicates that India already lost 90% of the area designated as hotspots for biodiversity. This loss has led to the extinction of 25 species in these regions. This situation emphasizes the urgent need for conservation measures to address the threats posed by habitat loss, wildlife trade, and other human activities (Centre for Science and Environment (CSE), 2021).

The long-term survival and fitness of species with low genetic diversity become more uncertain due to the risk of inbreeding, and this can have cascading effects on the ecosystems they inhabit. A diverse gene pool is necessary, as it leads to increased genetic diversity and resistance to diseases. Species with limited genetic variation are particularly vulnerable to adapting to climate change and emerging diseases. These cascading effects also compromise ecosystem resilience, and other drastic changes, ultimately threatening the ecosystem services and reducing carbon sequestration (Fig. 7.1).

The illegal wildlife trade involves the poaching and trafficking of animals and plants for various commercial purposes, including their use in traditional medicine, production of ornamental products, meeting the demand for exotic pets, and crafting luxury items (UNODC, 2024). The illegal harvesting of many endangered plant species including sandalwood and red sander trees in India not only threatens the survival of these species but also disrupts the ecosystems that depend on them. Additionally, the poaching of iconic species such as tigers, leopards, and elephants

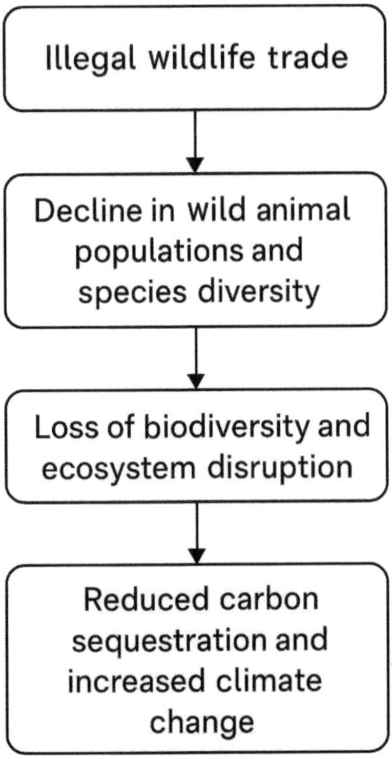

Fig. 7.1 Pathways linking illegal wildlife trade to ecosystem disruption and climate change vulnerability

for their skin, bones, and ivory has devastating effects on the wider ecosystem, as these animals are vital for maintaining ecological balance. These illegal activities not only disrupt food chains but also reduce genetic variability, and lead to loss of many iconic and keystone species.

7.1.2 Scientific Approaches and Conservation Strategies

As Illegal wildlife trade is a major threat to many wildlife and ecosystem across the globe. The effective management of species that are heavily traded or trafficked is often hindered by a critical gap—lack of reliable baseline data on their status in the wild. Understanding the native origin and region-specific genetic signatures of these species is essential for precisely identifying their origin and formulating targeted, science-based conservation strategies. For instance, a recent study conducted by Sahoo et al. (2024) on highly trafficked Indian Star Tortoise (IST), revealed the existence of two deeply divergent genetic lineages of the Indian star tortoise, namely northwestern and southern populations (Sahoo et al., 2024). The authors proposed recognizing these as distinct subspecies: *Geochelone elegans stellata* (northwestern group) and *Geochelone elegans elegans* (southern group). This refined taxonomy not only enhances our scientific understanding but also plays a vital role in the proper release and rehabilitation of seized individuals, ensuring they are returned to their correct native habitats. Ultimately, such insights are invaluable for guiding effective conservation and management efforts. Similarly, the illegal exploitation of the reptile species, monitor lizard (*Varanus bengalensis*), for its hemi penis, which is being sold under the false name Hatha Jodi in India. It has been frequently utilized for a variety of religious purposes, including witchcraft, black magic, and mysticism to satisfy superstitious beliefs. By impeding the products' originality, these items are freely offered on online platforms (Sharma et al., 2019). As the populations of *V. bengalensis* are declining from their distribution ranges, hence understanding the genetic state, origin, and targeted lineages of monitor lizards is vital. The population genetic structure of *V. bengalensis* was investigated by Gautam et al. (2023), who found two genetically different lineages: one from the Himalayan foothills and the other from mainland Indian populations (Gautam et al., 2023). These studies are useful for identifying species and their body components, but they are also useful for figuring out where seizures originate.

Considering the most profitable crime, IWT has become a major driver for biodiversity decline. Thus, this chapter focuses on multi-faceted topics by emphasizing the how IWT effect ecosystem mainly focused to decline in keystone species, the effect of alien species, the pet trade, and the commercial exploitation of wildlife that has disrupted the ecological balance. By understanding the specific case studies of species in per trade, we highlight the challenges that are faced in local biodiversity conservation. Herein, we aimed to address the importance of collective measures in conserving biodiversity and ensuring the continuity of vital ecosystem systems for sustainable development.

7.2 Cascading Effect of Wildlife Trade on Keystone Species

Keystone species are vital organisms which have a significant impact on ecosystems based on their distribution pattern. The term keystone species firstly define by Naturalist Robert T. Paine refer to those individuals' population whose presence refer the integrity of community structure and their stability, are determined by their abundances and activities (Paine, 1969). These have essential impact on ecological processes and the structures of communities which plays crucial role in maintaining a stable ecosystem.

7.2.1 Types of Keystone Species and Examples

The wide range of keystone species are broadly classified into a few categories' predators, ecosystem engineers, prey, plants and keystone mutualists (Budnukaeku et al., 2021). For instance, the removal of predator keystone species, Sea Star

(*Pisaster ochraceus*) from the intertidal zone resulted in the disruption of mussel's populations and effect the Pacific Northwest ecosystem stability. Moreover, an apex predator Sharks play an important role as keystone species by maintaining a balance marine ecosystem by controlling prey population such as capturing the weak fish, turtles and dolphins.

Keystone mutualists like Hummingbirds forms a mutual beneficiary relationship for survival with plant species by aiding in pollination. In Indian rivers, the Asian River dolphin (*Platanista gangetica*) regulates the population of mesopredators and helps in maintaining the stability of fresh water ecosystems (Braulik et al., 2021). Similarly, an ecosystem engineering, the North American beaver (*Castor canadensis*) builds dams, canals and dens that leads to the development of new flora and fauna species in that ecosystem. The Saguaro cactus (*Carnegiea gigantea*) act as a keystone host in deserts, providing resources and habitat to variety of species (Foley et al., 2021). In tropical forests, fig and palm trees plays the role of keystone mutualists; by supporting a vast number of animal species. These plants form a mutual relationship with fig wasps, which act as pollinators for the tree while the tree acts as a habitat for the wasp. This research introduces the idea of keystone species as the species which regardless of its biomass have a significant impact on the functioning and stability of its ecosystem. So, key stone species helps in maintaining the stability of ecosystem, regulation of species populations and enhancing their natural surroundings.

7.2.2 Ecological and Societal Cascades from Wildlife Trade

The large-scale wildlife trade includes both legal and illegal activities has far reached impacts on removing the species from their native habitat. Its consequences waves through ecosystems, triggering a cascade of ecological disruptions and deep into societies in complex, often overlooked. In case, where keystone species decline, the results is not just a local decline but systemic disentanglement of ecosystem from their services. High number of wildlife trade persist across a range of spatial scales, supported by wide-ranging socio-cultural and economic drivers. For instance, in certain cultural like Buddhist tradition of merit release huge number of birds are capture and released as part of religious believes. Research have shown that in just two temples in Phnom Penh, Cambodia approximately 680,000 birds were sell each year, that generate annually worth US $235,000 in net profit to sellers (Gilbert et al., 2012).

7.3 Invasive Alien Species (IAS)

Sometime pets become pests. The international pet's market has led to the creation of more than hundred invasive species with harmful economic and ecological impacts, which scale from individual species to entire ecosystems. Invasive species or the term exotic used for the species whose introduction and/or spread threatens the natural biological diversity of the region/habitat (CBD (Convention on Biological Diversity), 2002). Their introduction and invasion often disrupt natural and native species and their ecosystems, causing significant loss of biodiversity, the emergence of zoonotic diseases, and considerable economic loss. For example, IAS can threaten native fish populations by competing for the same food sources. In agriculture and forestry, they often damage crops, lower yields, and drive up the cost of control and management efforts. These wide-ranging effects highlight the necessity of strong monitoring and management strategies in place to reduce the harm caused by invasive species (Shastri et al., 2018). Recent, studies have reported that the invasive species impacted natural biodiversity globally, resulting in 39% of the species facing extinction during the past hundreds of years (Vyas, 2019). Sandilyan et al. (2018) compiled a consolidated list of 169 invasive alien species (IAS) in India (Sandilyan et al., 2018). As per the report, IAS occupied different ecosystems, including aquatic ecosystems (32.5%), terrestrial plants (31.4%), agricultural ecosystems (27.8%), and island ecosystems (8.3%). It highlights the widespread presence of IAS in different ecosystems and the necessity for scientific-based management in vulnerable

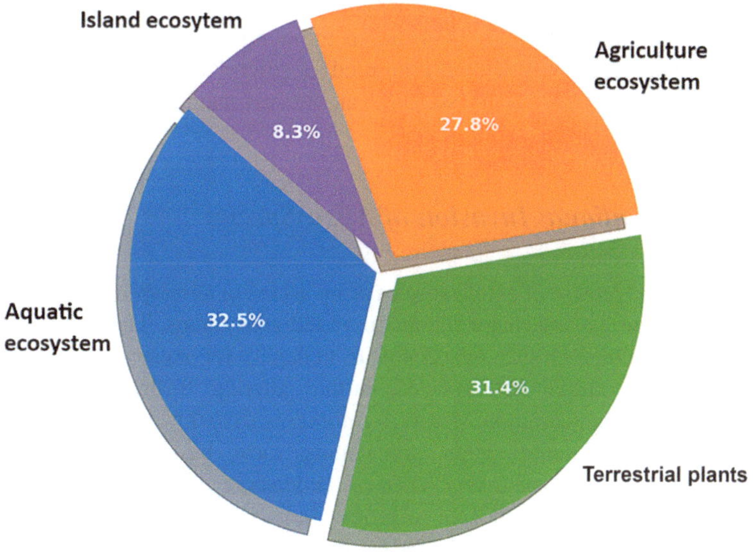

Fig. 7.2 Graph showing the distribution of invasive alien species (IAS) across major ecosystems in India

ecological zones (Fig. 7.2). The pathways for their introduction can be broadly categorized into intentional and unintentional means (Mack et al., 2000).

7.3.1 Intentional Invasion of Alien Species

The invasion of IAS are a worldwide issue. The human activities have caused accelerated movements and translocation of approx. 37,000 alien species across globe, with indication of negative impacts for more than 35,000 species categorised as invasive. India biodiversity significantly posed the threat by IAS, with approx. 66% of native diversity affected by this. The estimates of economic loss due to invasion IAS in India is high as $182.6 billion. The harmful effect of biological invasion of IAS are now widely recognised and estimated at different scale like at local, regional, national and international. There is many research are ongoing in many parts of countries to reduce present and possible future impacts (Reshi & Khuroo, 2012).

The intentional introduction of IAS in India has led to significant ecological disruption, often resulting in biodiversity loss and changes in ecosystem dynamics. A notable example is *Lantana camara*, which was introduced for ornamental purposes. This species has become an aggressive invasive plant, outcompeting native vegetation and displacing local flora in forest ecosystems. Similarly, *Eichhornia crassipes* (water hyacinth), initially introduced to enhance water bodies' aesthetic value, has clogged rivers and lakes, disrupted aquatic ecosystems and reduced biodiversity. The introduction of exotic animals like *Trachemys scripta* (Red-eared Slider turtle) for the pet trade has led to the decline of native fresh water biodiversity including wild turtle populations through resources competition. These intentional invasions of IAS highlight the need for stricter regulations and awareness to prevent such ecological harm (Shastri et al., 2018; Chandra & Chatterjee, 2007).

7.3.2 Unintentional Invasion of Alien Species

The unintentional release of exotic species or invasions of alien species have become a significant issue, posing major threats to varied ecosystems. The unintentionally introduction of *Oreochromis mossambicus* (Tilapia) for aquaculture has led to competition with native fish species, further altering freshwater ecosystems. Among the birds, the Common Myna (*Acridotheres tristis*) has become an aggressive competitor to native birds, threatening local bird populations. These unintentional invasions of many species underscore the need for enhanced biosecurity and vigilant monitoring to protect India's biodiversity and ecosystems (Shastri et al., 2018; Sharma et al., 2016). In India, till date around 31 species of aquaculture, 600 ornamental varieties and 3 larvivores fish species have been identified as exotic (Mahapatra & Mohanty, 2023). Inland Invasive Fish species that causing huge threat are African catfish (*Clarias gariepinus*), Nile tilapia (*Oreochromis niloticus*), Red-bellied piranha (*Pterygoplichthys nattereri*), and Alligator gar (*Atractosteus spatula*) are now dominate in many parts of Indian States due to aquarium trade. Now these species have been posing a threat to indigenous varieties, appears to be making its way across the country.

7.4 Zoonoses Unleashed

Zoonotic diseases are those that can be transmitted from animals to humans and are often linked to the introduction and spread of invasive species. These diseases can have serious public health implications, particularly when non-native animals act as intermediaries. This is especially true because people are coming into contact with wildlife more and more. A lot of these infections are linked to the movement or trade of invasive species. For example, the coronavirus associated with SARS has been connected to various animals, including dogs, pigs, and Himalayan palm civets, highlighting the significant role of wildlife trade and human-animal interactions in the spread of diseases (Fang et al., 2024). Similarly, Ebola hemorrhagic fever—believed to be transmitted through the consumption of apes as bushmeat—emphasizes the dangers associated with hunting and trading exotic species.

Moreover, health experts have repeatedly warned of the substantial risks posed by wildlife trade to both human and animal health. These include the direct transmission of diseases from live animals, plants, and wildlife products such as bushmeat, along with broader consequences for biodiversity, ecosystems, and global food security systems. Human activities—such as handling bushmeat, trading non-native animals, and disrupting natural habitats—greatly increase the likelihood of zoonotic disease emergence.

7.4.1 Impact on Public Health

Research has shown that approximately 61% of human infectious diseases are zoonotic in origin (Borah et al., 2025). Even more concerning is that pathogens originating from animal hosts are twice as likely to cause emerging infections in humans (van Doorn, 2021). For such pathogens to establish themselves in human populations, they must overcome a series of biological barriers—between species, within human hosts, and among individuals—to enable sustained human-to-human transmission. Since 1940, scientists have identified global hotspots where these emerging diseases are more likely to appear, often in regions with intense human-wildlife interaction or domestic animal exposure, and where drug-resistant and vector-borne pathogens are on the rise.

Together, this growing body of evidence underscores the urgent need to regulate wildlife trade, reduce human-wildlife contact, and strengthen our understanding of the interconnected risks at the human-animal-environment interface to prevent future outbreaks.

7.5 Conservation Legislation

The Wildlife (Protection) Act of 1972 stands as a foundation of India's legislative efforts to conserve its diverse and fragile natural heritage. Enacted by the Parliament as Act No. 53 of 1972, this law was designed to offer comprehensive protection to wild animals, birds, plants, and the ecosystems they inhabit. Recognizing the growing threats posed by poaching, habitat loss, and illegal trade, the Act laid the groundwork for regulating human interaction with wildlife and preserving ecological balance across the country.

Since its inception, the Wildlife Protection Act has been revised multiple times to adapt to emerging conservation needs and challenges. The first amendment came in 1982, followed by updates in 1986, 1991, 2003, and 2006—each strengthening the legal framework and enforcement mechanisms. These amendments reflect the Indian government's evolving commitment to biodiversity preservation, incorporating scientific advancements and responding to new threats over the decades.

A major development came with the most recent amendment in 2022, which officially came into force on April 1, 2023. This update brought about a significant rationalization of the Act's Schedules—classification lists that determine the level of protection granted to various species. Where the Act previously featured six Schedules, it now contains four streamlined categories, aiming to simplify enforcement and ensure clearer prioritization of conservation efforts. One of the most notable features of this amendment is the introduction of a dedicated Schedule for species listed under the Convention on International Trade in Endangered Species of Wild Fauna and Flora (CITES). This not only enhances India's compliance with international wildlife trade regulations but also demonstrates a strong alignment with global biodiversity conservation standards.

Additionally, the 2022 amendment introduced new provisions to address the increasing threat posed by invasive alien species, which can severely disrupt native ecosystems. Through better categorization and regulatory mechanisms, the revised Act now empowers authorities to implement more targeted and efficient management strategies for non-native species that pose ecological risks.

7.6 Case Study: Pet Trade

Illegal wildlife trafficking affects over 100 million plants and animals every year, contributing significantly to the rapid decline of species across the globe. Human-driven activities—such as overexploitation and widespread habitat destruction—

have pushed many species to the threatened category, leading to decline in population and even local extinctions in some regions (Mozer & Prost, 2023).

Turtles are the most ancient creatures on Earth, known for their unique adaptations that allow them to thrive in a different range of ecosystems—from land to freshwater, wetlands, and marine. In Asia, turtle biodiversity is especially rich, mainly in regions of Ganga and Brahmaputra River basins, mainland Southeast Asia, the South China coastal zone, and New Guinea. This area, considered a biodiversity hotspot, and home to at least 100 native species of freshwater turtles and tortoises (van Dijk et al., 2000).

Now a days, nearly 80% of turtles face a high risk of extinction, primarily due to unsustainable harvesting for commercial and subsistence purposes (Stanford et al., 2020). In this context, illegal pet keeping often poses a significant threat to natural biodiversity. We highlight the case of the Red-eared Slider, RES (*Trachemys scripta elegans*), whose introduction into non-native environments has led to serious ecological risks. This case highlights the urgent need for stronger regulations and increased public awareness to prevent further harm. The RES is a freshwater turtle native to the Mississippi River and the Gulf of Mexico. This species is easy to recognize because of it tiny size and two striking red patches on either side of its head. As per the report of Global Invasive Species Database.

According to the Global Invasive Species Database reported by Lowe et al. (2000), the introduction of RES into non-native habitat began posing a global threat between 1989 and 1997, during which more than 52 million individuals were shipped to various countries by local traders (Lowe et al., 2000). RES mature faster than native species, display more aggressive behaviour, and their reproduction rate is also high. This combination of traits has led to their recognition as one of the top 100 most invasive species on the IUCN list, and considered as the most invasive turtle in the world (Lowe et al., 2000). Native freshwater ecosystems in India have been significantly impacted by the species' due to its quick spread (Vyas, 2019). Hatchlings are frequently sold as pets in markets. Many hatchlings that survived, grew larger, or lost their owners' interest were eventually released back into the wild. Both deliberate and accidental releases have led to the successful establishment of RES populations in a number of Asian nations. In many regions of India, especially in water bodies such as lakes, ponds, and rivers, RES have successfully developed its breeding populations over time.

7.7 Illegal Trade and Conservation Initiatives

In India, many state and central enforcement agencies have been taking various actions to combat illegal wildlife trade. In December 2018, customs officials at Chennai International Airport intercepted two travelers from Bangkok attempting to smuggle 7200 juvenile RES for illegal sale in the Southern part of India. A year later, the Directorate of Revenue Intelligence seized more than 4500 turtles at Trichy International Airport that arrived from a Sri Lankan Airlines flight. The turtles were concealed in 30 carton boxes, underscoring the scale of the illegal pet trade and its impact on native ecosystems (Krishnan, 2021). Again, in the year 2024, within 1 week, Chennai Airport witnessed two major wildlife trafficking attempts involving exotic turtles. On April 20, 2024, customs officers caught almost 5000 hatchlings of red-eared sliders, that were illegally transported from Malaysia. Moreover, on April 12 customs officers had already stopped a shipment of 484 red-eared slider turtles and nine African spurred tortoises, which were all transported secretly in passengers' luggage (Times of India, April 20, 2024 by Venkadesan S report). These cases, which are being repeated, show the intensifying danger of illegal wildlife trade that is being done through international transit points. The quantity of illegal confiscations found indicates a risk of new invasions that is extremely high and a risk of new disease introductions that is also very high. Therefore, in order to protect the endemic and native species, it is necessary to launch mass awareness campaigns and take proactive steps to stop the further spread of the alien species. The initiatives should come from the grass-roots and community involvement. This way, the impact will be long-lasting and the changes sustainable. If India adopts such proactive measures, it will be able to reduce the impact of the invader and will also be able to protect its native biodiversity.

References

Borah, P., Das, P., Bordoloi, R., Begum, R., Hazarika, G., Saikia, D. P., et al. (2025). Prevalence of zoonotic diseases in the Northeastern region, one health perspective. *Animals and Zoonoses, 1*(2), 178–187.

Braulik, G., Atkore, V., Khan, M. S., & Malla, S. (2021). *Review of scientific knowledge of the Ganges river dolphin*. WWF, commissioned by the World Bank.

Budnukaeku, A. C., Ichendu, C. B., & Irene, D. (2021). Human impacts on keystone species within the ecological food web. *MOJ Ecology & Environmental Sciences, 6*(5), 182–186.

CBD (Convention on Biological Diversity). (2002). *Decision VI/23 (Annex, footnote): Alien species that threaten ecosystems, habitats and species* (Document UNEP/CBD/COP/6/23). Convention on Biological Diversity Secretariat, Montreal, Canada. Available from https://www.cbd.int/decision/cop/?id=7197. Accessed 11.01.15.

Centre for Science and Environment (CSE). (2021). *State of India's environment 2021: In figures*. Centre for Science and Environment. https://www.cseindia.org

Chandra, K., & Chatterjee, A. (2007). Invasive alien species in India: A review of their impacts and management. *Indian Journal of Forestry, 30*(4), 375–379.

Fang, R., Yang, X., Guo, Y., Peng, B., Dong, R., Li, S., & Xu, S. (2024). SARS-CoV-2 infection in animals: Patterns, transmission routes, and drivers. *Eco-Environment & Health, 3*(1), 45–54.

Foley, T., Swann, D. E., Sotelo, G., Perkins, N., & Winkler, D. E. (2021). Asynchronous flowering patterns in saguaro cacti (Carnegiea gigantea). *Ecosphere, 12*(12), 1–14, e03873.

Gautam, K. B., Kumar, A., Das, A., & Gupta, S. K. (2023). Himalayan upliftment and Shiwalik succession act as a cradle for divergence in Bengal monitor lizard Varanus bengalensis (Reptilia: Varanidae) in India. *Cladistics, 39*(5), 382–397.

Gilbert, M., Sokha, C., Joyner, P. H., Thomson, R. L., & Poole, C. (2012). Characterizing the trade of wild birds for merit release in Phnom Penh, Cambodia and associated risks to health and ecology. *Biological Conservation, 153*, 10–16.

Krishnan, M. (2021, June 3). What's dangerous about these cute turtles which are popular pets in India? *Mongabay-India*. https://india.mongabay.com

Lowe, S., Browne, M., Boudjelas, S., & De Poorter, M. (2000). *100 of the world's worst invasive alien species: A selection from the global invasive species database* (Vol. 12, p. 12). Invasive Species Specialist Group.

Mack, R. N., Simberloff, D., Lonsdale, W. M., Evans, H., Clout, M., & Bazzaz, F. A. (2000). Biotic invasions: Causes, epidemiology, global consequences, and control. *Ecological Applications, 10*(3), 689–710.

Mahapatra, B. K., & Mohanty, B. P. (2023). Invasive Alien fish species in India are threat to the inland aquatic ecosystem. *Advances in Fish Research, VIII*, 153–162.

Mozer, A., & Prost, S. (2023). An introduction to illegal wildlife trade and its effects on biodiversity and society. *Forensic Science International: Animals and Environments, 3*, 100064.

Nellemann, C., Henriksen, R., Kreilhuber, A., Stewart, D., Kotsovou, M., Raxter, P., Mrema, E., & Barrat, S. (2016). *The rise of environmental crime: A growing threat to natural resources peace, development and security*. UN Environment Programme.

Paine, R. T. (1969). A note on trophic complexity and community stability. *The American Naturalist, 103*(929), 91–93.

Reshi, Z. A., & Khuroo, A. A. (2012). Alien plant invasions in India: Current status and management challenges. *Proceedings of the National Academy of Sciences, India Section B: Biological Sciences, 82*, 305–312.

Sahoo, S., Kumar, A., Rai, J., & Gupta, S. K. (2024). Unlocking the genetic enigma: A paradigm shift towards evidence-based conservation of Indian star tortoise (Geochelone elegans). *Biological Conservation, 290*, 110421.

Sandilyan, S., Meenakumari, B., Babu, C. R., & Mandal, R. (2018). *Invasive alien species of India*. National Biodiversity Authority, Chennai.

Sharma, G., Chakraborty, P., & Ghosh, A. (2016). Impact of invasive alien plant species on biodiversity of India: A review. *Environmental Science and Pollution Research, 23*(19), 18933–18947. https://doi.org/10.1007/s11356-016-6874-x

Sharma, C. P., Kumar, A., Vipin, Sharma, V., Singh, B., Kumar, G. C., & Gupta, S. K. (2019). Online selling of wildlife part with spurious name: A serious challenge for wildlife crime enforcement. *International Journal of Legal Medicine, 133*(1), 65–69.

Shastri, R., Gajbhiye, S., & Ramaswamy, S. (2018). Invasive species in India: An overview. *Indian Journal of Ecology, 45*(1), 109–115.

Stanford, C. B., Iverson, J. B., Rhodin, A. G., van Dijk, P. P., Mittermeier, R. A., Kuchling, G., et al. (2020). Turtles and tortoises are in trouble. *Current Biology, 30*(12), R721–R735.

UNODC. (2024). *World Wildlife Crime Report 2024: Trafficking in protected species*. United Nations Office on Drugs and Crime.

van Dijk, P. P., Stuart, B. L., & Rhodin, A. G. J. (Eds.). (2000). *Asian turtle trade: Proceedings of a workshop on conservation and trade of freshwater turtles and tortoises in Asia* (Chelonian research monographs no. 2) (pp. 15–23). Chelonian Research Foundation.

van Doorn, H. R. (2021). The epidemiology of emerging infectious diseases and pandemics. *Medicine, 49*(10), 659–662.

Vyas, R. (2019). Distribution of invasive red-eared sliders, Trachemys scripta (Testudines: Emydidae), in the wetlands of Gujarat State, India. *Reptiles & Amphibians, 26*(2), 145–150. https://doi.org/10.17161/randa.v26i2.14389

Chapter 8
Wildlife Trade as a Vector for Zoonotic Diseases

Preeti Rani, Harsh Gulati, and Deepak Rai Babbar

Contents

P. Rani · H. Gulati (✉)
Department of Zoology, School of Bioengineering and Biosciences, Lovely Professional University, Phagwara, Punjab, India
e-mail: harsh12@kuk.ac.in

D. R. Babbar
Department of Zoology, Kurukshetra University, Kurukshetra, Haryana, India

A. Poonia et al. (eds.), *Wildlife Trade in India*,
https://doi.org/10.1007/978-981-95-3694-8_8

Abstract The global wildlife trade, encompassing both legal and illegal activities, and plays a crucial role in the emergence, transmission of zoonotic diseases, posing severe risks to biodiversity, public health, and global economies. The trade involves millions of live animals and their derivatives, spanning local markets to international networks, often operating with minimal regulation. Wildlife markets, especially those selling live animal as well as operating wet markets, bring different species into close contact, creating favorable conditions for pathogen to transmit from animals to humans. This has been associated with major zoonotic disease outbreaks, such as SARS, MERS, and COVID-19.

The Illegal Wildlife Trade (IWT), classified as an environmental crime, exacerbates these risks by bypassing health and safety regulations, further increasing the chances of disease transmission. The overexploitation of wildlife, driven by high economic demand for exotic pets, traditional medicine, bushmeat consumption, and luxury goods, has led to biodiversity loss and habitat destruction, which in turn heightens human-wildlife interactions. Deforestation, land-use changes, and urbanization are example of anthropogenic pressures that upset ecosystems and bring wildlife closer to the human populations and domestic animals, which rises the risk of zoonotic spillover.

Scientific research has shown that zoonotic pathogens, particularly RNA viruses, can mutate and adapt rapidly, making them highly capable of jumping between species. The emergence of novel pathogens, including coronaviruses and hemorrhagic fever viruses, underscores the need for an evidence-based approach to managing wildlife trade. Regulatory frameworks like the Convention on International Trade in Endangered Species of Wild Fauna and Flora (CITES) aim to mitigate these risks, but enforcement remains inconsistent across regions. Strengthening international collaboration, enhancing biosecurity measures, and implementing a One Health approach-integrating human, animal, and environmental health-are critical steps in preventing future pandemics.

Furthermore, raising public awareness about the dangers of wildlife trade, promoting sustainable alternatives, and strengthening legal frameworks to combat illegal trade are essential to reducing the risks associated with zoonotic disease transmission. By implementing targeted interventions, improving surveillance systems, and fostering global cooperation, it is possible to mitigate the threats posed by wildlife trade while ensuring biodiversity conservation and safeguarding human health.

Keywords Wildlife trade · Zoonotic diseases · Illegal wildlife trade · Emerging infectious diseases · Biodiversity loss · Habitat destruction · One health approach · Pathogen spillover · International collaboration · Public health risks

8.1 Introduction

Wildlife trading across borders has been going on for a long time and could bring in money and other benefits. But the Convention International Trade in Endangered Species of Wild Fauna and Flora (CITES) came into effect in 1975 as a global policy tool to ensure sustainability. Because this policy is recognized that unsustainable trade can have a negative impact on source populations (Tittensor et al., 2020). The wildlife trade encompasses the sales and exchanges of wild animal and plant resources, forming a critical component of biodiversity conservation and sustainable development. It includes the trade of living organisms and an array of products valued by humans, such as skins, medicinal ingredients, food. This trade often supports economically disadvantaged communities and generates significant revenue at a national level (Nijman, 2010). Primarily driven by economic incentives, the trade ranges from localized small-scale operations to profit-driven ventures. While the majority of wildlife trade takes place locally or remains confined within national boundaries, a significant portion transcends borders and operates on an international scale. The supply chain involves numerous intermediaries, including experts in marketing, transport, handling, storage, manufacturing, industrial production, retail businesses and exportation, and operating on both domestic as well as international levels.

Every year, billions of individual animals, representing a wide array of species, are hunted, captured, or killed to supply both the global legal and illicit wildlife trade markets. Mounting evidence links zoonotic diseases (e.g., COVID-19) and biodiversity loss to the overexploitation of wildlife, which adversely affects the livelihoods of people living in proximity to these animals. However, compassion and consideration for the welfare of the wildlife impacted by these trades are rarely prioritized. Both legal and illegal wildlife trades highlight the exploitation of animals for human use, often with minimal regard for animal welfare and even less acknowledgment of the rights.

Despite the growing global recognition and concern for iconic species, there is still widespread ignorance, denial and disregard for the intense and routine suffering endured by countless wildlife individuals caught up in both legal and illegal trade practices. Scientific studies increasingly affirm that animals are able to feel both suffering as well as delight; leading to protective legislation for some species against unnecessary suffering. Nonetheless, the brutal realities of the wildlife trade are extensively documented: Rhinos often cry out in pain when poached for their horns, which are sometimes gruesomely hacked off while they are still alive; calves may endure the horrific severing of their spines with machetes or forcibly taken into captivity. Snakes are forcibly pumped with water and left to starve for few days prior being flayed to meet demands of the reptile leather trade. Birds endure the horrific practice of having their eyelids sewn shut and are crammed into tight tubes, often dying from exposure, dehydration or starvation, all to supply markets for exotic food, fur, ornamental items and traditional medicine (Wyatt et al., 2022). Fatalities are frequent during transport, especially within the illegal wildlife trade. Survivors intercepted by law enforcement often endure inadequate housing conditions while awaiting legal resolution, and in some cases, they are euthanized due to a lack of suitable alternatives or resources for their care. Those who survive and enter captivity weather as pets, in aquariums, farms, laboratories, or the entertainment industry are unlikely to have their behavioral and physiological needs adequately met, leading to suffering. In some cases, abandoned animals may become invasive species and are often eradicated inhumanely as pests.

The sixth mass extinction is the term is used to describe the current rate of biodiversity loss and this is greater than it has ever been in human history. The wildlife overexploitation is one of the main causes of this decline including unsustainable harvesting, and the second biggest threat to biodiversity worldwide. Moreover, the vital services that nature provides to people are at risk due to this loss. According to World Wildlife Fund's (WWF) Living Planet Report 2020 (WWF, 2020) in 1970 and 2016 average loss of reptiles, fishes, mammals, and amphibians is 68% population monitored globally. These are the major changes in species populations indicates a fundamentally broken relationship between humans and the natural world. Zoonotic diseases can have a devastating effect on health linked to wildlife trade (Karesh et al., 2005; Nijman, 2010).

Annually there are thousand species are traded for pets, food, curios, fashion, traditional medicines invest millions of individuals, contrary to the context of faster decreasing biodiversity. These species need attention in different modes of observing. Socioeconomic inequality frames the global pattern of wildlife trade, with a network of trade linking lower-income exporting countries to wealthier importing nations (Nijman, 2010; Liew et al., 2021).

When efficiently managed, wildlife trade gives to biodiversity conservation, sustain ecosystem functions, and also provide long term support for regional maintenance. This ensures that the future generations can continue to gain a living without effecting decreases in wild populations (Cooney et al., 2017).

8.2 Legal Wildlife Trade and Illegal Wildlife Trade

Legal wildlife trade involves the authorized buying, selling, and exchange of wild animals and plants, along with their parts and derivatives conducted in accordance with national and international regulations. This trade is governed by regulatory frameworks such as the Convention on International Trade in Endangered Species of Wild Fauna and Flora (CITES), which ensure that global trade does not harm the existence of species in the wild.

Illegal wildlife trade (IWT) categorized as a green crimes or environmental crimes, encompasses activities that harm the environment while benefiting individuals, groups or corporation through the illegal exploitation, trade, damage, or theft of natural resources. It involves various actions such as acquiring, plunder, capturing, processing, importing, gathering, exporting, possessing along with utilization of wild fauna, flora and fungi whether water-dwelling or land-dwelling, living or

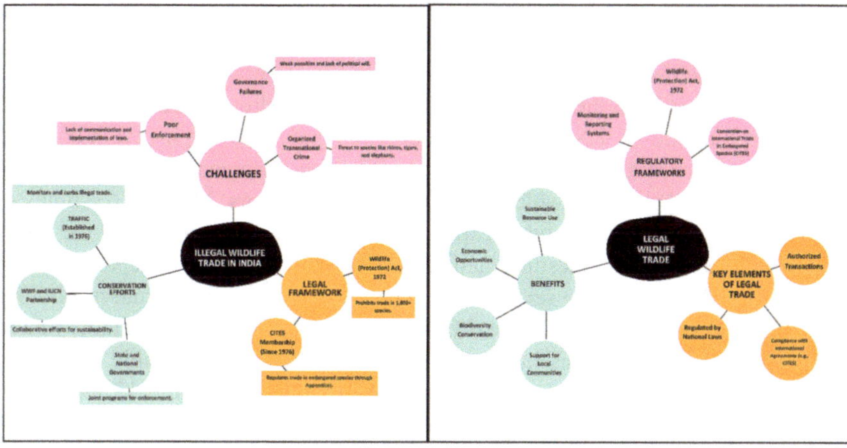

Figs. 8.1 and 8.2 Illegal and legal wildlife trade: challenges, frameworks, and conservation implications

deceased, including compounds, components and goods. These activities are regulated and protected by national and global laws (Mozer & Prost, 2023).

The overexploitation of natural resources, including illegal wildlife trade (IWT), is greater contributor in current biodiversity decrease than climate change (Maxwell et al., 2016). Considering all these factors, IWT has moved beyond an emerging issue but has been classified as a major crime by, e.g. the International Criminal Police Organization (Nellemann et al., 2014) and the United Nations Office on Drugs and Crimes (United Nations Office on Drugs and Crime, 2020) based on this any radiation chart or other (Figs. 8.1 and 8.2).

Animal-borne diseases are that kind of conditions or infestation that can be transfer among animals (domesticated and non-domesticated) and humans (Rodríguez, 2021). They have been implicated in numerous recent global health and economic crises, such as outbreaks severe acute respiratory syndrome (SARS), Ebola, Severe Acute Respiratory Syndrome Coronavirus 2 (SARS-CoV-2), Middle East Respiratory Syndrome (MERS), the virus responsible for COVID-19 (Intergovernmental Panel on Climate Change (IPCC), 2013).

Over the past two decades, the emergence of three novel coronaviruses has raised significant global health concerns due to their distinct genomic characteristics and potential for widespread epidemics. First of Severe Acute Respiratory Syndrome (SARS), was caused by SARS-CoV and emerged in 2003 in Guangdong, China. The virus rapidly spread globally, facilitated by air travel and human-to-human transmission. SARS-CoV primarily targets the ACE-2 receptors present in bronchial epithelial cells and type II pneumocytes in the lungs, leading to severe respiratory illness, atypical pneumonia, and respiratory failure. The mortality rate for SARS was around 10%, but it reached as high as 50% in elderly individuals, highlighting its potential for severe outcomes. The global response, coordinated by the World Health Organization (WHO), successfully contained the outbreak, which resulted in 8700 reported cases and 744 deaths across 33 countries. Early containment measures, including case isolation, contact tracing, and management, were critical in halting the spread of the disease.

Just a few years later, the Middle East Respiratory Syndrome (MERS) emerged in 2012, caused by the MERS-CoV, a zoonotic virus that likely originated in bats and was transmitted to humans through contact with infected dromedary camels (Contini et al., 2020). Unlike SARS, which binds to ACE-2 receptors, MERS-CoV binds to the DPP4 receptor, which is highly expressed in various tissues, including the lower bronchi, kidneys, small intestine, and liver. This difference in receptor binding contributes to MERS's more severe manifestations, which include acute respiratory failure, pneumonia, and renal failure. MERS has a notably higher case fatality rate of 35–40% significantly surpassing SARS, but it is less transmissible, requiring prolonged exposure for human-to-human transmission. Between 2012 and 2019, 2499 confirmed cases and 858 associated deaths were reported globally, with the majority of cases occurring in Saudi Arabia. MERS also exhibits gastrointestinal symptoms, which are less common in SARS, and while its transmission routes remain incompletely understood, it is primarily associated with close contact with camels or their products.

Both SARS and MERS underline the significant public health risks posed by novel coronaviruses. These outbreaks highlight the need for early detection, rigorous epidemiological surveillance, and effective containment strategies to prevent widespread transmission. The global response to both viruses, particularly the WHO's coordinated efforts in limiting the spread of SARS, demonstrates the critical importance of international collaboration in managing emerging infectious diseases. As both viruses continue to evolve, the possibility of future outbreaks remains a concern, necessitating continued research into coronavirus biology, transmission mechanisms, and vaccine development (Contini et al., 2020).

The relationship between wildlife trade and the emergence zoonotic diseases is a critical area of research that require a science-driven and evidence-based approach to analyze the multifaceted factors influencing disease transmission. A comprehensive understanding of mechanisms underlying zoonotic diseases transmission is vital before exploring strategies to mitigate the risks posed by the wildlife trade. This trade encompasses the capture, transport, and sale of live animals, animal products, and plants all of which have significant implications for zoonotic diseases emergence. Poorly regulated and illegal wildlife trade operations often involve the handling of diverse species in close quarter, creating optimal conditions for zoonotic infections to transfer from animals to humans (Saleem et al., 2023). Additionally, factors such as globalization, habitat destruction, and increased human-wildlife interactions further elevate the risk of zoonotic diseases outbreaks.

8.3 Wildlife Trade as a Driver of Zoonotic Disease Emergence

The wildlife trade serves as a pathway for dissemination, threatening human health along with endanger livestock, global commerce, rural economies, indigenous wildlife species, and ecological stability. Accurately measuring the scale of the international wildlife trade is extremely challenging due to its wide range-from local swab systems to major global networks, and the prevalence of illegal or unofficial channels. Evaluation suggests that approximately 40,000 live primates, four million live birds, 640,000 live reptiles, and 350 million live tropical fish are traded worldwide per annum.

In Guangzhou, China, a wide range of live wildlife including Asian palm civets, ferret mustelids, forest deer, feral pigs, spiny mammals, wild canids, tree-dwelling rodents, burrowing rodents, desert rodents, various species of reptiles and endangered small spotted cats are commonly traded alongside domestic animals like canids, domestic felids, and lagomorphs. Later in 2003 epidemic of severe acute respiratory syndrome (SARS), authorities across Guangzhou reports confiscated 838,500 wild animals from these markets. Wild mammalian, reptilian and avian taxa move daily through commercial trade hubs, in which they come into interaction with humans and numerous different population prior being sold locally, transported to other markets, or, in some cases released into the wild as a component of cultural or spiritual practice such as merit release, due to they are no longer wanted as pets.

For instance, Annual marketplace of North Sulawesi, in Indonesia, report selling 90,000 mammals. A survey conducted over 25 weekends at a single market in Thailand recorded the sale of over 70,000 avians, from 276 species. Similarly in 2001 assessment of four marketplaces in Bangkok 36,537 birds were observed, with solely 37% being indigenous to Thailand and remaining 63% consisting of invasive species.

Due to insufficient trade data, we made conservative estimate in which tens millions of wild animals are shipped annually within East and Southeast Asia as well as from parts of the world, for food or traditional medicine purposes. In central Africa, estimated annual trade and consumption of wild animal meat exceeds 1 billion kilograms. Similarly, in Amazon Basin, consumption is estimated to range between 67 and 164 million kilograms annually. Solely the consumption of mammals, approximately 6.4–15.8 million individual animals. In Central Africa, approximately annual animal consumption by human vary in number of animals utilized by humans fluctuates, with 579 million proposed.

The same parasite species that are harbored by imported animals are responsible for the transmission of numerous diseases. For instance, between November 1994 and January 1995, the US Department of Agriculture staff inspected 349 reptile shipments from 22 countries containing 117,690 animals. Of these, 54,376 animals were infested and 97 shipments were cleared of ticks (Karesh et al., 2005). Many ticks borne diseases threaten both people and animals, including heartwater disease, Lyme disease and babesiosis. Due to human activities like bushmeat handling, trade in exotic animals and destroying natural habitat, there is a growing risk of emergent pathogens spreading between humans and animals. Among 1415 documented human-infecting microorganisms, 61% classified as zoonoses, and pathogens with multiple hosts are twice as likely to be associated with human emergent pathogens. Additionally, other host species share 77% of the pathogens identified in livestock.

8.4 The Scope of Wildlife Trade

McNeely et al. (2009) identified key challenges to preserving biodiversity in Asia, as recognized by the Society for Conservation Biology's Asian Section. They highlighted that the rapid economic development and urban expansion in countries like India and China, which are driving significant infrastructure investments. They identified two main challenges: reducing the trade of threatened species as well as using conservation biology to enhance deeper understanding of the transfer zoonoses, both of which are closely linked to wildlife trade. Southeast Asia, including parts of China and Indonesia, is recognized as a "wildlife trade hotspot," where trade poses a disproportionately large threat to biodiversity. Wildlife trade encompasses the sale or exchange of wild animals and plants, which plays a critical role in both biodiversity conservation and sustainable development. In this trade include live animals and plants, as well as a variety of products such as skins, medicinal ingredients, and food, often providing income for economically disadvantaged communities and generating substantial national revenue. The primary driver of wildlife trade is economic, spanning from subsistence earning to high-volume commercial revenue. Although trading of wildlife occurs nationally as well as locally, significant volumes are traded internationally. The trade involves various intermediaries, including specialists in storage, transport, handling, retail, and manufacturing who function both locally and internationally (Traffic, 2008). With the rise of economic growth and globalization, the demand for wildlife has surged, leading to an expansion in the scale and scope of wildlife trade. The growing population, expanding consumer power, and globalization have driven the demand for exotic wildlife in both developed and developing countries. Without effective regulation and driven by the potential for substantial financial gain, this demand persists, exerting significant pressure on wildlife populations. Whereas precise trade volumes are often unclear, it is evident that significant numbers of species are traded annually, with conservation-sensitive species often being the focus of the most detailed data, though data quality varies. Both state and non-state organizations carefully track the collection of certain species from the wild, and trade in many of these species is controlled and consistently recorded under the Convention on International Trade in Endangered Species of Wild Fauna and Flora (CITES) (Nijman, 2010). This framework allows for retrospective assessments of actual trade levels. Although rare plants and animals are typically traded in smaller quantities, particularly when trade limits are set from conservation standpoint as it may be more valuable to focus trade analysis on species or species groups that are highly dependent on conservation efforts.

8.5 The Diversity of Wildlife Trade: Live Animals and Their Byproducts

Animal derived materials are commonly used in fashion items such as belts, leather purses, shoes and fur coats, making fashion manufacturers significant consumers of illegal wildlife trade (IWT), whether knowingly or unknowingly. While mostly 90% of fur constitutes of the legal wildlife trade in the ergonomic clothing industry, whereas illegal trade is primarily targets on exotic reptile hides for luxury products. Reduced role of illicit fur trade is attributed to the growth of fur farming, which has decreased the demand of illegal fur illicit. There are legal reptile farms, among Kenya for Nile crocodiles (*Crocodylus niloticus*), but in mostly countries, the annual export of python skins (*Pythonidae*) surpasses the breeding facility capability. Skins from large reticulated pythons (*Python reticulatus*) are unlikely to come from farms, considering it takes several annum for them to expand such sizes, making wild capture a more cost-effective option. Thus, wild capture likely supplies a significant portion of python skins, especially those imported into regions like the EU (Mozer & Prost, 2023). While the species trafficked for fashion are relatively few, Sosnowski and Petrossian's study of IWT seizures in the US from 2003 to 2013 showed that six genera (Varanus, Python, Homalopsis, Crocodylus, Caiman and Alligator) accounted for more than 80% of fashion-associated onfiscations, all of which are known for their use in leather goods. Python species were confiscated more frequently than any other genera. Despite this, trade of wildlife-related fashion goods has declined, while seizures of live animals for the pet trade have grown.

8.5.1 *Zoonotic Diseases Associated with Wildlife Trade*

The wildlife trade plays a significant role in the transmission of zoonotic diseases, posing threats to the health of both human and animals. The intimate connection between peoples and wild creatures during capture, transport, and trade increases the risk of pathogen spillover. Various zoonotic diseases, such as tuberculosis, brucellosis, anthrax, and Ebola hemorrhagic fever, have been linked to wildlife trade and consumption. The following Table 8.1 provides an overview of major zoonotic diseases, their causative agents, modes of transmission, affected hosts, symptoms, and preventive measures.

This Table 8.1 highlights the diverse pathogens, including bacteria, viruses, and parasites, that can be transmitted through wildlife trade. Effective biosecurity measures, vaccination programs, and hygiene practices are crucial in mitigating the risks

Table 8.1 Major zoonotic diseases: causative agents, hosts, transmission, and prevention

Diseases	Causative agents	Vector/ Transmission	Host	Symptoms	Treatment/Prevention
Tuberculosis	Bacteria: *Mycobacterium tuberculosis, M. bovis*	Aerosol, contaminated feed	Human primates, non-human primates, canids, felids	Weight loss, respiratory distress	Euthanasia; biosecurity and monitoring
Brucellosis	Bacteria: *Brucella spp.*	Ingestion of infected prey	Canids	Infertility, abortion, enlarged lymph nodes	Antibiotics, isolation of suspected cases
Anthrax	Bacteria: *Bacillus anthracis*	Contaminated feed	Felids	Sudden death, hemorrhage, difficulty breathing	Avoid contaminated food sources
Salmonellosis	Bacteria: *Salmonella spp.*	Fecal oral route	Songbirds, waterfowl, primates	Diarrhea, septicemia, lethargy	Hygiene at feeding stations, clean feeders
Plague	Bacteria: *Yersinia pestis*	Fleas, contact with infected rodents	Felids	Fever, lymphadenopathy, sudden death	Quarantine, flea control, isolation of exposed animals
Rabies	*Lyssavirus*	Bite, saliva contacts with broken skin	Canids, felids	Paralysis, aggression, salivation	Vaccination (inactivated for animals, prophylaxis for humans)
Herpesvirus	*Cercopithecine herpesvirus type 1 (B virus)*	Bite, scratches, mucosal exposure	Non-human primates (macaques, gibbons)	Vesicles, conjunctivitis, encephalitis	Avoid exposure, isolate infected animals
Rubeola (Measles)	*Measles virus*	Human contact	Non-human primates (marmosets, macaques)	Rash, fever, respiratory distress	Vaccination; avoid human interaction
Monkeypox	*Monkeypox virus*	Direct contact with infected animals	Non-human primates	Maculopapular rash, pustules	Supportive care; zoonotic risk awareness
Ebola hemorrhagic fever	*Ebola virus*	Direct contact with infected fluids	Non-human primates	Fever, rash, bleeding, diarrhea, hepatosplenomegaly	No licensed vaccine; supportive care
Kyasanur Forest Disease (KFD)	*KFD virus (Haemaphysalis ticks)*	Tick bites, contact with dead monkeys	Langurs, bonnet monkeys	Fever, headache, myalgia, high monkey mortality	Tick control, formalin-inactivated vaccine (humans), annual boosters

associated with these diseases. Additionally, raising awareness about zoonotic risks in communities involved in wildlife trade can contribute to reducing the spread of infectious diseases.

8.6 Zoonotic Diseases and Their Wildlife Origins

Emerging infectious diseases (EIDs) are defined as pathogens newly introduced or reintroduced into a population, leading to an increased incidence or expanded geographic range. Over the past two decades, novel pathogens have consistently emerged, driven by environmental changes such as globalization, human and animal movements, other factors that enable infectious agents to adapt to new ecological niches. Anthropogenic activities such as habitat degradation, species overexploitation, global pollution, climate change and the legal and illegal wildlife trade (IWT) are major contributors to EIDs (Rush et al., 2021). IWT in particular is a significant driver due to its threats to public health, biodiversity and ecosystem integrity. It facilitates the transmission of zoonotic and epizootic diseases, especially when wildlife is transported outside their endemic regions, increasing the risk of outbreaks.

Zoonotic diseases, which originate from animals, account more than 60% of newly arising infectious diseases among humans, with approximately 75% of these pathogens stemming from wildlife. Many zoonotic and epizootic viral infections involve single-stranded RNA viruses, such as coronaviruses, which can cause severe infections in both humans and animals. For instance, SARS-CoV-2, liable for the COVID-19 pandemic, thought to have originated from wildlife, emphasizing the role of animal-human interactions in the emergence of infectious diseases.

Zoonotic pathogens can be classified into two main types:

1. Pathogens of animal origin that rarely infect humans but, once transmitted, can occasionally sustain temporary human-to-human transmission, leading to potential outbreaks (e.g., HIV, SARS-CoV-2, and Ebola virus).
2. Pathogens primarily transmitted from animals to humans either directly or through vectors such as insects or rodents, which can cause significant public health risks, including diseases like Zika virus, Dengue virus, and Hantavirus (Fig. 8.3).

Although zoonotic pathogens can be transmitted through various routes, arboviruses, which are spread via arthropod borne-vectors such as mosquitoes, have lower probability to cause widespread global outbreaks due to their intricate life cycles. These pathogens require both an insect vector and a human or animal reservoir host to complete their transmission cycle. Additionally, the geographic spread of arboviruses is often limited by climate conditions and the availability of suitable vectors (Naicker, 2011). Most zoonotic pathogens infect humans as "dead-end hosts," meaning they are either non-transmissible or have limited human to human transmission (e.g., Rabies, Rift Valley fever). However, a small proportion of these

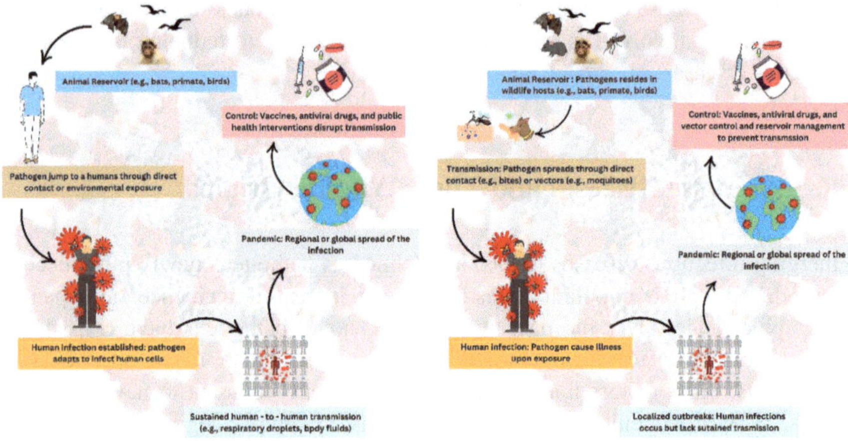

Fig. 8.3 Zoonotic pathogen cycles: From rare animal-origin infections with occasional human-to-human transmission (e.g., HIV, SARS-CoV-2) to direct or vector-mediated diseases (e.g., Zika, Dengue), both pose significant public health risks. (**a**) Illustrates pathogens of animal origin that rarely infect humans but can occasionally lead to human-to-human transmission, resulting in potential outbreaks (e.g., HIV, SARS-CoV-2, Ebola). (**b**) Shows pathogens primarily transmitted from animals to humans, either directly or through vectors like insects or rodents, causing significant public health risks (e.g., Zika, Dengue, Hantavirus)

pathogens (around 10%) have the ability to spread efficiently between humans (e.g., Mycobacterium tuberculosis, measles) or establish sustained human-to-human transmission after an initial zoonotic introduction (e.g., certain influenza strains, SARS coronavirus).

Live and wet markets (LWMs), commonly found in low- and middle-income countries, pose significant risks for the emergence of zoonotic diseases (Naguib et al., 2021). In these markets, where humans frequently come into close contact with live animals and animal products, conditions are conducive to the transmission of diseases. These markets have been directly linked to the outbreaks of coronaviruses and avian influenza, emphasizing the urgent need to address the threat related with human-wildlife interactions. Newly arising infectious diseases not only threaten global public health but also undermine economic stability, biodiversity, and environmental security, highlighting the critical need for proactive strategies to prevent their emergence.

One study categorized known pathogens and found that over 60% (Woolhouse & Gowtage-Sequeria, 2005) of emerging infectious diseases in humans are zoonotic, with the majority (72%) (Jones et al., 2008) originating from wildlife. Another estimate suggests that 75% of emerging pathogens are zoonotic. Although estimates differ, there is general agreement that most emerging viruses stem from animals. Many emerging zoonotic and epizootic viral infections involve single-stranded RNA viruses, such as coronaviruses, which can lead to severe infections in both animals and humans.

8.7 Case Studies

8.7.1 SARS-CoV

December 2019 saw the emergence of novel coronavirus, SARS-CoV-2, was identified from a cluster of a typical pneumonia cases among Wuhan, China, and rapidly escalated into a global pandemic with profound health, economic, and ecological impacts. Phylogenetic analysis indicates that SARS-CoV-2, along with related SARS-related coronaviruses (SARSr-CoVs), evolved in Old-World bats of the Rhinolophidae family. Although no direct evidence links SARS-CoV-2 transmission from bats to humans, its nearest identified counterparts, sharing 96.2% genome resemblance, was found in the Intermediate Horseshoe Bat (*Rhinolophus affinis*) in Yunnan, China, in 2013 (Zhou et al., 2020). The timing of virus's spillover to humans and the role of intermediate hosts remains unclear. While the United States has the highest number of confirmed COVID-19 cases, concerns have emerged regarding the potential transmission of SARS-CoV-2 from humans to wildlife, particularly bats.

The pandemic poses multiple risks to North American bat populations. SARS-CoV-2 could infect over 40 native bat species of temperate-zone, potentially causing disease and mortality. The virus might also establish reservoirs within these bat populations, similar to the way bat rabies lyssaviruses persist in the New World, thereby posing a recurring risk of human infection. Furthermore, persistent infection in bats could lead to viral evolution or recombination with endemic pathogens, potentially resulting in strains more infectious or pathogenic to humans and other animals. These scenarios could exacerbate public fears, transforming bats from perceived beneficial wildlife into vectors of unacceptable disease risks, undermining conservation and education efforts (Olival et al., 2020).

Multiple spillover events to mammals have led to numerous human SARS-CoV-2 infections. Although SARS-CoV-2 can infect a wide range of wildlife species, white-tailed deer are the only known wild species to have contracted the virus. Most of cases have been found in captive animals. The extent of SARS-CoV-2 exposure in wildlife communities and the factors influencing the risk of wildlife transmission remain unknown. Goldberg et al. (2024) investigated the impact of urbanization and human use on seropositivity while analysing 23 species of wildlife for SARS-CoV-2. Between May 2022 and September 2023, the study documented six species including common opossum, common raccoon, groundhog, Eastern cottontail, and Eastern red bat had positive SARS-CoV-2 RNA detections throughout Virginia and Washington, D.C., USA. It was discovered that seroprevalence was three times higher in areas with high human activity than in areas with little human presence. SARS-CoV-2 genomes from nine individuals of six species were classified to seven Pango lineages of the Omicron variant. At least seven recent human-to-animal transmissions are suggested by the strong match to variants that were present in humans at the time. These results support the idea that wildlife populations have been exposed to SARS-CoV-2 on a large scale and suggest that human-populated areas could serve as cross-species transmission points.

Although there is ample evidence of SARS-CoV-2 transmission to confined wildlife (Drozdz et al., 2021; Fritz et al., 2020; Murphy & Ly, 2021) detection in natural wildlife populations is still limited to a few species, such as white-tailed deer (*Odocoileus virginianus*) (Hale et al., 2022; Kuchipudi et al., 2022), feral mink (*Neovison vison*) (Aguiló-Gisbert et al., 2021) and Eurasian River otters (*Lutra lutra*) (Rahman et al., 2020). Many wildlife species may be suitable hosts for SARS-CoV-2, according to experimental infections and modelling of the functional SARS-CoV-2 receptors, angiotensin-converting enzyme 2 (ACE2) (Martínez-Hernández et al., 2020; Meekins et al., 2021; Griffin et al., 2021; Fagre et al., 2021; Francisco et al., 2022; Caserta et al., 2023). However, it remains unclear whether a variety of wildlife species are exposed to the virus in their natural environments, where exposure to SARS-CoV-2 is likely indirect and occurs at a lower dose.

8.7.2 COVID-19

At the end of December 2019, occurrences of pneumonia triggered by a novel coronavirus emerged in Wuhan, Hubei Province, China. The World Health Organization (WHO) later designated the virus as coronavirus disease 2019 (COVID-19). The virus quickly spread across Hubei Province and throughout China, with daily confirmed cases rising sharply between January 16 and 24, a surge likely linked to large-scale population movement during the Chinese New Year.

COVID-19 cases peaked on February 12, 2020, with 15,153 new cases reported in a single day. From January 23 to February 4, the rapid increase in daily cases highlighted the virus's expanding spread. However, from February 5 onward, the daily case numbers began to decline, reflecting the effectiveness of quarantine measures and travel restrictions. The extension of the Spring Festival holiday further reduced population mobility, significantly limiting the virus's transmission compared to previous years. By February 14, new confirmed cases outside Hubei Province had steadily decreased, with daily new cases in other regions dropping to single digits. This marked a turning point in controlling the outbreak within China (Wang & Su, 2020).

Live and wet markets (LWMs), common in the rapidly growing urban areas of low- and middle-income countries (LMICs) across Asia and Africa, play a critical role in supporting livelihoods and ensuring food security for many communities. These markets, which sell live animals and animal products, including ready-to-eat foods, serve millions of customers daily in tropical and subtropical regions. However, the close interaction between humans-both vendors and buyers-live animals, food items, and even wild or peri-household pet significantly elevate the risk of zoonoses disease spread. The 2002–2003 SARS outbreak, caused by the coronavirus SARS-CoV, is a striking example, as it originated from an LWM in southern China, highlighting the serious public health threats these markets can pose (Naguib et al., 2021).

8.7.3 *Nipah Virus: Associated with Bats and Pig Farming*

In 1998–1999, a novel highly transmissible respiratory and neurological disease affected swines in Malaysia, causing a simultaneous viral encephalitis outbreak among farm workers. A new paramyxovirus, unique from the Hendra virus, was identified as the Nipah virus, with fruit bats (Pteropus species) serving as its natural reservoir. The virus caused high mortality in pigs (up to 40%) and 265 human cases, with a 38% fatality rate. The epidemic was controlled by culling affected pigs, costing Malaysia nearly 45% of its pig population (Bengis et al., 2004).

Nipah virus emerged due to ecological changes, including deforestation, drought, and the expansion of pig farming. Deforestation displaced fruit bats, leading them to move into agricultural areas with pigs, increasing the risk of virus transmission from bats to pigs and humans. Human Nipah infections were also reported in Bangladesh in 2001, 2003, and 2004, with fruit bats again identified as the source, possibly through contaminated fruit consumption.

8.8 Role of Wildlife as Reservoirs for Pathogens

Wildlife serves as critical reservoirs for various pathogens, playing a pivotal influence on the spread of diseases to humans and domesticated animals. Pathogens in wildlife can either naturally persist in their hosts without causing significant disease (endemic infections) or spill over to new hosts under specific conditions, often facilitated by ecological and anthropogenic factors.

8.9 Natural Reservoirs of Zoonotic Pathogens

Bats are recognized as reservoirs for a wide range of viruses, including rabies and filoviruses, such as Ebola and Marburg viruses. Although many viruses have been detected in bats, the specific reservoir hosts for most remain unidentified. For instance, while experimental infections of certain bat species-like the Angola free-tailed bat and Wahlberg's epauletted fruit bat-have shown replication of Ebola Zaire virus, the broader role of bats as natural reservoirs for Ebola remains unclear. Recent studies detected Ebola virus RNA in fruit bats in Central Africa, including species like the hammer-headed fruit bat and Franquet's epauletted bat, but these findings are not yet definitive (Calisher et al., 2006). While Ebola virus-specific antibodies were found in some of these bats, no direct evidence of virus shedding or transmission has been established, making these results intriguing yet inconclusive.

Similarly, while rabies virus transmission has been clearly demonstrated in bats, particularly vampire bats, only a small proportion of human rabies cases are associated with bat variants, highlighting the relatively rare role of bats in human rabies transmission. The origin of rabies variants affecting carnivores is thought to stem from bat-associated strains, with genetic studies suggesting a divergence in these viruses as far back as the seventeenth century.

The complexity of zoonotic transmission involving bats is further illustrated by the filoviruses, specifically Ebola and Marburg viruses. These viruses belong to the Filoviridae family, with Ebola species such as Zaire, Sudan, and Reston viruses, and Marburg virus identified as major concerns for human health. Despite detecting Ebola virus RNA in various mammalian species, including bats and rodents, the natural reservoir host remains unidentified. Some theories suggest that a non-pathogenic form of Ebola might exist, potentially mutating into a pathogenic variant. Experimental studies on bats and other wildlife are necessary to clarify the full extent of their role in the transmission of these deadly viruses, but current evidence still leaves these questions open.

8.10 Drivers of Pathogen Transmission

Anthropogenic activities often exacerbate the role of wildlife as pathogen reservoirs: Land-use change, driven by activities like road building and logging, increases human-wildlife contact, fostering pathogen spillover and the emergence of zoonotic diseases. Habitat fragmentation and conversion into agricultural or urban areas expand human-wildlife interfaces, raising spillover rates, especially in transitional zones (Rodríguez, 2021). Deforestation has been linked to Ebola outbreaks, with a delay of 2 years between environmental change and disease emergence.

Cultural practices, such as wildlife trade and farming, also elevate zoonotic risks by intensifying human-animal interactions. While wildlife farming may reduce wild animal consumption, it increases pathogen spillover risk. Targeted interventions like regulating high-risk wildlife markets and improving surveillance are crucial.

Increased livestock production intensity further leads to zoonosis by creating densely populated environments can act as reservoirs for pathogen transmission between animals and humans. For instance, avian influenza, swine flu, and MERS have originated from livestock, with their transmission further intensified by transportation and agriculture-wildlife mixing (Intergovernmental Panel on Climate Change (IPCC), 2013). Moreover, recreational activities, such as wildlife watching, can also raise the risk of exposure to wildlife-borne pathogens. Mitigating the hazards of zoonotic diseases require addressing factors such as land-use change, agricultural intensification, and wildlife trade.

8.11 Contributing Factors to the Risk

8.11.1 Environmental Factors

In tropical areas, climatic changes have driven the extensive rise of zoonotic virus carriers and vectors, increasing virus transmission to animal hosts. Climate factors, including changes in temperature and precipitation, can trigger genetic mutations in viruses. For example, West Nile Virus in North America have facilitated beneficial mutations at higher temperature. As well as, studies on MERS-CoV show that 56.1% of cases occurred between April and August, when high temperatures and low humidity enhanced the virus's viability.

Rising global temperature, projected to increases by 3 °C by 2050 (Intergovernmental Panel on Climate Change (IPCC), 2013), are likely to alter the distribution of vectors and reservoirs to new regions. A significant number of emerging zoonotic are linked to arthropod vectors and animal reservoirs (Mishra et al., 2021). In Arctic regions warning promotes the spread of pathogens and vectors, changes the density and distribution of reservoirs, which increases pathogen transmission risks. Rising temperatures also lengthen vector survival and expand habitat ranges, enhancing zoonotic pathogen spread.

Climate change effect ecosystems both directly or indirectly through altered precipitation, temperature, and humidity, creating conditions for pathogen transmission. Vectors and reservoirs may also spread via trade and migration, as evidenced by the spread of dengue in southern Europe. Understanding these dynamics is beneficial for managing disease ecology and spillover risks.

8.11.2 Extreme Weather and Climate Conditions

Extreme climatic events, such as floods and droughts, have become significant drivers of pandemics, with a marked increase in these events over the past 50 years, largely due to climate change (UNCCD and FAO, 2020). Periods of droughts followed by wet conditions increase the risk of disease transmission by bringing hosts and vectors, such as mosquitoes, into closer contact, as observed with West Nile virus (WNV) in Southern Florida. By 2050, severe droughts could lead to threefold increase in WNV cases in high-risk regions. Droughts also led to increased reliance on bushmeat and heightening zoonotic risks.

Floods exacerbate surge of vector-borne illnesses such as Dengue, WNV, and malaria, with examples in Romania, Italy and Southeast Asia. Disasters like floods complicate pandemic responses, as seen during COVID-19, where healthcare systems were disrupted. Relief camps further amplify disease risks due to overcrowding, poor hygiene, and limited resources. Improved forecasting and preparedness are crucial for mitigating disease spread during extreme weather events.

8.11.3 Pets

Zoonotic diseases from companion animals often spread through close contact with owners. Salmonellosis in humans is linked to exotic reptiles, while petting zoos have led to *Escherichia coli* O157:H7 outbreaks. The rise of zoonotic leishmaniasis in Europe is partly due to domestic dogs being primary reservoirs, with Leishmania infantum causing cutaneous and visceral forms. This disease spreads through travel and the importation of infected dogs. Domestic cats are primary reservoirs for *Bartonella* spp. leading to conditions like cat-scratch disease and bacillary angiomatosis, which are expected to persist or increase (Naicker, 2011).

8.11.4 Biodiversity Loss

Biodiversity loss increases the risk of zoonotic diseases by disrupting ecosystems and elevating human exposure to pathogens. Changes in species diversity affect host-pathogen dynamics, with reservoir and vector species playing key roles. The COVID-19 pandemic highlights how biodiversity loss creates ecosystems vulnerable to pathogen spillover. Factors like high human population density, habitat destruction, and wildlife diversity contribute to disease emergence, as seen with the Nipah virus outbreak. Fragmented habitats bring wildlife into closer contact with humans, while reduced diversity among small mammals increases hantavirus risk. Deforestation, human-wildlife conflicts, and wildlife trade exacerbate these risks, leading to novel pathogens. Disruption of migratory species' routes and the consumption of their meat in wet markets also heightens transmission risks. Tracking migratory routes is vital to monitoring emerging threats and protecting global health (Mishra et al., 2021).

8.11.5 Urbanization and Socioeconomic Impacts

Urbanization, particularly in slums and peri-urban areas, has exacerbated environmental and health hazards, fuelling pandemics. High population density, poor sanitation, and inadequate living conditions make infectious diseases, like COVID-19, more likely to spread. Rapid urbanization influences pandemic timing and recurrence, with slums becoming hotspots due to poor infrastructure and healthcare. Lockdowns caused mass migration, spreading COVID-19 and worsening socio-economic conditions (Mishra et al., 2021). The pandemic highlighted vulnerabilities, especially among marginalized groups such as indigenous communities in Mexico, facing obstacles like limited access to water and healthcare. Overcrowded, polluted urban areas with poor healthcare access contributed to higher infection and mortality rates. Air pollution worsened disease severity, and improper medical waste disposal further increased transmission risks.

8.12 Impacts of Wildlife Trade-Driven Zoonotic Diseases

Internationally, a variety of species and ecosystems are involved in the wildlife trade, both legally and illegally. As a result of strong interaction between humans, domestic animals, and wildlife, this multibillion-dollar trade offers a high risk for the transmission of zoonotic diseases. Emergence of infectious diseases is when pathogens cross species barriers and invade human populations due to these interactions. The wildlife trade has significant implications for the establishment of zoonotic diseases. Due to habitat disruption and patch isolation brought on by trade-related activities, people, domestic animals, and wildlife have been brought into closer proximity, which makes it easier for zoonotic diseases to spread. Furthermore, high-density trading strategies, which are frequently associated with unregulated and illegal wildlife trade, can result in congested and unhygienic conditions, offering the perfect conditions for the spread of infectious diseases among various species and humans. The trading of wildlife requires special consideration in certain domains. The trade of exotic pets, bush meat, wild animal meat consumption, and wet markets, where live animals are bought and slaughtered right there, all contribute significantly to the transmission of animal borne illnesses. The likelihood of animal born illnesses spillover incidents is increased by the fact that these traditional activities frequently include subpar hygiene standards and the mixing of several species.

Numerous case studies demonstrate the impact of wildlife trading on the emergence of zoonotic diseases. For instance, we might look at well-known instances of zoonotic diseases that have been transmitted through the trade in wild animals (Rahman et al., 2020). The severe acute respiratory syndrome (SARS) outbreak, which took place between 2002 and 2003, brought attention to the link among wildlife commerce as well as animal borne illness. Similarly, bush meat intake has been connected to Ebola virus epidemics in Central and West Africa, highlighting the threats of this kind of trade in wildlife. The ongoing COVID-19 pandemic, which is being caused by the SARS-CoV-2 virus, has also raised awareness of the possible effects of wildlife trade on an international level. The importance of understanding and addressing the risks posed by the trade of wild animals to protect public health is evident from these examples.

8.13 Prioritization of Zoonotic Diseases

Preventing, detecting, and responding to zoonotic diseases in resource-limited settings is challenging due to competing public health priorities. Effective mitigation requires multisectoral collaboration across human, animal (domestic and wildlife), and environmental health sectors, which can take time to establish. A crucial step is for these sectors to work together identify and prioritize zoonotic diseases of greatest concern, fostering partnerships and resource sharing. Early involvement of diverse sectors foster collaboration, program ownership, and enables the utilization of established systems for priority diseases to effectively respond to other emerging health threats.

The CDC's One Health Zoonotic Disease Prioritization tool supports countries in identifying high-priority zoonotic diseases through a collaborative, semi-quantitative approach. This tool involves multidisciplinary teams from government ministries, international organizations (e.g., WHO, FAO, OIE), and other stakeholders. Workshops, facilitated by trained in-country professionals, develop country-specific criteria to rank diseases of national concern. Minimizing external facilitator roles ensures objectivity and decision-making by host country representatives, promoting ownership and sustainability of the process (Sokolow et al., 2019).

8.14 Targeted Interventions

Zoonotic diseases are reduced through the implementation of precise measures that are based on risk assessment. Authorities can implement particular measures, such as increased biosecurity, potent laws and focused surveillance, to lessen the likelihood of disease spillover by identifying high-risk wildlife species and trade networks. The efficacy of interventions is increased by this targeted approach since resources are directed to the areas that require them the most (Saleem et al., 2023).

8.15 Ecological and Conventional Approaches

The process of zoonotic spillover involves a pathogen overcoming a series of ecological barriers to transfer from a vertebrate host to a spillover at a specific time and location. While conventional interventions such as population management, immunization, and pesticide application (e.g., drugs, insecticides, disinfectants) aim to temporarily reduce the number of susceptible, infected, or recovered individuals, these actions often have short-term effects, logistical challenges, and potential adverse consequences like environmental damage or resistance evolution. For instance, disruptions in vaccination efforts during the 2014 Ebola epidemic in West Africa led to increased measles risk, highlighting the transient nature of such measures.

Ecological interventions, however, aim to address the underlying transmission processes for longer-term impact. Strategies such as managing reservoir population size, pathogen survival in the ecosystem, or vector proliferation, and altering the spatial and temporal alignment of reservoirs and spillover hosts, can effectively limit spillover. For example, rehabilitating the habitats of natural predators can lead to lasting changes in mortality rates of host species, while enhancing genetic diversity in reservoir hosts may reduce long-term susceptibility.

These strategies require a comprehensive understanding of ecological systems and typically address multiple transmission pathway instead of focusing on a single factor. Case studies reveal how ecological interventions can control reservoir populations, affect pathogen survival, or reduce contact risk, providing sustainable approaches to manage zoonotic spillover (Sokolow et al., 2019).

8.15.1 *Public Awareness and Education*

Rising awareness and fostering active community participation are crucial for effective management of zoonotic diseases. Educating the public about zoonotic diseases, their transmission routes, and preventive strategies empowers communities to take better prevent and control these infections. Promoting the One Health approach, which highlights the interconnectedness of people, fauna, and environmental well-being, play a key role in reducing zoonotic spillover risks. Strengthening collaboration among key stakeholders, including government agencies, non-governmental organizations, and medical bodies, strengthens coordination and enhances disease control initiatives. Delivering tailored messages to specific audiences and utilizing diverse communication platforms ensure effective outreach. Engaging communities builds trust, strengthens partnerships, and enables individuals to play a proactive role in preventing and controlling zoonotic diseases, ultimately protecting public health (Waheed et al., 2023).

8.15.2 *International Collaboration*

Effectively managing zoonotic diseases risks associated with wildlife trade requires a unified global approach and cooperative risk assessment. This collaborative effort, stretching beyond national boundaries, promote mutual learning through the dissemination of risk assessment frameworks, data repositories, and research insights. In the complex global landscape where wildlife trade and zoonotic diseases pose significant threats, international cooperation becomes essential. By pooling intellectual resources under this shared umbrella, countries can engage in collective problem-solving to address the challenges posed by high-risk wildlife species and their commercial networks.

In conclusion, mitigating the risks of zoonotic disease transmission requires comprehensive risk assessment to identifying high-risk animal species and trading networks. This approach facilitates early detection, resource allocation, informed decision-making, focused interventions, public awareness, and global collaboration. By conducting thorough, evidence-based risk assessments, we can effectively manage and reduce the risks associated with wildlife trading effectively, ultimately protecting both human health and biodiversity conservation (Saleem et al., 2023).

References

Aguiló-Gisbert, J., Padilla-Blanco, M., Lizana, V., Maiques, E., Muñoz-Baquero, M., Chillida-Martínez, E., Cardells, J., & Belló-Pérez, M. (2021). First description of SARS-CoV-2 infection in two feral American mink (*Neovison vison*) caught in the wild. *Animals, 11*(5), 1422.

Bengis, R. G., Leighton, F. A., Fischer, J. R., Artois, M., Morner, T., & Tate, C. M. (2004). The role of wildlife in emerging and re-emerging zoonoses. *Revue Scientifique et Technique-Office International des Epizooties, 23*(2), 497–512.

Calisher, C. H., Childs, J. E., Field, H. E., Holmes, K. V., & Schountz, T. (2006). Bats: Important reservoir hosts of emerging viruses. *Clinical Microbiology Reviews, 19*(3), 531–545. https://doi.org/10.1128/CMR.00017-06

Caserta, L. C., Fernandes, A. M., Wurtz, M., Webb, M., Robinson, K., Smith, J., Hall, R. J., McGraw, L., Nelson, T. M., & Langlois, R. A. (2023). White-tailed deer (*Odocoileus virginianus*) may serve as a wildlife reservoir for nearly extinct SARS-CoV-2 variants of concern. *Proceedings of the National Academy of Sciences of the United States of America, 120*(12), e2215067120.

Contini, C., Di Nuzzo, M., Barp, N., Bonazza, A., Vaccarezza, M., Gasperetti, A., Marcolini, G., & Rubino, S. (2020). The novel zoonotic COVID-19 pandemic: An expected global health concern. *Journal of Infection in Developing Countries, 14*(3), 254–264. https://doi.org/10.3855/jidc.12671

Cooney, R., Roe, D., Dublin, H., Phelps, J., Wilkie, D., Keane, A., Travers, H., Skinner, D., Challender, D. W. S., Allan, J. R., & Biggs, D. (2017). From poachers to protectors: Engaging local communities in solutions to illegal wildlife trade. *Conservation Letters, 10*(3), 367–374. https://doi.org/10.1111/conl.12294

Drozdz, M., Schulz, J., Machnik, G., Zieba, P., & Zieba, T. (2021). Current state of knowledge about role of pets in zoonotic transmission of SARS-CoV-2. *Viruses, 13*(9), 1149.

Fagre, A., Lewis, J., Eckley, M., Zhan, S., Rocha, S. M., Sexton, N. R., Burke, B., Geiss, B. J., & Kading, R. C. (2021). SARS-CoV-2 infection, neuropathogenesis and transmission among deer mice: Implications for spillback to new world rodents. *PLoS Pathogens, 17*(3), e1009275.

Francisco, R., Marinho, V., Azevedo, F., Santos, C., Ferreira, J., Carvalho, C., Cardoso, P., & Almeida, C. (2022). Experimental susceptibility of North American raccoons (*Procyon lotor*) and striped skunks (*Mephitis mephitis*) to SARS-CoV-2. *Frontiers in Veterinary Science, 8*, 715307.

Fritz, M., Rosolen, B., Krafft, E., Becquart, P., Elguero, E., Vratskikh, O., Denolly, S., Boson, B., Vanhomwegen, J., Gouilh, M. A., Kodjo, A., Chirouze, C., Rosolen, S. G., Legros, V., & Leroy, E. M. (2020). High prevalence of SARS-CoV-2 antibodies in pets from COVID-19+ households. *One Health, 11*, 100192.

Goldberg, A. R., Hoyt, J. R., et al. (2024). Widespread exposure to SARS-CoV-2 in wildlife communities across Virginia and Washington, D.C., USA. *Nature Communications, 15*, Article 5643. https://doi.org/10.1038/s41467-024-49891-w

Griffin, B. D., Chan, M., Tailor, N., Mendoza, E. J., Leung, A., Warner, B. M., He, S., Fuss, J., Zheng, J., Garnett, L., Tran, K. N., Lindsay, L. R., Jardine, C. M., & Strong, J. E. (2021). SARS-CoV-2 infection and transmission in the North American deer mouse. *Nature Communications, 12*, 10.

Hale, V. L., Dennis, P. M., McBride, D. S., Nolting, J. M., Madden, C., Huey, D., Ehrlich, M., Grieser, J., Winston, J., Lombardi, D., Twenhafel, N., Jenkins-Moore, M., Killian, M. L., Lantz, K., Tell, R. M., Torchetti, M. K., Robbe-Austerman, S., Nelson, M. I., Faith, S. A., & Bowman, A. S. (2022). SARS-CoV-2 infection in free-ranging white-tailed deer. *Nature, 602*, 481–486.

Intergovernmental Panel on Climate Change (IPCC). (2013). Climate change 2013: The physical science basis. In T. F. Stocker, D. Qin, G.-K. Plattner, M. Tignor, S. K. Allen, J. Boschung, A. Nauels, Y. Xia, V. Bex, & P. M. Midgley (Eds.), *Contribution of working group I to the fifth assessment report of the intergovernmental panel on climate change*. Cambridge University Press. https://doi.org/10.1017/CBO9781107415324

Jones, K. E., Patel, N. G., Levy, M. A., Storeygard, A., Balk, D., Gittleman, J. L., & Daszak, P. (2008). Global trends in emerging infectious diseases. *Nature, 451*(7181), 990–993.

Karesh, W. B., Cook, R. A., Bennett, E. L., & Newcomb, J. (2005). Wildlife trade and global disease emergence. *Emerging Infectious Diseases, 11*(7), 1000.

Kuchipudi, S. V., Surendran-Nair, M., Ruden, R. M., Yon, M., Nissly, R. H., VandeWoude, S., Saif, L. J., & Kapur, V. (2022). Multiple spillovers from humans and onward transmission of SARS-CoV-2 in white-tailed deer. *Proceedings of the National Academy of Sciences of the United States of America, 119*(6), e2121644119.

Liew, J. H., Kho, Z. Y., Lim, R. B. H., Dingle, C., Bonebrake, T. C., Sung, Y. H., & Dudgeon, D. (2021). International socioeconomic inequality drives trade patterns in the global wildlife market. *Science Advances, 7*(19), eabf7679. https://doi.org/10.1126/sciadv.abf7679

Martínez-Hernández, F., Correa-Cárdenas, C. A., Tavares, M. C. H., Martineau, Y., Viñuela, E., & Alves, P. C. (2020). Assessing the SARS-CoV-2 threat to wildlife: Potential risk to a broad range of mammals. *Perspectives in Ecology and Conservation, 18*(4), 223–234.

Maxwell, S. L., Fuller, R. A., Brooks, T. M., & Watson, J. E. M. (2016). Biodiversity: The ravages of guns, nets and bulldozers. *Nature, 536*(7615), 143–145. https://doi.org/10.1038/536143a

McNeely, J. A., Kapoor-Vijay, P., Zhi, L., Olsvig-Whittaker, L., Sheikh, K. M., & Smith, A. T. (2009). Conservation biology in Asia: The major policy challenges. *Conservation Biology, 23*(4), 805-810. https://doi.org/10.1111/j.1523-1739.2009.01284.x

Meekins, D. A., Gaudreault, N. N., & Richt, J. A. (2021). Natural and experimental SARS-CoV-2 infection in domestic and wild animals. *Viruses, 13*(10), 1993.

Mishra, J., Mishra, P., & Arora, N. K. (2021). Linkages between environmental issues and zoonotic diseases: With reference to COVID-19 pandemic. *Environmental Sustainability, 4*(3), 455–467.

Mozer, A., & Prost, S. (2023). An introduction to illegal wildlife trade and its effects on biodiversity and society. *Forensic Science International: Animals and Environments, 3*, 100064.

Murphy, H. L., & Ly, H. (2021). Understanding the prevalence of SARS-CoV-2 (COVID-19) exposure in companion, captive, wild, and farmed animals. *Virulence, 12*(1), 2777–2786.

Naguib, M. M., Li, R., Ling, J., Grace, D., Nguyen-Viet, H., & Lindahl, J. F. (2021). Live and wet markets: Food access versus the risk of disease emergence. *Trends in Microbiology, 29*(7), 573–581.

Naicker, P. R. (2011). The impact of climate change and other factors on zoonotic diseases. *Archives of Clinical Microbiology, 2*(2), 1–6.

Nellemann, C., Henriksen, R., Raxter, P., Ash, N., & Mrema, E. (Eds.). (2014). The environmental crime crisis: Threats to sustainable development from illegal exploitation and trade in wildlife and forest resources.. United Nations Environment Programme.

Nijman, V. (2010). An overview of international wildlife trade from Southeast Asia. *Biodiversity and Conservation, 19*(4), 1101–1114.

Olival, K. J., Cryan, P. M., Amman, B. R., Baric, R. S., Blehert, D. S., Brook, C. E., Epstein, J. H., Gilbert, A. T., Hayman, D. T. S., Jenkins, R. K. B., Johnson, C. K., Kading, R. C., Kingston, T., Lorch, J. M., Mendenhall, I. H., Peel, A. J., Plowright, R. K., et al. (2020). Possibility for reverse zoonotic transmission of SARS-CoV-2 to free-ranging wildlife: A case study of bats. *PLoS Pathogens, 16*(9), e1008758.

Rahman, M. A., Zaman, N., Asyhari, A. T., Al-Turjman, F., Bhuiyan, M. Z. A., & Zolkipli, M. F. (2020). Data-driven dynamic clustering framework for mitigating the adverse economic impact of COVID-19 lockdown practices. *Sustainable Cities and Society, 62*, 102372.

Rodríguez, C. M. (2021). Nature and humanity. *PARKS, 27*, 13.

Rush, E. R., Dale, E., & Aguirre, A. A. (2021). Illegal wildlife trade and emerging infectious diseases: Pervasive impacts to species, ecosystems, and human health. *Animals, 11*(6), 1821.

Saleem, M. I., Ramzan, F., Nazar, M., Khan, M. S., Mahfooz, A., Zaka, F., & Shakir, A. H. (2023). Mitigate the risk of zoonosis. In *Zoonosis* (Vol. 1, pp. 643–658). Unique Scientific Publishers.

Sokolow, S. H., Nova, N., Pepin, K. M., Peel, A. J., Pulliam, J. R., Manlove, K., Plowright, R. K., Rejmanek, D., Switzer, C., & De Leo, G. A. (2019). Ecological interventions to prevent and manage zoonotic pathogen spillover. *Philosophical Transactions of the Royal Society B, 374*(1782), 20180342.

Tittensor, D. P., Harfoot, M., McLardy, C., Britten, G. L., Kecse-Nagy, K., Landry, B., Outhwaite, W., Price, B., Sinovas, P., Blanc, J., Burgess, N. D., & Malsch, K. (2020). Evaluating the relationships between the legal and illegal international wildlife trades. *Conservation Letters, 13*(5), e12724. https://doi.org/10.1111/conl.12724

Traffic. (2008). *What's driving the wildlife trade?* The World Bank.

UNCCD & FAO. (2020). *The impact of disasters and crises on agriculture and food security*. Retrieved from https://catalogue.unccd.int/1688_FAO_cb3673en.pdf

United Nations Office on Drugs and Crime. (2020). *World wildlife crime report 2020: Trafficking in protected species*. United Nations.

Waheed, T., Ali, K., Asad, M., Baber, Z. U., Riaz, A., Rasheed, M., & Saleem, M. (2023). Awareness and community engagement in zoonotic disease management. In *Zoonosis* (Vol. 1, pp. 470–485). Unique Scientific Publishers.

Wang, Q., & Su, M. (2020). A preliminary assessment of the impact of COVID-19 on the environment: A case study of China. *Science of the Total Environment, 728*, 138915.

Woolhouse, M. E., & Gowtage-Sequeria, S. (2005). Host range and emerging and re-emerging pathogens. *Emerging Infectious Diseases, 11*(12), 1842–1847. https://doi.org/10.3201/eid1112.050997

WWF. (2020). Living planet report 2020. In R. E. A. Almond, M. Grooten, & T. Petersen (Eds.), *Bending the curve of biodiversity loss*. WWF.

Wyatt, T., Maher, J., Allen, D., Clarke, N., & Rook, D. (2022). The welfare of wildlife: An interdisciplinary analysis of harm in the legal and illegal wildlife trades and possible ways forward. *Crime, Law and Social Change, 77*(1), 69–89.

Zhou, P., Yang, X. L., Wang, X. G., Hu, B., Zhang, L., Zhang, W., et al. (2020). Addendum: A pneumonia outbreak associated with a new coronavirus of probable bat origin. *Nature, 588*(7836), E6.

Chapter 9
Impact of Wildlife Trade on Marine Fauna

Parvati Sharma, Monika Jangra, and R. Urmila

Contents

Abstract The marine animals trade is a serious concern throughout the World and the trade of marine species such as corals, turtles and seahorses, has a significant impact on ecological system, economic, and conservation initiatives. The frequent poaching of seahorses for its use in traditional medicine, decorative purposes and the aquarium industry through the World. The excessive commercial and ornamental traffic of coral leads to reef degradation and decline marine biodiversity of corals and its dependent species. Sea turtles are also categorized in endangered species due to excessive poaching for their eggs, meat, and shells. One international agreement that attempts to limit and control overexploitation is the Convention on International Trade in Endangered Species of Wild Fauna and Flora (CITES) is addressed the

P. Sharma (✉) · M. Jangra · R. Urmila
Department of Zoology, Chaudhary Bansi Lal University, Bhiwani, Haryana, India
e-mail: parvati.zoology@cblu.ac.in

A. Poonia et al. (eds.), *Wildlife Trade in India*,
https://doi.org/10.1007/978-981-95-3694-8_9

marine issues. Nevertheless, issues including ongoing illegal trade, weak enforcement, and the destruction of natural regions persist despite the efforts of CITES. To deal with issues, requires a combined project of locally driven, implementation of successful conservation tactics, and sustainable trading practices to oversee the survival of these species. In the regard, this chapter explores the effects of the trade in marine species, assesses the work of the current conservation measures, and discussed about adequate ways to decrease its detrimental effects.

Keywords Marine trade · Overexploitation · Seahorses · Turtle conservation · Coral trade · Biodiversity loss

9.1 Introduction

Seahorses, corals, and turtles are the marine species that are to be characterised under endangered species due to overexploitation and became a serious concern for biodiversity conservation worldwide (Dee et al., 2014). Their sustainability is in danger due to unsustainable harvesting practices of these species arise by the growing demand for these marine species in international market, which is execrated by the industries of traditional medicine, decorative crafts, and tourism. For example, the main reason for the decline seahorse populations was their widespread collection for use in aquariums and traditional Chinese medicine (Vincent, 1996).

In spite of sluggish regeneration and critical biological roles of coral reefs in marine ecosystems, these species are regularly poached without any fear for decorative and commercial uses (Kramer, 2003). Similarly, the great demand for their meat, eggs, and shells of turtles are also responsible for pushing into the verge of extinction by both complying and non-compliant markets (Chelonia, 2021).

International laws and regulations which are exist in the Convention on International Trade in Endangered Species (CITES), have been in use to prevent overexploitation in order to counteract the dangers related to these species. However, the complication of international trade networks and enforcement issues frequently negotiate the efficiency of such regulations (Pritchard, 2008). The undue ecological pressure arises due to the marine species trade, the alteration of socioeconomic factors, and the functions of international agreements to protect these endangered species are all discussed in this chapter. The chapter highlight the practices that can maintain a balance between conservation efforts and commercial interests for the community as well as marine species.

Effects of Marine Species Trade on Economy

The advancement in Technology and globalization have made the trade easier, but ecological issues underscore the need for stricter laws and more meticulous on market demand (Howse et al., 2005). A study of 2024 revealed that the trade of sea turtles, seahorses, and coral species is heavily affected the economic factors as it fluctuates on the demands of the consumers (Sasipriyan et al., 2024).

The substantial market value of these marine species propels global commerce and yields significant economic benefits, particularly for communities residing on islands and along coastal regions. In locales where alternative sources of income are scarce, the extraction of these marine species emerges not only as a crucial means of subsistence but also as a vital revenue stream for countless individuals within these communities.

9.1.1 Demand in Market and Choice of Consumers

The robust consumer interest in seafood, ornamental fish, and products derived from marine sources has notably enhanced international trade dynamics. The growth of global trade agreements has further eased the export of marine species, especially from developing countries to more affluent nations. China, reigning as the largest importer and exporter of seafood worldwide, plays a pivotal role in shaping the patterns of global marine trade (FAO, 2022). Rising incomes across Asia are contributing to an increasing demand for premium seafood products (FAO, 2022). The trade in marine species is exceptionally lucrative, spurring both legal fisheries and illegal fishing activities (Sadovy et al., 2018). In 2020, the value of the global seafood market surpassed $160 billion, with rare and high-demand species commanding even greater prices (OECD, 2021). This profitability has also facilitated the growth of black markets and illegal, unreported, and unregulated (IUU) fishing, which continues to persist due to its enticing financial rewards (Sumaila et al., 2020). The uses for trading various marine species, such as coral, seahorses, and turtles, are examined below for different purposes:

Construction Purpose
In many countries such as East Africa, South Asia, Southeast Asia, and the Pacific, Corals and its fragments are used for building materials (Cement for new construction), limestone and road fill or bricks. Corals are mined for calcium supplements also therefore it is known as "calcium mining."

Collection for Ornament Purposes and Memorabilia Tourist and exporters are interested to buy jewellery made up corals and mementos of corals are also famous. When buying coral to sell abroad, a jeweler should confirm that the vendor has the proper CITES permission for the coral and check with USFWS before importing or exporting any animal goods.

The overexploitation of some fish species and destructive fishing methods are having a major impact on reef ecosystems and coral reef fish and other organism populations. For the fish food trade, around 25,000 metric tons of live reef fish are collected annually, Unfortunately, the most popular way to catch these fish is by cyanide fishing, which is currently practised by at least ten major exporting nations.

By poisoning and killing non-target creatures, including corals, cyanide, one of the most lethal poisons known, ruins the environment of coral reefs while typically just stunning fish. To catch reef fish alive, other chemicals are often utilized, such as plant poisons and quinaldine (Bruckner, 2000). The Mediterranean and the Pacific reasons has broadly harvested due to its high demand.

Trade for Aquarium Live corals are highly useful in in the marine aquarium industry because it is used in captivating and ecologically diverse tank settings. Another reason to trade of corals as these are favored by hobbyists, collectors cater to aquarium enthusiasts and public aquariums and various species of corals as well as sea horse are collected from natural resources for this purpose.

Demand for Traditional Medicine Sea horse (Hippocampus spp.) is used in various herbal medicine in China due to its high medicinal properties. As local population have belief that sea horse is useful in curing various remarkable disease. In numerous cultures, the meat and eggs of sea turtles are very demanding as it regarded as delicacies that provide essential and economic support to local people and these practices are very common in different areas despite of various ongoing conservation efforts.

Trading of Coral Shell The shells of various sea turtle species particularly the hawksbill turtle (Eretmochelys imbricata), are used in making jewellery, ornaments, and luxury goods which are in huge demand. Therefore, the population of sea turtle has led to drastic declines in hawksbill populations over the time.

These economic drivers illustrate the intricate interplay of cultural traditions, consumer preferences, and the necessity for marine species conservation. Addressing these pressing issues necessitates a comprehensive approach involving sustainable harvesting methods, enhanced trade regulations, and the promotion of alternative livelihoods to alleviate pressure on these vulnerable species.

9.2 Impact on Ecosystem Due to Overexploitation

Many species have experienced ecological extinction as a result of overexploitation in the marine environment. The way marine ecosystems functioned and provided services was significantly changed by this loss (Estes et al., 2011; Lotze et al., 2006; Worm et al., 2006). Among the most abundant yet most endangered Mediterranean environments are coralligenous outcrops, where the Mediterranean red coral Corallium rubrum (Linnaeus, 1758) is regarded as an engineer species (Jones et al., 1994). People have been using red coral for thousands of years.

Clark's Theory: Overexploitation Economics Clark's 1973 landmark article, which linked discount rates to fisheries exploitation, shown that even in non-open access fisheries, the sole owner of a fishery resource may find it economically advantageous to overexploit, and in extreme situations, drive it to extinction.

The outcomes of testing Clark's theory were then contrasted with the exploitation statistics of Newton et al. (2007), who declared 49 island nations and territories' reefs to be either fully exploited, underexploited, overexploited, under/fully exploited, or fully/overexploited. We categorized nations that were deemed to be "under/fully exploited" or "fully/overexploited" as fully exploited. The authors also evaluated the ecological impact of each country's reef reefs to determine if they were viable or not for the five nations that were not included in Newton et al. (2007).

Overexploitation is main reason for the loss of biodiversity whatever it is i.e. it may be terrestrial or aquatic and decline of biodiversity is harmful for constant propagation of ecosystem along with the well-being of human societies. If it's an argue to protect our earth for upcoming generation, it's a need of an immediate action through sustainable practices, conservation approaches, policy improvements, community education, and responsible resource management. As harvesting of marine species are economically beneficial but inspite of that protection of environment should be on priority.

The marine sp. such as Seahorses has a very low reproductive rate and continuous degradation of its habitat stands this species in vulnerable category. It has observed that in certain areas, seahorse populations have been declined upto 75% within only 5 years and such a huge drastic decline can disturb marine ecosystems because seahorses play a significant role in marine food chain to maintain ecological equilibrium as it feed up on small crustaceans. Corals, often referred to as the "rainforest of the sea," are foundational to maintaining marine biodiversity.

The removal of coral causes the loss of habitat of numerous marine species as it provides shelter and breeding habitat to them. The degradation of coral reefs leads to loss of biodiversity and reef structures becomes week which is highly vulnerable to degrade. Furthermore, the coral reef protects the coastline from strong storm and erosion due to its inherent capacity to shield coastal areas. When turtle numbers decrease, their essential ecological functions like the maintenance of healthy seagrass beds and coral reefs are affected. As green turtles play a significant role in maintaining seagrass growth which is essential to stop excessive overgrowth that could be harmful for other marine habitats. The reported species of Sea turtle which are in trade given in Table 9.1. When destruction of natural resources occurs beyond the limit then it affects marine as well as terrestrial ecosystem such as forests and agricultural lands etc. and ultimately it affects ecosystem disruption and adverse effects on human livelihoods. To escape from its effect, there is an urgent need to take effective conservation initiatives along with strict regulations to prevent unsustainable practices, poaching, and habitat loss to ensure the longevity of such kind of ecosystems. But to implicate these efforts various economic factors such as demand in global market and consumer preferences become a hinderance to slow down the trade of marine species such as corals, sea turtles and seahorses. The substantial market worth of these marine species stimulates international commerce and generates significant economic gains, particularly for communities situated on islands and along coastal areas. Global there are many reasons where alternative resources for income are very less and trade of these marine species act as the main sources or a vital revenue stream for their live hood.

Table 9.1 Species composition of reported trade in freshwater turtles (percentage individuals)

Species	Indonesia 1998 Quota (%)	Malaysia Sumatra99 Traded (%)	Thailand (1999) (%)	Vietnam (1999) (%)	('94–'99) (%)
Wild-collected					
Amyda cartilaginea	n/incl.	3	0.36	–	n/incl.
Dogania subplana	9.2	3	–	–	–
Pelochelys cantorii	–	–	0.17	–	–
Cuora amboinensis	61.3	47	18.49	–	37.6
Cuora galbinifrons	–	–	–	–	14.3
Cyclemys dentata	14.6	–	–	–	8.8
Heosemys grandis	–	–	13.17	–	5.5
Heosemys spinosa	4.9	17	–	–	–
Malayemys subtrijuga	–	–	–	–	7.3
Notochelys platynota	–	1	0.50	–	–
Orlitia borneensis	0.3	9	0.89	–	–
Pyxidea mouhotii	–	–	–	–	18.4
Siebenrockiella crassicollis	–	14	5.47	–	4.2
Farmed					
Pelodiscus sinensis	–	–	50.77	100	–
Trachemys scripta	–	–	10.18	–	–

Note: Columns do not necessarily add to 100% since certain species may be traded but are not included in this table (Samedi & Djoko, 2000; Shepherd, 2000; Sharma & Tisen, 2000; Hendrie, 2000)

9.2.1 Loss of Biodiversity and the Degradation of Ecological Services

The trade of marine species involving corals, seahorses, and sea turtles are not only responsible for extinction for these remarkable species but also destroy the services provided by these species to the ecosystem. This action leads to ecological ramifications extensively which includes a marked reduction in biodiversity and the creation of socio-economic challenges (IUCN, 2021). Since the 1870s, the world's live coral cover has diminished by nearly 50%, with the decline intensifying in recent decades as a result of climate change and human actions. Coral reefs are essential habitats for approximately a quarter of all marine species; they bolster fisheries, shield coastlines from the ravages of erosion, and play a vital role in the tourism industry. However, the combined threats of overharvesting and climate-related degradation severely undermine these invaluable services, jeopardizing both food security and the economic stability of coastal communities.

Due to excessive demand for traditional medicine and the aquarium trade of Sea horse in Global market, overexploitation of Seahorses occur and it leads to disturb the balance of marine ecosystem. As Seahorses serve as natural predators of small crustaceans, helping to regulate these populations. The loss of seahorses can lead to ecological imbalances, consequently threatening the stability of marine environments. Furthermore, details regarding the prohibitions on the trade of wild, dried seahorses from countries that utilize them in traditional Chinese medicine and the dried seafood sector are displayed in Table 9.2.

Sea turtle has a significant role in the vitality of seagrass beds and coral reefs and decline in the population of this species create a deep-rooted ecological implication. Excessive illegal practices, unintentional capture and habitat destruction is responsible for serious threats of Sea turtles and even all seven species of sea turtle currently categorized as either threatened or endangered. A massive decline in the numbers may cause rampant seagrass growth as it checks excessive growth of sea grass and coral reefs overwhelmed by sponge species which leads to decrease biodiversity and altered habitat structures (Pauly & Zeller, 2016).

The overexploitation of natural resources has not only affected marine life but also it affects detrimentally on different crucial ecosystem services such as water purification, pollination, and carbon sequestration. It also effects on ocean health and contributing to the collapse of fisheries through disruption of marine food web and overfishing (FAO, 2020). To effectively deal with these pressing challenges, there is need to strongly implement coordinated conservation efforts, enforce stricter trade regulations and adopt sustainable management practices, and, thereby mitigating long-term ecological and economic repercussions.

9.2.2 Socioeconomic Consequences

Natural resources are the only income source for numerous communities and they depend upon for their economic livelihood but the excessive exploitation of these resources can lead to their destruction and resulting in economic instability with heightened food insecurity. For instance, the survival of coastal communities will detrimental when fish populations collapse as a direct consequence of overfishing. This situation not only interferes with their income sources but also risks food availability for the people which effects the overall nutritional health (FAO, 2020).

Initiatives such as the advancement of sustainable agricultural techniques, the enforcement of anti-poaching legislation and the creation of marine protected areas, can significantly support in conservation of biodiversity and to preserve ecological balance. Additionally, international treaties such as the Convention on International Trade in Endangered Species (CITES) are involved actively in regulating the overexploitation of endangered species, thereby the long-term sustainability of our natural resources can be ensured (CITES, 2021).

Table 9.2 The existence and, where relevant, nature of bans on wild, dried seahorse trade from countries/jurisdictions that traditional Chinese medicine and dried seafood traders reported as sources of dried sea horses (Foster, 2016)

Reported source	Ban?	Year	Species	Origin of ban
Argentina	NO			
Australia	NO			
Brazil	NO			
Bangladesh	NO			
China	YES	2011	All species	Self-imposed when RST initiated–"The export of wild Hippocampus spp. was banned on 01/01/2011 until further notice"
Egypt	NO			
Hong Kong	NO			
India	YES	2001	All species	Self-imposed–"Prohibited to acquire, receive, keep in his control, custody or possession, sell, offer for sale, or otherwise transfer or transport"
Indonesia	YES			
Japan	NO			
Malaysia	YES	2009	All species	Self-imposed when RST initiated-"an administrative suspension of all seahorse exports"
Maldives	NO			
Mexico	YES	1994	All species	Self-imposed–"The commercialization of cultured and incidentally-caught seahorses was legal in Mexico…the intentional capture of wild seahorses and their trade was illegal"
Philippines	YES	2004	All species	Self-imposed in response to CITES listing–Republic Act (RA) 8550 Section 97 imposed a blanket prohibition on the fishing or collecting of all CITES-listed species without any distinction among the Appendices.
Senegal	YES	2016	*H. algiricus*	CITES imposed–Parties recommended to suspend trade in *H. algiricus* from Senegal for failure to meet RST recommendations by deadlines.
Singapore	NO			
South Africa	NO	2016	All species	Self-imposed in response to RST recommendations–the Department of Fisheries issued "interim measure to cease the issuance of export permits for seahorses (Hippocampus spp.) …. Shall come into force starting from 1 January 2016 onwards, until further notification"
Thailand	YES			
Taiwan	NO			
Vietnam	YES	2013	*H. kuda*	CITES imposed–Parties recommended suspending trade in *H. kuda* from Vietnam for failure to meet RST recommendations by deadlines.

9.3 Regulatory Efforts for Conservation

The international trade of seahorses, sea turtles and corals, has become a considerable ecological concern, prompting to establish the various conservation and regulatory frameworks which ensures the sustainable management to protect these species. Some effective conservation initiatives have taken to safeguarding natural resources and escape from huge environmental degradation. These effective measures include the enforcement of habitat protection act, introduce the stringent regulations, and strictly follow the sustainable resource management practices which leads to a more balanced and sustainable future for societies. It is also essential that government as well as non-government organizations, and individual stakeholders work like a unit to ensure the success of these measures to protect biodiversity for upcoming generations.

9.3.1 International Conservation Efforts

International regulations play an important role to look after the trade of marine species. The CITES categorizes numerous marine creatures, including different species of seahorses, corals, and turtles, under its appendices, which necessitate to get the permit and the enforcement of sustainable harvesting practices in Ocean. Furthermore, the establishment of Marine Protected Areas (MPAs) which serves to safeguard for vital habitats and overcome exploitation pressures. But still challenges persist in enforcement, particularly in those regions that have scarcity of adequate resources to monitor and restriction on illegal trading activities in waterbodies.

9.3.1.1 Convention on International Trade in Endangered Species (CITES)

CITES, established in 1973, is a global treaty among 184 parties that aims to ensure that international trade including wild fauna and flora does not create any threat to their survival. Species of wild fauna and flora are categorized into specific appendices as per their conservation status (CITES, 2022) such as

Appendix I: include the species at risk of extinction, having the overall commercial trade generally banned.

Appendix II: include the species not currently threatened with extinction but must follow trade regulation to prevent population decline. Notably, seahorses (Hippocampus spp.) were categorized under Appendix II in 2002, while all sea turtle species including the green turtle (*Chelonia mydas*) and the hawksbill turtle (*Eretmochelys imbricata*) are grouped in Appendix I, thus prohibiting international commercial trade of these species. A wide range of coral species is also classified in CITES appendices, highlighting their susceptibility due to exploitation and habitat loss.

9.3.1.2 International Agreements

The following are the international agreements for protection of global environmental:

- The Paris Agreement (2015) was done to deal with climate change by reducing greenhouse gas emissions and promoting the sustainable development agenda (UNFCCC, 2015).
- The Convention on Biological Diversity (CBD) seeks to strengthen biodiversity conservation through strategic planning (CBD Secretariat, 2020).

9.3.2 National Legislation and Conservation Programs

The Countries around the world joined their hands for implementing specific laws and conservation initiatives to protect marine species. In the United States, the Endangered Species Act (ESA) protects sea turtles by intentionally harm, harassment, or killing, and for this purpose the Marine Turtle Conservation Fund provides essential financial and technical support to global sea turtle conservation endeavours.

In Mexico, the threats faced by the Pacific seahorse (*Hippocampus ingens*) and has listed it under special protection category to regulate its trade and exploitation.

9.3.3 Global Initiative of Non-Governmental Organizations (NGOs)

NGOs and various conservation groups are involved in the protection of marine species.

- Project Seahorse: This project focuses on the promotion of sustainably seahorses harvesting and encouraging responsible fishing practices by maintaining their international trade as per CITES.
- A certification system is established by the Marine Aquarium Council for aquarium fish and coral species which ensures that trade practices are sustainable and do not nave adversely impact on Aquarium species.

9.3.4 Regulatory Authority and Policy Frameworks

Several policies have been made to shape the trade of marine species and these policies play a significant role:

- Trade Regulations: CITES and other international agreements between different countries supervise the legal trade of marine species (CITES, 2022).

- Market Influences: Tariffs, quotas, and subsidies can significantly affect the competitiveness and pricing of marine products in global markets (OECD, 2021).
- Climate Change and Overfishing: the climate variation affects the ocean conditions which may cause availability of different species, consequently affecting supply chains of a particular species and thus trade dynamics alters (Pauly & Zeller, 2016).
- Sustainability Certifications: The Marine Stewardship Council (MSC) emphasize on eco-labelling to change consumer choices and support in promoting sustainable fishing methods (Marine Stewardship Council (MSC), 2021).

9.4 Challenges and Ongoing Efforts

The ongoing international conservation efforts of the illegal trade of marine species also faces a lot of challenges. Even though the strict regulation follows as per administration still Dried seahorses still available in illegal markets and sea turtles regularly to be hunted for their shells and meat. The regulatory measures and initiatives for Conservation poses numerous obstacles, including:

- Weak Enforcement: Due the lack of resources in developing countries affects effective monitoring and enforcement of trade restrictions.
- Financial Constraints: The financial condition is one of the main causes which hinders conservation programs and anti-poaching activities.
- Political Resistance: The economic interest of political parties can hinder the implementation of stricter environmental regulations to protect wild life.

To overcome these challenges, it is essential to strengthen international collaboration, funding for conservation projects, should be increased and involvement of community. To monitor illegal activity, technological advancements such as remote sensing and artificial intelligence should be added in future policies.

9.5 Case Studies

The trade of marine species carries extensive ecological and socioeconomic impact. The following case studies illustrate the complexities involved in the illegal trade:

9.5.1 Coral Trade in the Philippines

The Philippines is the major exporter of live corals used for aquarium and for uniqueness in market. The unsustainable mined practices have led to increase reef degradation. The government has required to implicate stricter regulations and enhanced community-based management programs for sustainable harvesting and reef conservation.

9.5.2 Coral Reef Species Trade in the United States

U.S. is one of the largest importers of coral reef species for the purpose of aquarium, the U.S. Coral Reef Task Force indicating that high demand of coral reef leads to overexploitation. The different recommendations such as promoting aquaculture, implementing certification system programs, and raising public awareness program to overcome adverse effects.

9.5.3 Seahorse Trade in Vietnam

The Global seahorse trade is very profound in Vietnam, basically for traditional medicine and for aquarium industry. A study in 2006 revealed that the most seahorses were caught as bycatch in non-selective fishing practices leads to population decline. In response, recent initiatives emphasize on promoting sustainable fishing methods and regulating illegal trade to support seahorse conservation.

9.5.4 Sea Turtle Trade in Indonesia

Indonesia has suffered with the illegal trade of marine turtles specially for their shells and eggs since long time. A 2018 assessment by TRAFFIC enlightened the ongoing issue and emphasized on the need for stricter law enforcement and community-driven conservation efforts.

9.6 Conclusion

To achieve both ecological preservation and economic sustainability, it is difficult but crucial to strike a balance between marine conservation and economic livelihoods (Elegbede et al., 2023). Protecting biodiversity and assisting populations that rely on marine resources requires strong governance structures, effective maritime management, and the investigation of alternate sources of income. To combat the detrimental effects of marine species exploitation, it is imperative to implement sustainable trading practices, enforce strict restrictions, and support conservation efforts (Mora et al., 2009).

Future policies must emphasize community engagement and innovative conservation approaches to ensure the protection of marine ecosystems while providing economic security for those reliant on the marine trade (Bennett et al., 2021). Governments, environmental groups, and local communities must work together to advance the cause in order to maintain marine biodiversity and ensure the long-term sustainability of marine-based economies. Overexploitation issues require a multipronged strategy that combines community-driven projects, sustainable behaviors, and regulatory frameworks (Sibanda, 2024). The only way to protect marine ecosystems and improve the welfare of future generations is to balance economic development with conservation (Visbeck et al., 2014).

References

Bennett, N. J., Katz, L., Yadao-Evans, W., Ahmadia, G. N., Atkinson, S., Ban, N. C., & Wilhelm, A. (2021). Advancing social equity in and through marine conservation. *Frontiers in Marine Science, 8*, 711538.

Bruckner, A. W. (2000). New threat to coral reefs: Trade in coral organisms. *Issues in Science and Technology, 17*, 1–6.

CBD Secretariat. (2020). *Convention on biological diversity*. Retrieved from https://www.cbd.int/

Chelonia, J. (2021). Global trade and conservation of sea turtles. *Marine Conservation Journal, 15*(2), 45–57.

CITES. (2021). *Convention on international trade in endangered species of wild fauna and flora*. Retrieved from https://www.cites.org

CITES. (2022). *Appendices I, II, and III*. Convention on International Trade in Endangered Species of Wild Fauna and Flora.

Dee, L. E., Horii, S. S., & Thornhill, D. J. (2014). Conservation and management of ornamental coral reef wildlife: Successes, shortcomings, and future directions. *Biological Conservation, 169*, 225–237.

Elegbede, I. O., Fakoya, K. A., Adewolu, M. A., Jolaosho, T. L., Adebayo, J. A., Oshodi, E., et al. (2023). Understanding the social–ecological systems of non-state seafood sustainability scheme in the blue economy. *Environment, Development and Sustainability, 27*, 2721–2752.

Estes, J., et al. (2011). Trophic downgrading of planet earth. *Science, 333*, 301–306.

FAO. (2020). *The state of world fisheries and aquaculture*. Food and Agriculture Organization of the United Nations.

FAO. (2022). *The state of world fisheries and aquaculture 2022*. Food and Agriculture Organization of the United Nations.

Foster, S. J. (2016). Seahorses (Hippocampus spp.) and the CITES review of significant trade. *Fisheries Centre Research Reports, 24*, 48. https://doi.org/10.14288/1.0319240

Hendrie, D. B. (2000). Status and conservation of tortoises and freshwater turtles in Vietnam. In P. P. van Dijk, B. L. Stuart, & A. G. J. Rhodin (Eds.), *Asian turtle trade: Proceedings of a workshop on conservation and trade of freshwater turtles and tortoises in Asia* (Chelonian Research Monographs, Number 2) (pp. 63–73). Chelonian Research Foundation.

Howse, R., Eliason, A., & Trebilcock, M. (2005). *The regulation of international trade*. Routledge.

IUCN. (2021). *The IUCN red list of threatened species*. Retrieved from https://www.iucnredlist.org

Jones, C. G., Lawton, J. H., & Shachak, M. (1994). Organisms as ecosystem engineers organisms as ecosystem engineers. *Oikos, 69*, 373–386.

Kramer, P. (2003). The impact of coral trade on reef ecosystems. *Environmental Studies, 12*(1), 23–30.

Lotze, H. K., et al. (2006). Depletion, degradation, and recovery potential of estuaries and coastal seas. *Science, 312*(5781), 1806–1809.

Marine Stewardship Council (MSC). (2021). *Sustainable seafood certification: Market impact report.*

Mora, C., Myers, R. A., Coll, M., Libralato, S., Pitcher, T. J., Sumaila, R. U., et al. (2009). Management effectiveness of the world's marine fisheries. *PLoS Biology, 7*(6), e1000131.

Newton, K., Cote, I., Pilling, G., et al. (2007). Current and future sustainability of Island coral reef fisheries. *Current Biology, 17*, 655–665.

OECD. (2021). *Fisheries and aquaculture in the global economy.* Organisation for Economic Co-operation and Development.

Pauly, D., & Zeller, D. (2016). Catch reconstructions reveal global fisheries trends. *Nature Communications, 7*, 10244.

Pritchard, P. (2008). Turtles and trade: A global perspective. *Marine Turtle Conservation, 16*(4), 132–145.

Sadovy, Y., et al. (2018). The trade in live reef fishes for food: A review and research agenda. *Reviews in Fish Biology and Fisheries, 28*(1), 107–125.

Samedi, & Djoko, T. I. (2000). Freshwater turtle and tortoise conservation and utilisation in Indonesia. In P. P. van Dijk, B. L. Stuart, & A. G. J. Rhodin (Eds.), *Asian turtle trade: Proceedings of a workshop on conservation and trade of freshwater turtles and tortoises in Asia* (Chelonian Research Monographs, Number 2) (pp. 106–111). Chelonian Research Foundation.

Sasipriyan, P., Shameer, T. T., & Manimozhi, A. (2024). *Compendium of species recovery plan with species focus on marine species-2024.* Project Completion Report. Submitted to Tamil Nadu Biodiversity Conservation and Greening Project for Climate Change Response (TBGPCCR). Advanced Institute for Wildlife Conservation (Research, Training and Education), Vandalur, 76.

Sharma, D. S. K., & Tisen, O. B. (2000). Freshwater turtle and tortoise utilisation and conservation status in Malaysia. In P. P. van Dijk, B. L. Stuart, & A. G. J. Rhodin (Eds.), *Asian turtle trade: Proceedings of a workshop on conservation and trade of freshwater turtles and tortoises in Asia* (Chelonian Research Monographs, Number 2) (pp. 120–128). Chelonian Research Foundation.

Shepherd, C. R. (2000). Export of live freshwater turtles and tortoises from North Sumatra and Riau, Indonesia: A case study. In P. P. van Dijk, B. L. Stuart, & A. G. J. Rhodin (Eds.), *Asian turtle trade: Proceedings of a workshop on conservation and trade of freshwater turtles and tortoises in Asia* (Chelonian Research Monographs, Number 2) (pp. 112–119). Chelonian Research Foundation.

Sibanda, N. (2024). *Eco-social contracting for environmental justice: Assessing the effectiveness of community-based natural resource management* (No. 2024-06). UNRISD Working Paper.

Sumaila, U. R., et al. (2020). Illicit trade in marine fish catch and its impact on ecosystems and economies. *Science Advances, 6*(9), 3801.

UNFCCC. (2015). *The Paris agreement.* United Nations Framework Convention on Climate Change. Retrieved from https://unfccc.int/process-and-meetings/the-paris-agreement/the-paris-agreement

Vincent, A. C. J. (1996). The seahorse trade: A conservation crisis. *Biological Conservation, 75*(3), 197–208.

Visbeck, M., Kronfeld-Goharani, U., Neumann, B., Rickels, W., Schmidt, J., Van Doorn, E., et al. (2014). Securing blue wealth: The need for a special sustainable development goal for the ocean and coasts. *Marine Policy, 48*, 184–191.

Worm, B., et al. (2006). Impacts of biodiversity loss on ocean ecosystem services. *Science, 314*(5800), 787–790.

Part IV
Socio-Economic Drivers and Impacts

Part IV

Chapter 10
Economic Drivers and Market Forces in India's Wildlife Trade

Monika Jangra and Parvati Sharma

Contents

Abstract Illegal wildlife trafficking (IWT) is a highly lucrative industry that poses a severe danger to ecosystems and biodiversity. The demand for wildlife products, including luxury fashion items, traditional medicines, and exotic pets, fuels the illegal trade in India and adds to the extinction of endangered species. Although laws are in place to prevent wildlife crime, organized criminal syndicates involved in other illicit operations, such as drug trafficking, often support the trade, which is

M. Jangra (✉) · P. Sharma
Department of Zoology, Chaudhary Bansi Lal University, Bhiwani, Haryana, India
e-mail: monika.jangra81@cblu.edu.in

A. Poonia et al. (eds.), *Wildlife Trade in India*,
https://doi.org/10.1007/978-981-95-3694-8_10

made possible by corruption and a lack of enforcement. India's wildlife trade is mostly driven by economic motivations; animals such as tigers, pangolins, and rhinoceroses are taken for their body parts, which are subsequently utilized in fashion, medicine, and ornaments. The exotic pet trade also plays a significant role in the trafficking of birds and reptiles. Major transit hubs that link domestic and international markets, like as Uttar Pradesh, Karnataka, West Bengal, and Assam, are frequently the sites of wildlife trafficking. The majority of the trade often occurs in areas with high biodiversity but insufficient law enforcement to stop illegal activity. The economic motivations underlying IWT are made worse by societal factors, as animal products are commonly utilized as status and wealth symbols. Additionally, the COVID-19 pandemic has exacerbated wildlife trafficking, especially for bush meat, which has raised demand for illegal animal goods. To combat IWT in India, a comprehensive approach is needed. Enhancing enforcement, raising public awareness, and strengthening legal frameworks are all necessary to reduce demand and halt illegal activities. Involving local communities and fostering international cooperation are also necessary to address the problem. By combining legal, social, and economic actions, India may begin tackling the complex problem of wildlife trafficking and safeguarding its biodiversity for future generations. It is essential to comprehend market and economic dynamics in order to create solutions that effectively counter this trade.

Keywords Illegal Wildlife Trade (IWT) · Poaching · Biodiversity · Conservation · Wildlife products · Ecological impact · Wildlife trade hubs · Wildlife trafficking · India · Organized crime

10.1 Introduction (Historical Context of Wildlife Trade in India)

The illegal wildlife trade (IWT) in India is a complex and persistent issue, influenced by a combination of economic drivers, market forces, weak law enforcement, and socio-cultural factors. Despite stringent laws and international conventions, illegal wildlife trade persists, posing significant threats to biodiversity and conservation efforts. Key economic forces fueling the trade include demand for traditional medicines, luxury goods, and exotic pets, with wildlife products often fetching high prices in both local and international markets. This illicit trade is exacerbated by financial incentives, cultural beliefs, and corruption, undermining efforts to curb the practice despite stringent national laws, such as the Wildlife Protection Act of 1972, and international agreements like CITES (Khalsa, 2023; Raja, 2024; Mohanty & Patra, 2021).

India serves as a major source and transit hub for illegal wildlife products, particularly in regions like Uttar Pradesh, Assam, and Madhya Pradesh, where porous borders and weak law enforcement facilitate smuggling to neighboring countries, including Nepal, China, and Myanmar (Singh et al., 2023; Kuswanda et al., 2023). Organized crime networks capitalize on these vulnerabilities, driving the illegal trade of endangered species such as tigers, leopards, rhinos, and pangolins, while also contributing to ecosystem disruption and biodiversity loss through harmful practices like poisoning and indiscriminate trapping (Raja, 2024; Morton et al., 2021). The growing global demand for wildlife products, coupled with international conomic inequality, further fuels the IWT, as wealthier nations acquire illegal wildlife items from poorer countries, thus strengthening the transnational criminal network (Liew et al., 2021). Despite the presence of legal frameworks, enforcement remains insufficient due to corruption, lack of penalties, and resource constraints, particularly in areas prone to wildlife trafficking (Khalsa, 2023; Uprety et al., 2021). Addressing the IWT requires not only strengthening legal and enforcement frameworks but also reducing consumer demand for wildlife products, which remains underfunded and under-researched (Veríssimo et al., 2020). Furthermore, India's role as a signatory to CITES necessitates stronger bilateral agreements with destination countries to address the global nature of wildlife trafficking and reduce demand (Raja, 2024; Challender et al., 2015).

The legal and illicit wildlife trade in India has a big impact on the nation's biodiversity and economy. The illegal wildlife trade (IWT) poses a serious threat to conservation efforts, driven by economic factors such as demand for traditional medicines, luxury goods, and exotic pets, among others (Khalsa, 2023; Raja, 2024). Despite the existence of stringent laws like the Wildlife Protection Act of 1972 and international conventions such as CITES, illegal trade persists, largely due to weak enforcement, corruption, and a lack of public awareness (Raja, 2024; Mohanty & Patra, 2021). While the trade provides livelihoods for some, its impact on biodiversity and ecological balance is profound, as it often leads to the extinction of keystone species and disrupts entire ecosystems (Raja, 2024; Hughes, 2021). Additionally, communities that rely on wildlife for their livelihood are affected by the trade, highlighting the urgent need for sustainable alternatives to reduce dependency on illegal activities (Raja, 2024). The illegal wildlife trade in India is driven by complex factors, including demand for traditional medicines, luxury goods, and exotic pets, as well as weak law enforcement and corruption. This trade has significant economic benefits, particularly in traditional medicine and decorative items, but also incurs substantial environmental and regulatory costs due to biodiversity loss (Khalsa, 2023; Nalayini, 2024). Poaching, closely linked to illegal wildlife trade, has been exacerbated by rapid economic development and increased demand for wildlife products. On a global scale, the wildlife trade is influenced by international wealth inequality, with wealthier nations typically importing more animals from poorer exporting countries (Liew et al., 2021). Addressing this issue requires a comprehensive approach, including strengthening legal frameworks, improving enforcement, enhancing public awareness, and reducing consumer demand for wildlife products (Khalsa, 2023; Nalayini, 2024).

India occupies a critical position in global wildlife trade networks, acting as both a source and transit country for various species (Singh et al., 2023). Its rich biodiversity makes it a target for illegal trade, further compounded by the country's porous borders and weak law enforcement. The economic disparities between wealthier and poorer nations also exacerbate the issue, as wealthier countries exert significant economic advantage over poorer exporting countries, influencing trade dynamics (Liew et al., 2021). In high-income markets, particularly in Asia, there is a strong demand for wildlife-derived luxury goods such as ivory, tiger pelts, and exotic pets. This demand is closely linked to traditional medicine, status symbols, and ornamental consumption, creating a lucrative market for wildlife trafficking (Khalsa, 2023; Raja, 2024). The economic incentives driving the illegal wildlife trade are intricately tied to poverty and socio-cultural factors. In regions where economic alternatives are scarce, individuals and communities often resort to illegal wildlife trade for income. While the demand for wildlife products continues to grow, fuelled by both domestic and international markets, organized crime networks exploit these economic vulnerabilities. These criminal networks facilitate the smuggling of endangered species such as tigers, leopards, rhinos, and pangolins, contributing to biodiversity loss and ecological disruption. This smuggling occurs predominantly in regions like Uttar Pradesh, Assam, and Madhya Pradesh, where weak law enforcement and porous borders make it easier to transport wildlife products to neighbouring countries like Nepal, China, and Myanmar (Singh et al., 2023; Kuswanda et al., 2023). In turn, this illegal activity leads to harmful practices like poisoning and indiscriminate trapping, which further threaten species and ecosystems (Raja, 2024; Morton et al., 2021).

The historical context of wildlife trade in India reveals a long-standing relationship with commercial exploitation. In the pre-colonial period, wildlife trade was largely subsistence-based, with local communities hunting for personal use and local exchange. However, with the arrival of British colonial powers, commercial exploitation of wildlife flourished as demand for exotic goods in Europe surged. This shift led to large-scale hunting of animals for trophies and the export of animal parts, resulting in the depletion of several species and altering the country's biodiversity (Khalsa, 2023; Raja, 2024). The British established structured trade networks that facilitated the export of wildlife products like ivory, skins, and feathers, setting a precedent for unsustainable wildlife exploitation. After India gained independence in 1947, industrialization and urbanization contributed to the continued demand for wildlife products, both domestically and internationally. The expansion of cities and increased human-wildlife conflict exacerbated pressures on wildlife populations, while conservation efforts began to take shape. An important step in stopping the illegal wildlife trade was the Wildlife Protection Act of 1972, which established legal frameworks to save endangered species and control hunting and trade (Raja, 2024; Mohanty & Patra, 2021). India's membership in CITES further bolstered international cooperation to combat illegal wildlife trade by offering a global legal framework for the protection of endangered species (Andersson et al., 2021). The Environment Protection Act of 1986 expanded the legal framework to encompass broader environmental protection measures, indirectly supporting wildlife conservation efforts by addressing habitat destruction and pollution (Khalsa, 2023; Raja, 2024).

Despite these legal advancements, enforcement remains weak, hindered by corruption, inadequate penalties, and limited resources, particularly in regions prone to wildlife trafficking (Khalsa, 2023; Uprety et al., 2021). The persistent illegal trade is further fuelled by the financial incentives of criminal networks and the cultural and economic factors driving demand, which continue to undermine conservation efforts despite the presence of strong legal frameworks. The complexity of the wildlife trade is highlighted by India's dual role as a supplier and a transit centre for illegal wildlife items. Financial motivations, cultural beliefs, and the lack of effective enforcement mechanisms further perpetuate the cycle of exploitation. The increasing global demand for luxury wildlife products, combined with international economic inequalities, strengthens transnational criminal networks involved in wildlife trafficking (Liew et al., 2021). Addressing this issue requires a multifaceted approach that not only strengthens legal frameworks and enforcement mechanisms but also reduces consumer demand for wildlife products. Furthermore, comprehensive strategies should include providing sustainable alternatives to communities that depend on wildlife trade for their livelihood, while tackling the socio-economic drivers and international inequalities that perpetuate the trade (Raja, 2024; Veríssimo et al., 2020). Through international cooperation, stronger enforcement, and demand-reduction efforts, India can better manage its role in the global wildlife trade and contribute to meaningful conservation outcomes (Challender et al., 2015).

10.2 Economic Drivers of Wildlife Trade in India

The economic drivers behind India's wildlife trade are deeply interconnected with both domestic and international demand for wildlife products, supply-side factors fueling illegal and legal trade, and the pricing dynamics that govern market forces. Despite stringent laws such as the Wildlife Protection Act of 1972, illegal trade persists due to the high demand for luxury goods, traditional medicine, exotic pets, and animal products linked to fashion and tourism. High-value wildlife products like ivory, skins, and trophies are sought after in both the global and domestic luxury markets for fashion, art, and status symbols (Raja, 2024; Symes et al., 2017; Prasad et al., 2022). Traditional medicine drives the demand for products like tiger bones and rhino horns, exacerbating poaching and trafficking (Veríssimo et al., 2020; Wyatt, 2018). The growing market for exotic pets also contributes to the trade, with rare and endangered species often illegally captured and sold for financial gain (Khalsa, 2023; Raja, 2024). Wildlife tourism, while promoting conservation, also contributes to the trade in animal products such as feathers, skins, and shells, often sold as souvenirs to tourists (Chavez et al., 2023; Ruenes et al., 2023). Additionally, the fashion industry's demand for animal-derived materials like leather, fur, and exotic feathers fuels wildlife exploitation, leading to the illegal trade of protected species (Veríssimo et al., 2020). These economic drivers, coupled with weak enforcement and consumer demand, highlight the need for better regulation, international cooperation, and targeted conservation efforts to mitigate the impact of wildlife trade on endangered species (Khalsa, 2023; Raja, 2024).

10.2.1 Domestic and International Demand and Consumption of Wildlife Products

In India, the demand for wildlife products is deeply embedded in cultural, religious, and traditional practices. Traditional medicine, particularly Ayurveda, utilizes animal parts like tiger bones, rhinoceros' horns, and bear bile, contributing to a domestic market for these products (Agarwal, 2016). Additionally, animals and animal parts are consumed in local food traditions (e.g., wild meat) and used for ornamental purposes, with exotic pets such as reptiles and birds in high demand among certain urban populations (Kumar, 2020). Despite legal frameworks, the WWF India (2020) reports that illegal demand persists at a local level, driving poaching and the trafficking of endangered species.

High-value goods like ivory, tiger hides, and rhino horns are transported to markets in China, Southeast Asia, and the Middle East, making India a source and a transit location for the wildlife trafficking on a global scale (UNODC, 2020). In these markets, wildlife products are highly valued for their medicinal properties in traditional Chinese medicine, where tiger bones and rhino horns are believed to cure various ailments (Duffy, 2018a). The luxury goods market further fuels demand, with elephant ivory carvings and snake skins being sold for substantial profits, creating a thriving black market that often circumvents regulatory efforts like CITES.

10.2.1.1 High Demand for Wildlife Products

The illegal wildlife trade is driven by strong economic incentives. The demand for different wildlife items is increased by markets like decorating, traditional medicine, fashion, and unusual/exotic pets, which results in overexploitation of species.

Demand for Fashion

The fashion industry is a significant consumer of wildlife products, with animals being exploited for items such as fur coats, leather purses, belts, and shoes. Reptile skins used to make leather items are the main source of the illegal trade, even if fur makes up a sizable amount of the legal wildlife trade in fashion. Because fur farming has become so popular, fur's place in the illicit trade has decreased. On the other hand, reptile skins, such those of pythons, continue to be in high demand, particularly for leather goods. A significant amount of python skin is obtained from wild

Table 10.1 Species associated with the illicit fashion industry wildlife trade

Species	Affected products	Implications of illegal trade	References
Fur-bearing animals	Coats, accessories, textiles	Decline in illegal trade due to established farming practices	Wyatt (2018)
Nile crocodile (*Crocodylus niloticus*)	Leather goods, bags, footwear	Legal farming exists, but illegal harvesting still prevalent	CITES (2019)
Reticulated python (*Python reticulatus*)	Leather garments, accessories	High breeding costs lead to sourcing from wild populations	Lyons and Natusch (2011)
Other python species (*Python* spp.)	Fashion items, belts	Wild capture exceeds breeding capacity for commercial demand	Kasterine et al. (2012)
Varanus (Monitor lizards)	Leather goods, wallets, straps	Commonly confiscated in fashion-related wildlife seizures	Sosnowski and Petrossian (2020)
Crocodylus (Crocodiles)	Bags, wallets, straps	Both farmed and illegally sourced for high-demand products	Sosnowski and Petrossian (2020)
Caiman (*Caiman* spp.)	Leather apparel, accessories	Illegal hunting and trade for luxury products	Sosnowski and Petrossian (2020)
Alligator (*Alligator* spp.)	Shoes, belts, bags	Significant seizures linked to illegal wildlife trafficking	Sosnowski and Petrossian (2020)

captures since the size of the reptile skin trade frequently surpasses the capacity of authorized breeding facilities. These skins are especially sought after in European markets, where trade regulations aim to prevent illegal sourcing. Research on confiscations in the US from 2003 to 2013 found that specific genera, such as pythons and crocodiles, were most frequently seized in relation to fashion-related wildlife crimes. Despite a general decline in wildlife trade for fashion, the demand for exotic skins persists, and illegal trafficking remains a significant problem (Table 10.1).

Demand for Exotic Pets

The growing demand for exotic pets, especially reptiles and birds, also contributes to the wildlife trade (IWT). The unlawful traffic in these animals, which are frequently sold at high rates on international markets, originates in India and passes across the country (Khalsa, 2023). This trade not only threatens the survival of species but also disrupts local ecosystems when species are removed from their natural habitats. Exotic pets, typically non-native or not regularly domesticated species kept for companionship or entertainment, are central to the IWT. The exotic pet trade is driven by various motivations, including financial investment, competition participation (such as song contests), and the desire for status symbols. While birds and reptiles are the most commonly trafficked animals, mammals and invertebrates like tarantulas are also frequently exploited. Smugglers often exploit newly discovered species by using scientific descriptions and social media to locate and sell these animals, which adds a layer of complexity to the conservation efforts. For instance, tigers are frequently found in unlicensed zoos in the USA, signifying riches and power, and cheetahs are kept as status symbols in Gulf states. These practices exacerbate the demand for endangered species and contribute to their exploitation.

Although nature tourism has the potential to benefit endangered species, when profit-driven, it can harm them instead. Carnivores like tigers, for instance, are frequently employed as tourist attractions for pictures, films, and rides, which encourages unethical behavior like drugging and animal mutilation. Additionally, trafficking on demand has been observed, where species are poached and smuggled specifically for the pet trade. Genetic methods have been employed to prevent illegal wildlife laundering, such as in the case of cheetahs, to identify whether animals in captivity are bred legally or illegally. This system helps ensure that trade regulations are enforced, though illegal breeding remains a pervasive issue. In Brazil, for instance, a study showed that a significant proportion of species bred in captivity were done so illegally, often by commercial breeders. Some species, such as the Barbary macaque, once hunted for research, are now part of a profitable pet market, further compounding the challenges posed by illegal wildlife trade (Table 10.2).

Demand for Traditional Medicines

Traditional medicine in India plays a critical role in driving the illegal wildlife trade (IWT), particularly for species like tigers, pangolins, and rhinos, whose body parts are poached for use in various remedies. Despite scientific studies demonstrating that many animal-derived substances lack proven medicinal benefits (Khalsa, 2023), the demand for these products continues to drive significant trade in illegal wildlife. Regions such as Kerala, Tamil Nadu, and Assam are home to small-scale producers who rely on wildlife-derived medicinal products, including tiger bones, pangolin scales, and deer antlers (Sinha & Patil, 2020). Traditional medicine systems like Ayurveda and Unani heavily depend on these wildlife products, contributing to the ongoing demand for endangered species (Mehta et al., 2018). The purported health benefits of these wildlife-derived chemicals promote their continued usage, despite the conservation costs. Species like tigers and pangolins, already endangered, face increased challenges due to the poaching caused by the demand for their body parts (Mehta et al., 2018). These species are listed as endangered by the International Union for Conservation of Nature (IUCN), and their decline is being accelerated by illicit poaching and trade (Bhatnagar & Rao, 2019). This ongoing demand not only undermines conservation efforts but also accelerates the depletion of already vulnerable species.

A collection of indigenous knowledge, abilities, and practices, traditional medicine is used to treat and maintain physical and mental health conditions (WHO, 2013). It is consumed worldwide, not only in Asia but also in regions such as Africa,

Table 10.2 Exotic pet trade and species exploitation: key species and issues

Species	Category	Details	References
Parrots (*Psittaciformes*)	Exotic Pet Trade and IWT	Captures for the local pet trade are the largest danger to parrot populations in the Neotropics, affecting 68% of the populations under study.	Berkunsky et al. (2017)
	Price Discrepancies	When compared to Latin American markets, the costs of certain parrot species are more than 1500 times higher in Europe.	Zimmerman (2003)
Tarantulas (*Theraphosidae*)	Illegal Trade in Invertebrates	Invertebrates like tarantulas are becoming more and more illegally traded, often as a result of scientific descriptions or social media advertisements for recently found species.	Law (2019)
Snowy Owls (*Bubo scandiacus*)	Movies and the Pet Trade	According to studies, there is little to no correlation between a higher volume of trade in the respective species and well-known films; for example, Snowy Owls in Harry Potter.	Nijman and Nekaris (2017)
Cheetahs (*Acinonyx jubatus*)	Exotic Animals as Status Symbols	Cheetahs are more in demand and exploited because they are trafficked as prestige symbols, especially in Gulf states.	Tricorache et al. (2018)
	Wildlife Laundering	Although the commercial international trade of wild cheetahs is prohibited and they are listed in CITES Appendix I, captive-bred individuals are identified and money laundering is avoided through the use of genetic techniques.	Magliolo et al. (2021)
Tigers (*Panthera tigris*)	Exotic Animals as Status Symbols	In the USA, tigers are frequently housed in unlicensed or private zoos and are utilized as status symbols, which raises demand for and abuse of these animals.	UNODC (2020), WWF (2014)
	Nature Tourism and Conservation	Tigers are frequently harmed (drugged, their teeth and claws removed), and their use as tourist attractions for photography, film, and riding may pave the way for the illegal wildlife trade.	Buckley et al. (2016)
Barbary Macaques (*Macaca sylvanus*)	Use of Wildlife for Research and Pet Trade	In order to conduct biomedical research, barbary macaques were harvested from the wild. Previously marketed as pets, surplus animals are now a lucrative business.	van Uhm (2016)

the Caribbean, and Latin America (Romero-Daza, 2002). Over 500 animal species are used globally for medicinal purposes, with 53% of reptiles used in traditional medicine listed as endangered (Cardoso et al., 2021). The illegal trade in medicinal plants is still poorly documented, although it is known that approximately 25% of modern medicines are adaptations of traditional herbal remedies (Margulies et al.,

2022; Phelps et al., 2016). In South Africa, 27 million people use traditional medicine, consuming approximately 20,000 tonnes of indigenous plant material annually, with only 0.25% sourced from cultivated plants (Mander et al., 2007). The IUCN has classified approximately 723 plant species used in traditional medicine as threatened, while an estimated 15,000 plant species are consumed for medicinal purposes (Antonelli et al., 2020). Animal-based items used in traditional medicine, such as pangolin scales (Pholidota), bear bile (Ursidae), and large cat products (*Panthera* spp.), frequently don't have any demonstrated therapeutic benefits (UNODC, 2020). In China, 88% of respondents agreed that it is unlawful to purchase or sell products purportedly containing tiger parts, and 43% of respondents said they have eaten such products. Nevertheless, despite their ineffectiveness, tiger-based products are still un use (Still, 2003). Additionally, a shift has been observed in the trade of tiger products from primarily medicinal use to being sought after as status symbols (CITES, 2014). The continued reliance on wildlife for traditional medicine not only threatens biodiversity but also undermines efforts to protect endangered species. While the traditional medicine trade provides livelihoods for some, its negative impact on wildlife populations calls for urgent action to regulate the trade and reduce consumer demand for these products (Table 10.3).

Demand for Wildlife Food/Meal

Wild food harvesting, which includes the consumption of bushmeat and other plant species, is an important part of many communities' diets and economies; for lower-income households, bushmeat is frequently the main source of protein, while for wealthier individuals, wild food is a luxury good or status symbol, which leads to its demand in high-end markets. Species like pangolins (*Pholidota*) and king cobras (*Ophiophagus hannah*) are particularly sought after for their perceived delicacy or medicinal properties, driving illegal trade and threatening biodiversity. Wild food has also become a symbol of wealth and status in business settings, and its consumption is culturally influenced, with the COVID-19 pandemic potentially reversing the decline in bushmeat consumption observed in central Africa between 1990 and 2005. The illegal wildlife trade is heavily impacted by the demand for luxury items such as caviar, with up to 90% of caviar coming from illegal sources. Unregulated harvesting of wild food species, both animals and plants, accelerates the decline of ecosystems and endangered species populations. For example, even though all orchid species are listed in CITES Appendix II, illicit harvesting has occurred in Tanzania due to rising demand for orchid root tubers (Kinaka). Furthermore, although public demand for wild-caught products is still a problem, raising wild food species has been suggested as a conservation technique to lessen over-exploitation.

Additionally, wild food harvesting often overlaps with the use of medicinal plants, as some species are utilized both for food and traditional medicine. This overlap further exacerbates the depletion of wild species and contributes to the pressures on ecosystems and biodiversity. While wild food harvesting is generally

Table 10.3 Impact of traditional medicine on wildlife species and conservation status in Illegal Wildlife Trade (IWT)

Wildlife species	Body parts used	Traditional medicine systems	Regions impacted	Conservation status	References
Tiger (*Panthera tigris*)	Tiger bones, skins, other body parts	Ayurveda, Unani, Traditional medicine, status symbols	Kerala, Tamil Nadu, Assam, China, Southeast Asia, India	Endangered (IUCN)	Khalsa (2023), Mehta et al. (2018), Bhatnagar and Rao (2019)
Pangolin (*Pholidota*)	Pangolin scales	Ayurveda, Unani, Traditional medicine	Kerala, Assam, Asia, Africa, South Asia	Endangered (IUCN)	Mehta et al. (2018), Bhatnagar and Rao (2019)
Rhino	Rhino horn	Ayurveda, Unani	Assam, other regions	Endangered (IUCN)	Khalsa (2023), Bhatnagar and Rao (2019)
Deer	Deer antlers	Ayurveda	Assam, Tamil Nadu	Vulnerable (IUCN)	Mehta et al. (2018)
Bear (*Ursidae*)	Bear bile	Traditional medicine	Asia, parts of Africa	Vulnerable/ Endangered	UNODC (2020)
Big Cats (*Panthera* spp.)	Bones, skins, body parts	Traditional medicine	Asia, Africa, parts of South Asia	Endangered (Various species)	UNODC (2016)
Various Reptiles	Reptile body parts (e.g., skin, bile)	Traditional medicine	Various regions	53% listed as endangered	Alves et al. (2008)
Various Animal Species	Various animal products	Various traditional medicine systems	Worldwide (Asia, Africa, Caribbean, Latin America)	Over 500 species used	Cardoso et al. (2021)

legal unless protected species are involved or regulations such as off-season hunting are violated, unregulated harvesting can lead to significant biodiversity loss.

This Table 10.4 is centered on Harvesting wild food is usually permitted, although it may be prohibited when it involves protected species or violates certain laws. Wild food intake frequently varies by income level, with some species being considered luxury goods or status symbols. Unregulated harvesting of wild foods, especially bush meat, can result in a considerable loss of biodiversity.

Table 10.4 Wild food harvesting: legal status and conservation efforts

Category	Details	References
Legality of Wild Food Harvesting	Wild food harvesting and eating are lawful unless they involve protected species or laws are broken (e.g., off-season hunting).	Fa et al. (2003)
Impact of Unregulated Harvesting	A significant reduction in biodiversity may result from uncontrolled exploitation.	Fa et al. (2003)
Role of Wild Meat (Bushmeat) in Diet	Bushmeat is either a luxury good for those with better incomes or a staple diet for lower-income households, where it's frequently the only source of protein.	Nasi and Fa (2015), Shairp et al. (2016)
Trend in Bushmeat Consumption (1990–2005)	Between 1990 and 2005, the majority of central African states saw a decline in bushmeat consumption.	Ziegler (2010)
Impact of COVID-19 on Bushmeat Consumption	COVID-19 might have reversed the trend of declining bushmeat consumption.	Funk et al. (2022)
Wild Food as Status Symbol	Wild food is consumed as a status symbol or delicacy, including species like pangolin, king cobra, and caviar.	Shairp et al. (2016)
Wild Food in Business Contexts	Bushmeat and other wild foods are eaten to show off affluence and prestige, frequently in professional contexts.	Shairp et al. (2016)
Bushmeat as Source of Income	Bushmeat is increasingly used as a source of income, with organized crime groups involved in exploiting the trend, especially in urbanized and high-income neighborhoods.	UNODC (2012)
Caviar and Illegal Wildlife Trade	It is estimated that up to 90% of caviar comes from illicit sources.	Dickinson (2022)
Farming as Conservation Tool for Wild Food	Although some consumers still prefer wild over farmed items, the introduction of wild food species farming was done as a conservation measure.	Musing et al. (2019)

Demand for Luxury Goods and Decorative Items

The trade in wildlife products for decorative and luxury items is another major economic driver. Ivory, tiger skins, and other exotic animal products are in high demand for use in luxury goods and ornaments. The fashion industry, in particular, has been a notable consumer of animal products, fueling further illegal poaching (Nalayini, 2024).

The demand for wildlife-derived luxury goods is another important market within India's illegal wildlife trade. Products such as ivory, exotic feathers, and tortoiseshell are highly sought after for ornamental purposes and social status (Bhatnagar & Rao, 2019). This segment of the market fuels the consumption of wildlife products, particularly in urban areas, where aesthetics and status symbols are key drivers of demand. The growing popularity of these goods complicates the regulatory

Table 10.5 Overview of decorations, accessories, and jewellery in Illegal Wildlife Trade (IWT)

Category	Details	Species name	References
Range of Items in IWT	Hunting trophies (such as antlers or skins), elephant footstools, turtle shell jewelry, talismans, and amulets are among the decorations, accessories, and jewelry used in IWT.	Elephant (*Elephantidae*), Turtle (*Testudines*), Various species	UNODC (2020)
Common Uses in IWT	Accessories, jewelry, souvenirs (such as pinned butterflies or coral), ivory carvings, and equipment like bowls or horn pipes are all produced from the parts of wild animals.	Butterfly (*Lepidoptera*), Coral, Various species	UNODC (2020)
Seizures of Ivory Products in IWT	Elephant ivory, which is mostly used for ornamental items like jewelry and accessories, is widely used. Almost one-third of all wildlife seizures are related to it. According to the UNODC World WISE Database, between 2009 and 2018, about one-third of all wildlife seizures were elephant goods, including ivory.	Elephant (*Elephantidae*)	UNODC (2020), Gao and Clark (2014)
Ornamental Plants in IWT	Orchids, succulents, and cactus are among the many wild plants that are traded for their aesthetic qualities. The Orchidaceae family comprises more than 85% of ornamental plants that are traded.	Cactus species, Succulents, Orchid species (*Orchidaceae*)	Margulies et al. (2022)
Illegal Logging and Trade	Illegal logging and trading, especially in the timber industry, cause enormous financial losses and play a major role in environmental crime.	Rosewood (*Dalbergia* spp.)	Nellemann et al. (2014)
Rosewood in IWT	Because of its high market value, particularly because of its reddish hue, rosewood is frequently sold for musical instruments and furnishings.	Rosewood (*Dalbergia* spp.)	Nellemann et al. (2016)
Whitewashing of Timber	The practice of passing off illegal goods as legal in order to get them into markets is known as "whitewashing" rosewood and other timber species.	Rosewood (*Dalbergia* spp.)	UNODC (2020)

landscape, as the ornamental trade is intertwined with traditional practices and modern luxury consumption. Additionally, the luxury wildlife trade has substantial economic implications, generating billions of dollars annually, often at the expense of biodiversity (IUCN, 2021).

With a focus on typical goods and products including jewellery, hunting trophies, ivory, and ornamental plants, this Table 10.5 lists the different wildlife and plant species that are exploited in the illegal wildlife trade (IWT). It also discusses "whitewashing"—the practice of smuggling lumber into legitimate markets—and illegal logging, particularly as it relates to rosewood.

At the end market, wildlife demand is driven by distinct clusters, each with specific characteristics that influence both legal and illegal trade. These demand clusters include food, medicine, fashion, ornamentals, investment, and pets. Each has its own trends and dynamics, and some operate independently of others. For instance, the demand for python skins is influenced by fashion trends, while python meat or live pythons as pets have different driving factors. The World Wildlife Crime Report categorizes these demand clusters, recognizing that some species may satisfy multiple markets (e.g., rhinoceros horn is used for both medicinal purposes and as a carving material).

Although this typology isn't exhaustive, it highlights the complexity of demand and trade dynamics. In these clusters, the nature of the wildlife product influences supply dynamics. Non-perishable items, like collectibles, can be stockpiled, whereas perishable goods, such as meat or fish, must be consumed quickly, limiting storage potential. Some products, such as foodstuffs, have ongoing demand, while niche items, like specialized collectibles, tend to attract one-off purchases. The demand scale varies between bulk products and those requiring novelty or exclusivity. Each cluster's dynamics are shaped by both legal and illegal supply, with illegal trade often linked to factors affecting the general demand for that product. Understanding these underlying drivers is essential for addressing the illegal trade effectively.

The wide range of sectors involved in wildlife trafficking and the related problems are depicted in this Table 10.6, which offers a thorough summary of the international trade in wildlife goods. In order to prevent trade from jeopardizing the survival of species, CITES restricts the international trade of endangered species (CITES, 2025). However, the illegal wildlife trade (IWT) remains a significant concern because, in addition to threatening biodiversity, it supports a variety of industries, including luxury goods, fashion, traditional medicine, the exotic pet trade, and wild food.

10.2.2 Supply-Side Factors in the Illegal Wildlife Trade in India

The illegal wildlife trade in India is primarily driven by poaching and illegal hunting, with high-value species such as tigers (*Panthera tigris*), elephants (*Elephas maximus*), and rhinoceros (*Rhinoceros unicornis*) being the primary targets due to the demand for their body parts, including tiger bones, rhino horns, and elephant ivory (Indian Ministry of Environment, Forest and Climate Change, 2021). The growing global and domestic demand for these products has fostered the development of organized criminal networks that exploit regions rich in wildlife, particularly areas like Madhya Pradesh, Odisha, and Assam, where wildlife abundance creates opportunities for poaching (Gupta & Dey, 2017). Despite the presence of legal

Table 10.6 Global wildlife trade: key sectors, trends, and conservation impacts

Wild product	Markets	Key countries involved	Trade characteristics	Key issues & practices	References
Exotic Pets	Pet Trade	Brazil, USA, Latin America, Europe, Gulf States	Trade of exotic animals for companionship, status, and investment	Parrots, reptiles, and cheetahs highly traded, smuggling, exotic pet market	Cardoso et al. (2021), Marker (2019)
Fashion (Fur, Leather)	Fashion Industry	EU, USA, Kenya, Southeast Asia	Legal and illegal trade of fur (mainly reptiles) and skins for leather goods	Python skins, reptile skins, and crocodile farming in Kenya, illegal reptile trade	Andersson et al. (2021), CITES (2019), Sosnowski and Petrossian (2020)
Traditional Medicine	Medicinal Products	China, Africa, Southeast Asia, Latin America	Animal and plant products used in traditional medicines	Bear bile, pangolin scales, tiger products, and misuse of animal products	WHO (2013), Cardoso et al. 2021)
Wild Food (Bushmeat)	Food Industry (Delicacies)	Central Africa, Southeast Asia, India, China	Consumption for status, or as a primary protein source	Pangolins, king cobras, and seafood (e.g., caviar) as delicacies; unregulated trade	Funk et al. (2022, Shairp et al. (2025)
Wild Plants	Ornamental Goods	USA, China, India, Tanzania, Brazil	Trade of plants for decorative and ornamental purposes	Orchids, cacti, succulents heavily traded, illegal logging and harvest of endangered species	Margulies et al., (2022), Ziegler (2025, Phelps and Webb (2015)
Jewellery & Accessories	Luxury Goods, Souvenirs	USA, China, India, Southeast Asia	Animal-derived items like turtle shell jewellery, ivory, and amulets	Ivory, rosewood, and tortoiseshell as valuable luxury goods, illegal wildlife trafficking	UNODC (2020, 2016, 2025)
Wildlife Breeding	Commercial Breeding, Illegal Breeding	Brazil, India, Southeast Asia	Illegal breeding of endangered species for trade	Laundering of wildlife into legal breeding systems, captive breeding of cheetahs, macaques	Magliolo et al. (2021), Hogg et al. (2018)

(continued)

Table 10.6 (continued)

Wild product	Markets	Key countries involved	Trade characteristics	Key issues & practices	References
Laboratory Animals	Biomedical Research	Global	Trade of animals for research purposes, some being illegal	Illegal capture of Barbary macaques for research; market for surplus animals	van Uhm (2016), Cardoso et al. (2021), Hayman and Brack (2002)

frameworks such as CITES, the persistence of illegal wildlife trade is largely due to weak enforcement, insufficient resources for monitoring, and the lack of deterrent penalties, which allow criminal syndicates to operate with relative impunity (UNODC, 2020). The National Tiger Conservation Authority (NTCA) reported at least 45 tiger poaching cases between 2018 and 2020, with over 20% linked to organized criminal groups, highlighting the high demand and value placed on these species in the illegal trade (National Tiger Conservation Authority (NTCA), 2020). Furthermore, corruption within local law enforcement and judicial delays further exacerbate the situation, as enforcement efforts are undermined, leading to continued exploitation of endangered species (Gupta & Dey, 2017). In response to rising demand, certain species are legally bred or farmed, particularly for industries like the exotic pet trade and sustainable products like crocodile leather. India's crocodile farming industry, for example, produces over 100,000 crocodile skins annually (Das, 2020). However, these legal operations are often exploited, with unregulated farms operating under the guise of sustainability, facilitating the illegal trade of wildlife under inadequate oversight (Sharma, 2021). Species such as Tokay geckos (*Gekko gecko*) and Indian star tortoises (*Geochelone elegans*), often bred or captured illegally for the exotic pet market, further complicate the issue, as they are trafficked both within India and internationally. The overlap between legal breeding operations and illegal trade networks fuels the demand for these high-value wildlife products, creating a significant challenge for wildlife conservation and regulation.

The role of intermediaries in the wildlife trade cannot be understated. These individuals, ranging from local poachers to international smugglers, act as critical facilitators in the supply chain of wildlife products. They bridge the gap between the source of wildlife and the final consumer, often utilizing sophisticated trafficking routes that bypass border controls and evade regulatory frameworks like CITES. Middlemen exploit porous borders, particularly through routes in neighboring countries such as Nepal, Bangladesh, and Myanmar, making enforcement efforts particularly challenging (Sharma, 2021; Duffy, 2018b). In 2021, the Wildlife Crime Control Bureau (WCCB) recorded over 100 instances of wildlife trafficking across these borders, underscoring the scale and complexity of the illegal trade and the role intermediaries play in maintaining its persistence (WCCB (Wildlife Crime Control Bureau), 2021). These intermediaries are adept at hiding illicit wildlife products within legitimate shipments or trafficking them through areas with weak enforcement, which further complicates the effective monitoring of wildlife trade chains. Their operations circumvent both national and international regulations, perpetuating the cycle of illegal wildlife trafficking. Overall, the combination of organized poaching, the exploitation of legal wildlife breeding operations, and the critical role of intermediaries contributes significantly to the ongoing illegal wildlife trade in India. This multifaceted supply-side dynamic not only undermines conservation efforts but also poses a serious threat to biodiversity, both at the national and international levels. To combat this issue effectively, there is a pressing need for stronger enforcement measures, improved international cooperation, and better regulation of wildlife trade operations, including stricter oversight of breeding farms and markets dealing in exotic pets and wildlife products. To ensure the preservation of endangered species throughout India and to stop the operations of organized crime networks, improved cross-border cooperation is crucial. This multifaceted supply-side dynamic contributes significantly to the persistence of illegal wildlife trade in India, affecting both conservation efforts and biodiversity protection at the national and international levels.

10.2.3 Price Dynamics and Profit Margins in the Wildlife Trade

The high prices for some species, especially tigers (*Panthera tigris*), elephants (*Elephas maximus*), and rhinoceroses (*Rhinoceros unicornis*), are major drivers of the illegal wildlife trade. These species, revered for their rarity and cultural significance, command exorbitant prices on both legal and illicit markets. For instance, the price of rhino horn can exceed $60,000 per kilogram, a factor that makes it highly lucrative for poachers and traffickers (UNODC, 2020). Similarly, tiger pelts and elephant ivory can be sold at substantial amounts, especially in black markets in Asia, Africa, and parts of the Middle East (Karanth, 2017). The demand for these products stems from various sectors, including traditional medicine, luxury goods, and ornamental purposes. As these species become increasingly scarce, their value in the illegal wildlife trade continues to rise, further driving poaching and trafficking activities. The price of wildlife products is shaped by several factors, most notably the rarity of the species and the demand driven by cultural and medicinal beliefs. For instance, rhino horn is overpriced because it is mistakenly thought to have therapeutic qualities, especially in traditional Chinese medicine (Sharma, 2021). The rarity of these animals in the wild increases their desirability, and as endangered species become rarer, their market prices escalate. Additionally, fluctuations in international demand for wildlife products can significantly influence their market prices. In 1989, the global ban on ivory trade led to a temporary spike in ivory prices as illicit markets adapted to circumvent the new regulations (Kumar, 2020). These price dynamics create a market imbalance that fuels the illegal wildlife trade, where traffickers are motivated by the immense financial rewards.

In addition to the illicit markets, the legal trade of wildlife products also responds to pricing dynamics, especially in cases where species are farmed or sustainably harvested. For example, India's crocodile farming industry, which produces over 100,000 crocodile skins annually for the leather market, exemplifies how legal wildlife industries can also be influenced by the profitability of high-value species (Das, 2020). However, the massive profitability of the illegal wildlife trade complicates enforcement efforts. Despite regulations such as the Wildlife Protection Act (1972) and CITES, traffickers often find ways to exploit loopholes in the legal system, creating a persistent tension between legal and illegal markets (Gupta & Dey, 2017). This overlap not only complicates enforcement but also hinders effective conservation efforts, as the incentive to poach and traffic remains strong due to the high financial returns. Furthermore, market-driven demand for wildlife products is increasingly influenced by globalization. International demand for exotic wildlife, particularly for luxury goods and traditional medicine, has shifted the focus of illegal wildlife trade to global markets in Southeast Asia, China, and even the Middle East (Duffy, 2018b). Products like elephant ivory, rhino horns, and tiger skins are highly sought-after in these marketplaces, and trafficking networks have grown to satisfy this demand, frequently crossing international borders (Sethi, 2018). Research by the World Wildlife Fund (WWF) highlights that demand from wealthy consumers, along with the rise of online markets, has made tracking and enforcement even more difficult, complicating efforts to curtail the illegal wildlife trade (WWF-India, 2020).

The intricate pricing system and its impact on networks involved in the legal and illicit wildlife trade highlight how important financial incentives are to the continuation of wildlife crime. Enforcement authorities have several difficulties because of the illegal wildlife trade's tremendous profitability and the overlap between the legal and criminal markets. More severe legislative frameworks that can narrow the gaps traffickers exploit, international cooperation, and better enforcement procedures are all urgently needed to effectively combat wildlife trafficking. The continued focus on the economic drivers of wildlife crime is crucial to understanding and combating the illicit wildlife trade on both national and global scales.

10.3 Market Forces Influencing Wildlife Trade

The market forces influencing wildlife trade are multifaceted, ranging from the globalization of trade networks to the role of digital platforms and technological advancements. These forces shape how wildlife products flow across regions and how government regulations interact with market demand. This section examines the key market forces-globalization and trade networks, technological and digital influences, and government policies and regulations-that affect the wildlife trade, especially in the context of India.

10.3.1 Globalization, Cross-Border Networks, and Digital Platforms: Dynamics of India's Wildlife Trade

Particularly in nations like India, which are both a source and a vital transit hub for illegal wildlife trafficking, globalization has had a substantial impact on the wildlife trade. International demand for endangered species, exotic pets, and luxury wildlife products drives India's central role in the global wildlife trade network. High-value wildlife products such as tiger skins, elephant ivory, rhino horn, and wildlife medicines are often sought after in international markets like China, Southeast Asia, and the Middle East (Duffy, 2018b; Bennett, 2020). The presence of well-organized transnational smuggling networks, exploiting weak regulatory enforcement at national borders, makes it increasingly challenging for law enforcement agencies to track and seize illegal wildlife products. Research by the United Nations Office on Drugs and Crime (2020) highlights India's central position in Asia's wildlife trafficking routes, exacerbating the challenge of addressing both domestic consumption and international trade flows of illicit wildlife goods.

India shares extensive land borders with neighbouring countries such as China, Bangladesh, Nepal, Myanmar, and Bhutan, which facilitates both legal and illegal cross-border trade of wildlife products. The trade routes between China and India are especially noteworthy since they are notorious for smuggling animal products including elephant ivory, tiger pelts, and rhino horn. Similarly, Southeast Asia, especially Thailand and Malaysia, is a major transit hub for wildlife products destined for global markets (Kumar & Gupta, 2018). According to UNODC (2020), wildlife trafficking along these routes has become increasingly sophisticated, with traffickers employing multiple layers of intermediaries to ensure discreet and secure transportation. For instance, ivory poached in India's north-eastern regions is often smuggled into China, where it is processed into luxury items and sold in high-end markets (India, 2021). The porous nature of borders, combined with weak enforcement mechanisms and endemic corruption, has made cross-border trafficking rampant, further complicating enforcement efforts (Sharma, 2021).

By placing restrictions on the trade of endangered species, international accords like the Convention on International Trade in Endangered Species of Wild Fauna and Flora (CITES) are essential in controlling the global wildlife trade. India, a signatory of CITES, is obligated to regulate the trade of species listed under the convention, such as tigers, elephants, and rhinos. However, the implementation of CITES is often undermined by weak enforcement, policy gaps, and the persistent existence of illegal markets. For example, while CITES restricts the trade of ivory, illegal ivory markets continue to thrive in parts of Asia and Africa, with India serving as both a source and transit country for smuggled ivory (Duffy, 2018b; Karanth, 2017). Although CITES has achieved some successes, such as reducing rhino horn trade internationally, challenges remain in terms of enforcement across different jurisdictions, highlighting the complex dynamics at play in the global wildlife trade (Bennett, 2020). The rise of e-commerce and digital platforms has added a new dimension to the illegal wildlife trade, particularly in India. Platforms such as

Alibaba, eBay, and social media sites like Facebook and Instagram have become popular venues for the sale of illegal wildlife products. A report by TRAFFIC (2021) found that social media platforms, especially Facebook, are frequently used to facilitate the sale of exotic pets, tiger skins, and live reptiles, despite regulations prohibiting such sales. The anonymity of online transactions, coupled with the ease of communication, allows traffickers to bypass law enforcement agencies, often operating with impunity. In India, illegal wildlife trade is further facilitated by social media groups and private messaging platforms, where wildlife products are marketed covertly to a select audience. The use of cryptocurrency and encrypted messaging apps further complicates law enforcement efforts, as it provides traffickers with secure, anonymous transactions (Gupta & Dey, 2017; CITES, 2020).

Additionally, digital marketing techniques contribute to driving demand for wildlife products. Exotic wildlife, often marketed as luxury items or rare collectibles, are increasingly promoted through targeted advertisements on social media platforms. Influencers and celebrities, often unknowingly, amplify this demand by showcasing wildlife products as symbols of exclusivity and prestige (Sharma, 2021). For example, advertisements for exotic pets, including endangered species, or products derived from wildlife (such as tiger bones or rhino horn for purported medicinal purposes) often circulate in private online groups. Despite some social media platforms, such as Instagram and Facebook, taking action against the sale of illegal wildlife products, enforcement remains inconsistent, leaving significant gaps for traffickers to exploit (CITES, 2020). The rapid growth of e-commerce and online platforms has created new avenues for illegal wildlife trade, complicating regulatory efforts and demanding more targeted digital interventions (Duffy, 2018b). The illegal wildlife trade in India has been greatly impacted by globalization, cross-border trading networks, and the growth of digital platforms. Wildlife protection now faces a complex issue due to the combination of growing global demand, lax border control, and the exploitation of digital technologies. Efforts to combat wildlife trafficking in India must consider the global nature of the issue, focusing on international cooperation, stronger enforcement mechanisms, and more effective use of technology to close the gaps that traffickers currently exploit. The growing role of digital platforms in facilitating wildlife trade also underscores the urgent need for innovative solutions that address both traditional and digital avenues for trafficking (Kumar, 2020; UNODC, 2020).

Poaching, endangered species trafficking, and the illicit trade in exotic animal items are among the major challenges highlighted in the Table 10.7, which lists numerous wildlife crime activities in India's main regions. Additionally, it lists the most often confiscated species, poaching and trafficking trends, notable instances, and the difficulties in stopping these illicit operations.

Table 10.7 Wildlife crime trends and key seizures in India (2014–2024)

Region/ State	Wildlife crime activities	Top seized species	Poaching/Trafficking trends	Key seizures/Incidents	References
Assam	Significant poaching of rhinos and trafficking of exotic species through international borders, along with wildlife crimes related to human-wildlife conflict, remain major challenges for conservation efforts.	Rhinoceros horns, Tokay geckos, Exotic birds (Macaws)	Poaching of rhinos, primarily driven by the demand for rhino horns, and the trafficking of exotic species like Tokay geckos are major wildlife crimes.	Poaching of 191 rhinos in Assam (2000–2021); seizure of exotic species.	Rajpoot et al. (2018), Vu and Nielsen (2018), Chitale et al. (2014)
				2022: Zero poaching of rhinos due to coordinated efforts, technology, and specialized protection forces.	Roe and Booker (2019)
Bihar	Sale of rare birds and illegal wildlife trade through Indo-Nepal border.	Gangetic River dolphin	Poaching and hunting of Gangetic dolphins for bait and illegal trade.	There are an estimated 3000 Gangetic River dolphins, half of which are located in Bihar and are constantly threatened by hunting, riverfront development, and pollution.	Singh et al. (2021), Kolipakam et al. (2020)
Chhattisgarh	Poaching of spotted deer, pangolins, and leopards.	Pangolin scales, leopard skin	Rise in illegal wildlife trade, especially pangolins and leopard skins. Corruption among officials.	Feb 2022: Seizure of 19 kg of pangolin scales, star tortoises, and leopard nails in Bastar division.	Duffy et al. (2015), Sharma et al. (2018)
Gujarat	Wildlife trafficking involving Asiatic lions and other endangered species.	Asiatic lions, Wild boar meat, Leopard skins	Smuggling of Asiatic lions and wildlife products for sorcery and medicine.	Large seizures of wildlife in Gir region; Trade in wild animal meat.	Hughes (2021)
Jharkhand	Poaching and smuggling of wildlife products through Indo-Nepal border.	Sloth bears, elephant ivory	Increase in wildlife crime, including poaching of sloth bears and illegal ivory trade through inter-state rackets.	Aug 2019: Rescue of 5 sloth bears from poachers transporting them through Indo-Nepal border. Nov 2022: Seizure of 10 pieces of elephant ivory from 5 poachers in West Singhbhum.	Singh et al. (2021)

(continued)

Table 10.7 (continued)

Region/ State	Wildlife crime activities	Top seized species	Poaching/Trafficking trends	Key seizures/Incidents	References
Karnataka	Poaching and trafficking of tigers, elephants, and other endangered species.	Elephant tusks, Tiger skins, Leopard skins	Poaching of elephants for tusks, with organized smuggling networks.	Frequent poaching of elephants in Western Ghats region.	Goswami et al. (2019)
Kerala	High incidents of wildlife trafficking related to elephants and tigers.	Elephant tusks, Tiger skins, Leopard skins	Poaching and illegal logging in Kerala's forests.	Seizures of elephant tusks and tiger bones.	Shaffer et al. (2019)
Madhya Pradesh	Poaching of tigers, leopards, blackbucks, and other wildlife species, along with increasing conflicts between humans and wildlife, pose significant threats to biodiversity conservation.	Tiger skins, Leopard bones, Rhino horns	Poaching is concentrated in tiger reserves like Kanha and Bandhavgarh, with a high number of cases involving blackbucks and tigers, while escalating human-wildlife conflict has led to the deaths of tigers and leopards	In 2022, poaching of tigers was reported from Pench Tiger Reserve, while in October 2021, six individuals were arrested for selling a leopard hide and nails, and in December 2022, poachers were apprehended for slaughtering a nilgai	Nittu et al. (2022)
		Blackbuck			Karanth et al. (2012), Gulati et al. (2021)
Maharashtra	High poaching of tigers and seizure of exotic species for trade.	Tiger skins, Shark fins, Leopard bones	Continued trafficking of tiger products, exotic animals for medicinal use.	665 exotic animals seized at Mumbai airport (2022).	TRAFFIC (2021)
Northeast India	Transnational trafficking, with states acting as a transit route for wildlife from Southeast Asia to China and other parts.	Tokay geckos, Exotic reptiles, Wild mammals	Increased poaching and smuggling due to porous international borders.	Seizure of 468 exotic species (2022) from Mizoram; Rhino horn smuggling.	Vu and Nielsen (2018), Barabanov et al. (2019), Lim (2017)
Odisha	Rampant smuggling of pangolins for their scales and meat.	Pangolin scales	High demand for pangolins in China for medicinal and aphrodisiac purposes. Illegal smuggling from Odisha and Jharkhand.	Five kilograms of pangolin scales were seized from Shamsuddin Khan's home in the Nayagarh District's Daspalla neighborhood in June 2018.	Singh et al. (2021), Kolipakam et al. (2020)

Rajasthan	Poaching activities related to leopards, tigers, and various exotic species.	Leopard skins, Tiger bones, Pangolins	In recent years, there have been a lot of wildlife crime instances.	High incidence of poaching in Sariska Tiger Reserve.	Hughes (2021)
Southern India	Wildlife crime and illegal trade in species like red sanders, pangolins, and sea cucumber.	Rose-ringed parakeet, pangolin	Increasing wildlife trade in southern states, with Bengaluru as the central hub. Significant trafficking of exotic species such as pangolins and sea cucumbers.	Nov 2021: Seizure of 571 star tortoises in Bengaluru. Aug 2021: Arrest of 3 people with 8.25 kg of ambergris.	Sharma et al. (2018), Chan et al. (2021)
Uttar Pradesh	Highest wildlife crime cases; prime hub for poachers.	Turtles, Tiger bones, Leopard skins, Ivory	High poaching rates, especially for turtles and tigers.	6000 endangered turtles seized in Amethi (2017).	Rajpoot et al. (2018)
West Bengal	Poaching and illegal trade of reptiles and birds, along with a rise in exotic species trafficking.	Snake venom, Hornbills, Exotic birds (Macaws, Falcon)	Increase in trafficking of snake venom and exotic birds.	Snake venom seizures worth ₹70 crore (2017); Exotic birds seized (2018).	Singh et al. (2014)
General (Across India)	Smuggling and illegal trade of exotic and invasive species, including plants and wildlife parts.	Exotic species, invasive plants	Rise in international trade of exotic species and invasive plants. Lax regulations on exotic species trade.	2022: Wildlife (Protection) Amendment Bill passed to include exotic species and enforce stricter penalties.	Hulme (2021), Anoop et al. (2021)

10.4 Economic Contribution and Hidden Costs of Wildlife Trade in India: Legal Versus Illegal Impacts

The legal wildlife trade in India, which includes eco-tourism, sustainable wildlife farming, and exports of wildlife products, significantly contributes a role in driving both local and national economic growth. Eco-tourism, one of the country's fastest-growing economic sectors, generates approximately ₹20,000 crores ($2.7 billion) annually, with popular wildlife tourism destinations like Ranthambore, Kaziranga, and Jim Corbett National Park, attract millions of domestic and international tourists, generating income for local communities through employment in tourism, hospitality, and the sale of wildlife-based crafts (Bennett, 2019). Sustainable wildlife farming, including the farming of crocodiles for leather products, represents a sustainable alternative to poaching. These farming practices reduce the pressure on wild crocodile populations, while simultaneously providing local employment opportunities and contributing to global markets for luxury goods (CITES, 2020). The legal wildlife trade also benefits from sectors like honey beekeeping and the lawful sale of medicinal herbs like ginseng, which guarantee a steady supply of goods that ease the burden on wild populations. The export of legally wildlife based products, including handicrafts made from sustainably sourced timber, herbal remedies, and ornamental fish, further bolster the Indian economy. According to India is a significant exporter of these wildlife products like crocodile skins and medicinal herbs, contributing substantially to its GDP and international trade, while providing local communities with sustainable livelihoods.

However, the illegal wildlife trade presents hidden costs that significantly undermine these economic benefits and create broader societal challenges. The illegal trade in wildlife often finances organized crime, including drug cartels, arms traffickers, and terrorist organizations, exacerbating national security risks. A UNODC (2020) report emphasizes the connection between wildlife trafficking and other illicit activities, such as money laundering and human trafficking, creating an environment of instability in regions affected by poaching. In India, for instance, smuggling rings operating in national parks increase security risks, diverting resources from essential law enforcement efforts. Beyond national security, illegal wildlife trade has devastating effects on conservation efforts. Poaching for highly valuable wildlife products, such as elephant ivory, tiger bones, and rhino horns, threatens the survival of endangered species, driving biodiversity loss. As noted by Karanth et al. (2012), India's tiger population has been severely impacted by poaching driven by demand for tiger bones and skin in illegal markets. The illegal ivory trade, primarily fueled by demand from China and Southeast Asia, has contributed to the rapid decline of elephant populations. Such illegal activities undermine long-term conservation goals, escalating the costs of wildlife protection and threatening the country's ecological stability. Furthermore, the illegal wildlife trade poses significant public health risks, facilitating the transmission of zoonotic diseases. The trade in live wild animals, as well as bushmeat, has been directly linked to several major disease outbreaks, including the SARS outbreak and the ongoing COVID-19 pandemic.

Zhou et al. (2016) report that wet markets in China and Southeast Asia, which often source wildlife from India, have been implicated in the transmission of diseases such as HIV, SARS-CoV, and Ebola. These zoonotic diseases pose serious health threats to human populations, leading to substantial economic losses due to healthcare costs, lost productivity, and the collapse of tourism industries. For example, the global response to the COVID-19 pandemic led to severe disruptions in international travel and tourism, which are key contributors to India's economy. The resultant economic fallout further exacerbates the burden of illegal wildlife trade, as resources meant for economic growth are diverted to address public health emergencies. In conclusion, while the legal wildlife trade contributes significantly to India's economy through eco-tourism, sustainable farming, and exports, the hidden costs of illegal wildlife trade-ranging from organized crime and national security threats to the destruction of biodiversity and public health risks-pose serious challenges. The cumulative effect of these costs undermines the positive economic contributions made by the legal trade, highlighting the urgent need for stronger enforcement of wildlife protection laws and international cooperation to combat illegal wildlife trafficking. This balance of legal and illegal trade necessitates a comprehensive approach to ensure both the sustainable use of wildlife and the protection of vital ecosystems for future generations.

10.5 Economic Consequences of Wildlife Overexploitation and Trafficking

The overexploitation of wildlife, driven by both illegal and unsustainable legal trade, has profound long-term economic consequences, particularly for local communities and biodiversity. As key species such as elephants, tigers, and rhinos are depleted, the availability of wildlife products diminishes, which significantly impacts industries reliant on wildlife, such as eco-tourism and wildlife-based crafts. A decline in species populations, especially apex predators like tigers, can lead to reduced tourist numbers at national parks. For instance, the famous Ranthambore Tiger Reserve in India has experienced a reduction in tiger sightings, which directly impacts tourism revenue, affecting local businesses and tour guides. Bennett (2019) notes that such declines in wildlife populations create a "domino effect" on local economies that heavily depend on wildlife for their livelihood. Similarly, decreased wildlife populations diminish the supply of wildlife-based products like crafts, which serve as an additional source of income for local communities. The loss of biodiversity further exacerbates economic challenges, particularly in agriculture. Predators like tigers and wolves play a crucial role in controlling herbivore populations, preventing overgrazing and protecting crops. When these apex predators decline, herbivore numbers can explode, causing damage to local crops and disrupting food security for farmers. For example, the depletion of elephant populations in several regions has led to a rise in human-wildlife conflict, including crop destruction and subsequent economic losses for farmers (Sharma, 2020). The economic impact of wildlife trafficking is immense, with significant losses in biodiversity and ecosystem services. The poaching of iconic species like rhinos, tigers, and elephants not only threatens species survival but also destabilizes ecosystems. For example, the absence of elephants, known as keystone species, severely disrupts forest and grassland ecosystems by impacting seed dispersal and maintaining plant biodiversity. Their loss negatively affects the livelihoods of local community's dependent on these ecosystems. Elephant numbers have decreased by almost 60% in the past 50 years, according to the International Union for Conservation of Nature (2021), which has exacerbated environmental imbalances and put many other species' future in jeopardy. An enforcement costs to combat wildlife trafficking add to this economic burden. In India, millions of dollars are spent annually on anti-poaching patrols, border security, and law enforcement operations (TRAFFIC, 2020). However, enforcement efforts remain inadequate due to resource limitations, corruption, and poor institutional coordination. A 2019 World Bank report estimated that global wildlife crime costs between $7 billion to $23 billion annually, with India being a significant source and transit point for trafficked species. This loss affects national economies, hindering sustainable development goals. Moreover, the disappearance of pollinators such as bees and bats-sometimes exploited in wildlife trade—further disrupts ecosystems and agricultural productivity. This, in turn, causes additional economic losses for farmers, as pollinator-dependent crops like fruits and vegetables become more difficult to grow (Kumar & Dey, 2019). As a clear reminder of the far-reaching effects of wildlife depletion, a 2020 research by the European Commission (2020) estimated that the loss of insect pollinators alone might cost the world economy more than $500 billion yearly. Illegal wildlife trafficking has far-reaching hidden costs, but legitimate wildlife commerce makes a substantial contribution to national economies, particularly through ecotourism, sustainable wildlife farming, and exports. These include not only biodiversity loss but also national security risks, public health crises (due to zoonotic diseases), and the strain on law enforcement resources. For example, the illicit wildlife trade was implicated in the 2013 West African Ebola outbreak (WHO, 2014). A well-regulated and sustainable wildlife trade industry can foster economic growth while simultaneously protecting ecosystems and local communities, but the illegal wildlife trade continues to undermine these efforts. As noted by Phelps et al. (2016), a coordinated approach focusing on both enforcement and community-driven conservation efforts is essential to mitigate the economic impacts of wildlife trafficking and overexploitation.

Tourism plays a significant role in many national economies, contributing to job creation and GDP growth. Over 230 million jobs, approximately 8% of the global workforce, are directly or indirectly linked to tourism (Bolwell & Weinz, 2008). For example, Rwanda's Parc National des Volcans brings in over 140 million USD annually, which makes up 3.7% of the nation's GDP and 80% of its tourism earnings (Maekawa et al., 2013). Poaching jeopardizes not only the tourism sector that depends on iconic species like the Komodo dragon (*Varanus komodoensis*) in Indonesia's Komodo National Park, but also the species itself (Walpole & Goodwin, 2001). As demonstrated by the closure of Virunga National Park in 2018 following the kidnapping of two visitors, poaching-related crimes in national parks, such as those in Borneo's Lambir Hills National Park, where 90% of its protected species have gone extinct (Harrison, 2011), can discourage tourists due to safety concerns (Barnes, 2018). On a broader scale, the economic losses due to illegal wildlife trade (IWT) have been estimated at 48–153 billion USD annually (Nellemann et al., 2014), a figure that far exceeds the funds allocated for ecosystem restoration and conservation, with only 261 million USD spent annually in Africa and Asia to combat IWT (World Bank, 2019). This imbalance reflects a global under-prioritization of IWT, with only a small fraction of restoration efforts funded by governments, such as in Brazil where 90% of rainforest conservation efforts were privately funded (Nellemann et al., 2014). Additionally, illicit trade distorts prices, undermining legitimate markets, and governments lose about $15 million USD in tax revenue each year as a result of IWT (World Bank, 2019). Furthermore, IWT-affiliated organized crime groups frequently use extortion, violence, and corruption, which can topple governments and deter foreign investment, hence impeding socioeconomic progress (Zimmerman, 2003).

10.5.1 Economic Consequences of Wildlife Overexploitation

The overexploitation of wildlife, driven by both illegal and unsustainable legal trade, results in profound long-term economic consequences, particularly for local communities and biodiversity. As key species such as elephants, tigers, and rhinos experience population declines, industries that rely on wildlife, such as eco-tourism and wildlife-based crafts, are significantly impacted. The depletion of apex predators like tigers can lead to a reduction in tourist numbers. For instance, the Ranthambore Tiger Reserve in India has experienced a decline in tiger sightings, directly affecting tourism revenue and local businesses, including tour guides and hospitality services. Bennett (2019) highlights how such declines create a "domino effect" on local economies that heavily depend on wildlife for their livelihoods. Furthermore, a decrease in wildlife populations limits the supply of wildlife-based products, like crafts, which are a significant income source for local communities. The loss of biodiversity also exacerbates economic challenges in agriculture. Predators like tigers and wolves play a vital role in controlling herbivore populations, which, if left unchecked, can lead to overgrazing and damage to crops. For example, the depletion of elephant populations in several regions has contributed to increased human-wildlife conflicts, including crop destruction, causing substantial economic losses for farmers (Sharma, 2020).

The economic impact of wildlife trafficking is far-reaching, with poaching of iconic species like rhinos, tigers, and elephants destabilizing ecosystems. Elephants, known as keystone species, play an essential role in seed dispersal and maintaining plant biodiversity. The forest and grassland ecosystems that local residents rely on for their livelihoods are seriously disrupted by their disappearance. Elephant populations have declined by about 60% over the past 50 years, according to the

International Union for Conservation of Nature (2021), aggravating environmental imbalances and endangering the existence of numerous other species. The financial burden is further increased by the expenses associated with implementing anti-poaching measures. In India alone, millions of dollars are spent annually on anti-poaching patrols, border security, and law enforcement operations (TRAFFIC, 2020). However, enforcement remains inadequate due to resource constraints, corruption, and poor institutional coordination. A World Bank report (2019) estimates that the global cost of wildlife crime ranges from $7 to $23 billion annually, with India being a major source and transit point for trafficked species. These losses affect national economies, hinder sustainable development goals, and underscore the need for greater resources and coordinated efforts in combatting wildlife crime.

10.5.2 Impact on National Economies and Ecosystem Services

With approximately 230 million jobs created globally, or 8% of the workforce, tourism is essential to the economic development of many nations. A major contributor to national income is wildlife-based tourism, which is focused on famous animals like the Komodo dragon (*Varanus komodoensis*) in Indonesia's Komodo National Park. According to Maekawa et al. (2013), the park brings in over 140 million USD annually, which represents 3.7% of Rwanda's GDP and 80% of its tourism-related income. Estimates of the annual economic losses resulting from the illegal wildlife trade (IWT) range from $48 to $153 billion (Nellemann et al., 2014). Only $261 million USD is invested annually across Africa and Asia to tackle IWT, which is significantly less than the value of lost resources. This amount dwarfs the funding allotted for conservation and restoration (World Bank, 2019). The disparity indicates that IWT is not given enough priority worldwide, and in some places, like Brazil, 90% of rainforest conservation initiatives are funded by private sources (Nellemann et al., 2014). Additionally, the illicit trade influences legal market prices, causing imbalances, and governments lose an estimated $15 million USD in tax revenue each year as a result of IWT (World Bank, 2019). IWT-affiliated organized crime gangs frequently use extortion, brutality, and corruption, which weaken governments and discourage foreign investment, hence impeding socioeconomic growth (Zimmerman, 2003).

The depletion of pollinators, such as bees and bats, also leads to economic losses in agriculture, as crops dependent on pollination, including fruits and vegetables, become harder to grow. Kumar and Dey (2019) note that the loss of these pollinators exacerbates the strain on food security. A report by the European Commission (2020) estimated that the global economic cost of the decline of insect pollinators could exceed $500 billion annually, underscoring the far-reaching consequences of wildlife depletion for both ecosystems and economies. While legal wildlife trade contributes significantly to national economies, especially through eco-tourism, sustainable wildlife farming, and exports, the illegal trade imposes hidden and devastating costs. These include biodiversity loss, national security risks, public health crises (such as zoonotic diseases), and the strain on law enforcement resources. For example, the illicit wildlife trade was implicated in the 2013 West African Ebola outbreak (WHO, 2014). The illegal wildlife trade continues to thwart efforts to safeguard local populations and ecosystems while promoting economic growth through a sustainable and well-regulated wildlife trade economy. According to Phelps et al. (2016), reducing the financial effects of wildlife trafficking and overexploitation requires a concerted strategy that prioritizes both enforcement and community-driven conservation initiatives.

10.6 Social and Cultural Drivers of Wildlife Trade and Conservation Challenges

Wildlife trade is deeply intertwined with cultural traditions, social values, and community structures, which play a significant role in shaping both the demand and supply sides of the trade. Cultural significance, community involvement, and traditional practices not only sustain certain aspects of the trade but also present opportunities for sustainable conservation practices if carefully addressed. The use of wildlife in traditional medicine, especially in Asia, is a major factor driving the wildlife trade. Traditional medicine has been a cornerstone of healthcare systems for centuries, especially in countries like China and India, where wildlife products are believed to possess medicinal or healing properties. For instance, tiger bones are used in traditional Chinese medicine (TCM) to treat ailments such as arthritis, and rhinoceros horn is valued as a cure-all for fevers, headaches, and even cancer (Bennett, 2018). Despite the lack of scientific evidence supporting the medicinal efficacy of these products, the cultural belief in their healing properties fuels high demand. The economic incentive generated by the sale of wildlife for medicinal use plays a pivotal role in the persistence of the illegal wildlife trade. A 2019 report by the World Health Organization (WHO) indicated that the global traditional medicine market is worth over $60 billion, with a significant portion attributed to wildlife-derived products, including those from endangered species. This demand for wildlife-based traditional medicines fuels the black market for species like tigers, rhinos, and pangolins, which are often poached and trafficked to meet the demand (WHO (World Health Organization), 2019). Additionally, traditional medicine practices in South Asia have been linked to the use of endangered species parts, with a study by Shrestha et al. (2016) highlighting that parts of tigers, elephants, and other threatened species are commonly used in South Asian healing practices, further contributing to the illegal trade.

Moreover, wildlife also plays an important role in religious and cultural ceremonies in various parts of the world, particularly in India, where cultural traditions perpetuate demand for certain wildlife products. Elephants, revered in Hinduism as symbols of wisdom and remover of obstacles (represented by the deity Ganesha), are highly regarded, and their tusks and ivory are used in religious rituals and festivals (Choudhury, 2020). Elephant ivory is still used in religious ceremonies even though the international trade in elephant ivory is prohibited by the Convention on International Trade in Endangered Species of Wild Fauna and Flora (CITES). This is especially true in rural areas where the custom is still very much ingrained in the local culture. Similarly, the use of peacock feathers in cultural rituals, especially during Indian weddings and festivals like Navratri, has contributed to the continued demand for these feathers, despite the Indian Peacock being legally protected (Choudhury, 2020). The cultural significance of these species creates challenges for conservation efforts, as these practices are seen as integral to local identity and tradition. According to a 2019 study by the Wildlife Protection Society of India (WPSI), the demand for wildlife products in religious ceremonies, particularly ivory and peacock feathers, has continued to support illegal wildlife trade in India, even with strong legal frameworks in place to prevent such practices.

The intersection of cultural practices and wildlife trade highlights a critical challenge in balancing conservation efforts with respect for cultural heritage. While these cultural beliefs and practices are deeply rooted in social traditions, they often undermine conservation efforts and contribute to the illegal wildlife trade. Addressing this challenge requires nuanced approaches that respect cultural practices while promoting sustainable and legal alternatives. For example, the promotion of synthetic ivory and alternative materials for religious ceremonies and cultural rituals has been suggested as a solution to reduce demand for elephant ivory (Houghton et al., 2018). Similarly, raising awareness about the importance of wildlife conservation and the role of endangered species in ecosystem stability can help change attitudes toward the use of wildlife products in traditional medicine and cultural ceremonies. The 2017 study by the United Nations Environment Programme (UNEP) emphasized the importance of community-based conservation initiatives that involve local communities in wildlife protection efforts, ensuring that cultural traditions are respected while encouraging the adoption of sustainable practices.

10.7 Economic Incentives and Community Engagement in Mitigating Illegal Wildlife Trade

The trade in wildlife provides significant economic opportunities for local communities, particularly in rural and marginalized areas where other sources of income may be scarce. Poaching and illegal wildlife trade often arise as financially viable options due to the high monetary value placed on wildlife products such as ivory, rhino horn, and tiger pelts. Baker and Miller (2019) emphasize that poachers are typically driven by the immediate financial returns offered by wildlife trade, often outweighing the long-term environmental and legal consequences. In rural regions of India, such as Madhya Pradesh, where communities face limited formal employment or economic development opportunities, illegal wildlife trade becomes a primary source of income, perceived as less risky than other forms of criminal activity (UNODC, 2020). The lack of educational opportunities and persistent poverty further exacerbate the vulnerability of these communities to engaging in illegal wildlife trade (WWF India, 2018). This illicit activity involves not just poachers but also middlemen who facilitate the transport and sale of wildlife products, with the profits from illegal trade attracting a larger pool of actors into the criminal economy. Increasingly sophisticated smuggling networks, which involve multiple actors across national borders, make it difficult for authorities to curb the trade (Lembo et al., 2018).

The involvement of local communities in conservation efforts offers a critical solution to reducing the prevalence of illegal wildlife trade. Research by Sharma and Kumar (2020) suggests that the introduction of sustainable economic alternatives such as eco-tourism, wildlife-friendly businesses, and community-based agriculture can reduce the economic incentives for participating in illegal wildlife trade. The success of eco-tourism programs is particularly notable, as demonstrated in Kaziranga National Park in Assam. Through the development of eco-tourism, local villagers benefit from employment opportunities and skill-building, providing long-term financial stability and reducing their reliance on poaching for survival (Choudhury, 2020). The integration of indigenous and rural communities in conservation efforts is critical, as these communities often possess rich traditional ecological knowledge (TEK) due to their proximity to wildlife habitats. TEK plays an integral role in managing biodiversity, with practices based on generations of experience ensuring sustainable use of natural resources (Gadgil et al., 1993). For instance, the Khasi Tribe in Meghalaya has effectively managed forest resources by promoting sustainable hunting practices and the collection of non-timber forest products (NTFPs), preventing overexploitation and maintaining biodiversity (Ramakrishnan, 2014).

Studies confirm that community-based conservation initiatives that empower local populations to manage natural resources tend to result in better conservation outcomes than top-down, government-led programs. A case study by Choudhury (2020) in the Sundarbans Mangrove Forest in West Bengal highlights how eco-tourism has successfully integrated local communities into conservation efforts. Revenue generated from tourism has supported community infrastructure, healthcare, and schools, while promoting sustainable practices that protect endangered species such as the Royal Bengal Tiger. Furthermore, by encouraging eco-friendly practices and sustainable resource use, government regulations like the Wildlife Protection Act of 1972 and the Convention on International Trade in Endangered Species (CITES) assist community-led conservation efforts. These policies offer financial and technical assistance to local communities, helping them build viable alternatives to illegal wildlife trade, thus enhancing their engagement in conservation. Furthermore, the provision of financial incentives and capacity-building opportunities has been shown to encourage communities to take ownership of conservation efforts and actively participate in the sustainable management of natural resources (Jadhav & Barve, 2020). Research also indicates that when communities benefit economically from conservation, the likelihood of illegal wildlife trade diminishes, as demonstrated in community-based programs in Africa, where the economic benefits of sustainable tourism significantly reduced poaching (Bennett, 2018).

10.7.1 Addressing Weak Enforcement, Corruption, and Rural Poverty: Persistent Barriers in Combating Illegal Wildlife Trade in India

The illegal wildlife trade in India faces considerable challenges, primarily stemming from weak enforcement, corruption, and the economic vulnerability of rural communities. Despite India's rich biodiversity and a comprehensive legal framework, the enforcement of wildlife protection laws is significantly under-resourced. The Forest Department, responsible for overseeing the nation's protected areas, is underfunded and understaffed, with vast regions of national parks and reserves-over 160 million hectares-lacking proper monitoring and patrolling infrastructure (Pradhan, 2021). The Wildlife Crime Control Bureau (WCCB), tasked with combating wildlife crime, struggles due to budget constraints and is dependent on local law enforcement, which often lacks modern tools like drones, night-vision devices, and GPS tracking systems (Pande, 2021; Pradhan, 2021). As a result, there is insufficient capacity to handle the scope of wildlife trafficking, particularly given that India is a significant hub for the illegal wildlife trade worldwide and that trafficking in species including elephants, tigers, rhinos, and pangolins is common (WWF-India, 2020). Corruption within enforcement agencies compounds the issue, as it allows traffickers to circumvent legal protections through bribery, intimidation, and political influence. Verma and Sinha (2019) estimated that corruption is responsible for at least 30% of the wildlife trafficking cases in India. For instance, wildlife criminals in Kaziranga National Park, one of India's most prominent tiger reserves, exploit corrupt forest officials to facilitate the poaching and trafficking of rhino horns (WPSI, 2019). The situation is exacerbated by rural poverty, as many communities living near national parks lack alternative livelihoods and are often driven to poaching as a means of survival. According to WWF-India (2020), impoverished rural areas often rely on illegal wildlife trade because it offers immediate financial returns that outweigh the perceived risks. This situation is particularly evident in regions like the Sundarbans, where local poachers target tigers for economic gain (Sarma, 2021). To address these challenges, providing alternative livelihoods is crucial. Research by Sharma and Kumar (2020) indicates that initiatives like eco-tourism and sustainable agriculture can offer significant economic benefits while reducing dependence on illegal wildlife trade. Programs that integrate local communities into conservation efforts-such as offering employment in wildlife-friendly businesses-have shown promising results in areas like Kaziranga National Park (Sharma & Kumar, 2020). Additionally, supporting the development of non-timber forest products and promoting sustainable farming methods can provide long-term income while preserving biodiversity (Gadgil et al., 1993).

Table 10.8 Challenges in combating illegal wildlife trade

Challenge	Impact	Examples	References
Poaching and Smuggling Networks	Widespread trafficking of endangered species through well-established criminal networks and key transhipment points.	Pangolin scales, elephant ivory, tiger pelts trafficked across India-China and Southeast Asia corridors.	UNODC (2020), Pande (2021)
Weak Enforcement and Corruption	Lack of resources and corruption within enforcement agencies hinder effective wildlife protection.	Rhino poaching in Kaziranga and the Sundarbans is fueled by corruption, bribery, and insufficient funding for wildlife protection, allowing illegal activities to thrive.	Pradhan (2021), Verma and Sinha (2019), WPSI (2019)
Rural Poverty and Lack of Alternatives	Economic vulnerability of communities leads them to engage in illegal wildlife trade for short-term financial gain.	Poachers in the Sundarbans, Madhya Pradesh, and Uttarakhand, driven by poverty and lack of alternative livelihoods, engage in illegal fishing and wildlife trade, further exacerbating conservation challenges.	WWF-India (2020), Sharma and Kumar (2020), Sarma (2021)

The Table 10.8 identifies three major obstacles to stopping the illegal wildlife trade: networks of poaching and smuggling that enable the trafficking of endangered species; a lack of enforcement and corruption that impedes efforts to protect them; and rural poverty that forces people into the illegal wildlife trade because they have few other options.

10.8 Strengthening Wildlife Protection: Enhancing Laws, Penalties, and Enforcement

The effectiveness of wildlife protection laws, such as India's Wildlife Protection Act (WPA) of 1972, is crucial for curbing illegal wildlife trade and protecting biodiversity. However, the Act's enforcement remains inconsistent due to resource constraints and outdated approaches. Pradhan (2021) underscores the need to integrate modern technology into law enforcement, such as satellite imagery, drones, and AI-based tracking systems, to monitor wildlife habitats in real time. Such advancements would enable proactive interventions and enhance the detection of illegal activities. Moreover, strengthening the Wildlife Crime Control Bureau (WCCB) would be critical in consolidating national efforts to combat illegal wildlife trade. Choudhury (2020) argues that cross-border legal frameworks should be enhanced, allowing for greater collaboration between countries to address transnational wildlife trafficking. Furthermore, as proposed by Duffy (2019), a stronger integration of national legislation with international conservation standards could support India's adherence to international accords such as the Convention on International Trade in Endangered Species of Wild Fauna and Flora (CITES). Alongside improving regulatory frameworks, the creation of stricter penalties and more efficient enforcement is essential. Verma and Sinha (2019) note that weak deterrence mechanisms in India's legal system contribute to the persistence of wildlife crime. Trafficking syndicates often exploit lenient sentencing, with reduced penalties or delays in prosecution. The World Bank (2021) recommends that India adopt the more stringent penalty systems used in countries like Kenya and South Africa, where higher fines and longer prison sentences have led to a decline in wildlife crime. To further disrupt criminal networks, a multidisciplinary approach that includes collaboration between law enforcement, financial intelligence units, and international agencies is crucial. Pande (2021) highlights the need for financial tracking systems to disrupt money laundering schemes used by traffickers to launder profits through international financial systems. Finally, specialized wildlife crime courts, as proposed by Pradhan (2021), could expedite the adjudication process, ensuring consistency in handling cases and streamlining the justice system's response to wildlife trafficking. Such reforms would contribute to a more efficient and equitable legal framework for wildlife protection, ultimately strengthening India's capacity to combat illegal wildlife trade.

10.9 Conclusion

In conclusion, the wildlife trafficking in India is a complex matter that is impacted by social, ecological, and economic factors. Particularly through exports, ecotourism, and sustainable wildlife husbandry, the legal wildlife trade stimulates the economy. However, the illegal wildlife trade (IWT), which is motivated by the desire for luxury products, traditional medicine, and cultural norms, poses a severe danger to the nation's biodiversity. Because poaching of endangered animals, such tigers, rhinos, and elephants, continues to devastate ecosystems and disturb natural balances, it has significant long-term impacts on local populations and the national economy. While rhino populations remain in serious danger, illegal activities have led the population of species like tigers to drop by over 95% over the past century, according to scientific study. Economic incentives, particularly the high demand for wildlife products in both home and foreign markets, are what fuel this illicit trade. Rural residents are more vulnerable to becoming involved in wildlife trafficking networks because they often live in poverty and have few economic options. This highlights the need for sustainable revenue streams that can reduce reliance on illegal behaviour. Despite strong statutory frameworks like the Wildlife Protection Act of 1972 and international agreements like CITES, the problem is nonetheless exacerbated by weak law enforcement, corruption, and a lack of public awareness. Smuggling networks are made possible by porous borders, endangering conservation efforts and exacerbating illegal trade. To solve these issues, a holistic approach is required. It is crucial to strengthen law enforcement, enhance public education, and stiffen penalties for poaching in order to curb illegal trade. Equally important is lowering the demand for wildlife products by increasing public knowledge and offering communities viable, sustainable alternatives. Government agencies, NGOs, and local residents must work together to implement effective conservation strategies. Studies show that communities that take part in conservation programs are more likely to adopt sustainable behaviours, which ultimately contribute to the long-term preservation of biodiversity. Ultimately, both community-based remedies and legislative changes are required to combat the illegal wildlife trade in India. By combining stronger regulations, enforcement, and sustainable livelihood options, India can protect its unique biodiversity and ensure that future generations can continue to enjoy its rich natural legacy.

10.10 Future Outlook for India's Wildlife Trade

If India's wildlife trade is to shift to sustainable methods, an integrated approach that balances economic development and wildlife protection is vitally needed. This policy must include stricter enforcement of wildlife protection legislation, more severe penalties for poaching and illegal trafficking, and promotion of sustainable alternatives such as wildlife farming and ecotourism. Public awareness campaigns and community-based conservation initiatives will be crucial to involving local populations and reducing their dependence on the illicit wildlife trafficking. Increased international cooperation in the fight against cross-border wildlife trafficking will also reinforce enforcement measures. When taken as a whole, these efforts have the potential to protect India's abundant biodiversity and promote long-term, sustainable economic growth. Establishing a framework that keeps environmental harm and species extinction from being the cost of economic expansion is challenging, especially in regions near animal habitats. If these problems are appropriately handled, it is possible to prevent the illicit wildlife trade, save endangered species, and promote sustainable livelihoods.

References

Agarwal, R. (2016). Traditional medicine and the demand for wildlife products in India: Cultural practices and conservation challenges. *Journal of Ethnobiology and Ethnomedicine, 12*(1), 27.

Alves, R. R. N., da Nóbrega, W. L., da Silva Vieira, G. G., & Santana, G. G. (2008). Reptiles used in traditional folk medicine: conservation implications. *Biodiversity and Conservation, 17*, 2037–2049. https://doi.org/10.1007/s10531-007-9305-0

Andersson, A. A., Tilley, H. B., Lau, W., Dudgeon, D., Bonebrake, T. C., & Dingle, C. (2021). CITES and beyond: Illuminating 20 years of global, legal wildlife trade. *Global Ecology and Conservation, 26*, e01455. https://doi.org/10.1016/j.gecco.2021.e01455

Anoop, N. R., Sen, S., Vinayan, P. A., & Ganesh, T. (2021). Native mammals disperse the highly invasive *Senna spectabilis* in the Western Ghats, India. *Biotropica, 2021*, 1–5. https://doi.org/10.1111/btp.12996

Antonelli, A., Smith, R. J., Fry, C., Simmonds, M. S., Kersey, P. J., & Pritchard, H. W. (2020). *State of world's plants and fungi 2020*. Royal Botanic Gardens (Kew). https://doi.org/10.34885/172

Baker, S. E., & Miller, L. (2019). The impact of illegal wildlife trade on local economies: The role of financial incentives in poaching. *J Wildlife Economics, 14*(3), 45–60.

Barabanov, V. M., Gulimova, V. I., Berdiev, R. K., & Saveliev, S. V. (2019). Individual features of play behavior in thick-toed geckos in weightlessness and normal gravity conditions. *Life Sciences in Space Research (Amsterdam), 22*, 38–46.

Barnes, T. (2018). *Congo wildlife park to remain closed 'until further notice' after kidnap of British tourists*.

Bennett, E. (2018). Traditional medicine and the demand for wildlife products: A global challenge. *Journal of Traditional and Complementary Medicine, 22*(4), 185–199.

Bennett, E. (2019). The domino effect: How declines in wildlife populations impact local economies. *Journal of Environmental Economics, 45*(2), 112–129.

Bennett, E. (2020). Global trends in wildlife trade: A historical perspective. *Environmental Science & Policy, 112*, 50–58.

Berkunsky, I., Quillfeldt, P., Brightsmith, D. J., Abbud, M. C., Aguilar, J., Alemán-Zelaya, U., Aramburú, R. M., et al. (2017). Current threats faced by neotropical parrot populations. *Biological Conservation, 214*, 278–287. https://doi.org/10.1016/j.biocon.2017.08.016

Bhatnagar, S., & Rao, M. (2019). Poaching and the illegal trade of endangered species in India: Impacts on tiger and pangolin populations. *Indian J Conserv Sci., 27*(3), 215–223.

Bolwell, D., & Weinz, W. (2008). *Guide for social dialogue in the tourism Industry*. International Labour Organization.

Buckley, R. C., Morrison, C., & Castley, J. G. (2016). Net effects of ecotourism on threatened species survival. *PLoS One, 11*, e0147988. https://doi.org/10.1371/journal.pone.0147988

Cardoso, P., Amponsah-Mensah, K., Barreiros, J. P., Bouhuys, J., Cheung, H., Davies, A., et al. (2021). Scientists' warning to humanity on illegal or unsustainable wildlife trade. *Biological Conservation, 263*, 109341. https://doi.org/10.1016/j.biocon.2021.109341

Challender, D., Harrop, S. R., & MacMillan, D. C. (2015). The role of international cooperation in combating illegal wildlife trade: Challenges and solutions. *Global Conservation Enforcement, 23*(4), 333–348.

Chan, D. T. C., Poon, E. S. K., Wong, A. T. C., & Sin, S. Y. W. (2021). Global trade in parrots: Influential factors of trade and implications for conservation. *Global Ecology and Conservation, 30*, e01784. https://doi.org/10.1016/j.gecco.2021.e01784

Chavez, A. S., Lee, T. M., Maldonado, O., & Nuñez-Regueiro, M. (2023). Wildlife tourism and its unintended contribution to the illegal wildlife trade: The case of souvenir demand. *Conservation Science and Practice, 5*(4), e1289.

Chitale, V. S., Behera, M. D., & Roy, P. S. (2014). Future of endemic flora of biodiversity hotspots in India. *PLoS One, 9*, e115264. https://doi.org/10.1371/journal.pone.0115264

Choudhury, A. (2020). Wildlife in religious ceremonies and cultural rituals in India: Challenges to conservation. *Indian Journal of Environmental Sciences, 28*(1), 102–117.

CITES. (2014). *Interpretation and implementation of the convention: Species trade and conservation: Asian big cats*. CITES Sixty-fifth meeting of the Standing Committee (SC65), Doc. 38.

CITES. (2019). *Harvest and ranching of Nile crocodiles in Kenya*. CITES & Livelihoods Case Study.

CITES. (2020). *Annual report on the implementation of CITES*. Convention on International Trade in Endangered Species of Wild Fauna and Flora.

CITES. (2025). *CITES and the regulation of international wildlife trade: Annual report 2025*. CITES Secretariat.

Das, A. (2020). Crocodile farming and its role in wildlife trade in India. *Journal of Wildlife Conservation, 31*(2), 112–118.

Dickinson, H. (2022). Caviar matter(s): The material politics of the European caviar grey market. *Political Geography, 99*, 102737. https://doi.org/10.1016/j.polgeo.2022.102737

Duffy, R. (2018a). The global wildlife trade: A review of trends and strategies for conservation. *Journal of Conservation Biology, 32*(6), 1172–1184.

Duffy, R. (2018b). The global demand for wildlife and its impact on conservation. *Environmental Science & Policy, 86*, 104–110.

Duffy, R. (2019). Global environmental governance and the illegal wildlife trade: A policy perspective. *Conservation and Society, 17*(3), 145–158.

Duffy, R., St John, F. A., Buscher, B., & Brockington, D. (2015). Toward a new understanding of the links between poverty and illegal wildlife hunting. *Conservation Biology, 30*(1), 14–22. https://doi.org/10.1111/cobi.12622

European Commission. (2020). *Pollinators and agriculture: A growing concern for the economy*. European Commission Press Release.

Fa, J. E., Currie, D., & Meeuwig, J. (2003). Bushmeat and food security in The Congo Basin: linkages between wildlife and people's future. *Environmental Conservation, 30*, 71–78. https://doi.org/10.1017/S0376892903000067

Funk, S. M., Fa, J. E., Ajong, S. N., Eniang, E. A., Dendi, D., Nasi, R., et al. (2022). Impact of COVID-19 on wild meat trade in Nigerian markets. *Conservation Science and Practice, 4*, e599. https://doi.org/10.1111/csp2.599

Gadgil, M., Berkes, F., & Folke, C. (1993). Indigenous knowledge for biodiversity conservation: Approaches and applications. *Ecological Applications, 3*(4), 900–918.

Gao, Y., & Clark, S. G. (2014). Elephant ivory trade in China: trends and drivers. *Biological Conservation, 180*, 23–30. https://doi.org/10.1016/j.biocon.2014.09.020

Goswami, V. R., Yadava, M. K., Vasudev, D., Prasad, P. K., Sharma, P., & Jathanna, D. (2019). Author correction: Towards a reliable assessment of Asian elephant population parameters: The application of photographic spatial capture-recapture sampling in a priority floodplain ecosystem. *Scientific Reports, 9*, 1. https://doi.org/10.1038/s41598-019-51389-1

Gulati, S., Karanth, K. K., Le, N. A., & Noack, F. (2021). Human casualties are the dominant cost of human-wild life conflict in India. *Proceedings of the National Academy of Sciences of the United States of America, 118*, 8.

Gupta, S., & Dey, S. (2017). The Role of Organized Crime in Wildlife Trafficking in India: Challenges and Solutions. *Indian Journal of Environmental Law, 14*(2), 53–70.

Harrison, R. D. (2011). Emptying the forest: hunting and the extirpation of wildlife from tropical nature reserves. *Bioscience, 61*, 919–924. https://doi.org/10.1525/bio.2011.61.11.11

Hayman, G., & Brack, D. (2002). *International environmental crime: The nature and control of environmental black markets*. Royal Institute of International Affairs.

Hogg, C. J., Dennison, S., Frankham, G. J., Hinds, M., & Johnson, R. N. (2018). Stopping the spin cycle: genetics and bio-banking as a tool for addressing the laundering of illegally caught wildlife as 'captive-bred'. *Conservation Genetics Resources, 10*, 237–246. https://doi.org/10.1007/s12686-017-0784-3

Houghton, R., Inoue, H., & Singh, A. (2018). Alternative materials for ivory: Promoting sustainable practices in cultural ceremonies. *Global Conservation Review, 14*(3), 64–76.

Hughes, M. (2021). Wildlife trade: Biodiversity loss and the ecological impacts of illegal wildlife trade in India. *Environment Conservation Review, 34*(2), 112–130. https://doi.org/10.1016/j.cub.2021.08.056

Hulme, P. E. (2021). Unwelcome exchange: international trade as a direct and indirect driver of biological invasions worldwide. *One Earth, 4*(5), 666–679. https://doi.org/10.1016/j.oneear.2021.04.015

Indian Ministry of Environment, Forest and Climate Change. (2021). *Status of wildlife crime and illegal trade in India: Annual report 2021*. Government of India. https://moef.gov.in

IUCN. (2021). *The impact of illegal wildlife trade on endangered species in India*. International Union for Conservation of Nature. Available from: www.iucn.org

Jadhav, S., & Barve, A. (2020). Building community resilience: How economic incentives and capacity building support conservation efforts. *Environmental Management for Sustainability, 18*(1), 98–110.

Karanth, K. U. (2017). Conservation and poaching: Examining the struggles for India's iconic species. *Journal of Asia-Pacific Biodiversity, 45*(4), 346–360.

Karanth, K. K., Gopalaswamy, A. M., DeFries, R., & Ballal, N. (2012). Assessing patterns of human-wildlife conflicts and compensation around a central Indian protected area. *PLoS One, 7*, e50433. https://doi.org/10.1371/journal.pone.0050433

Kasterine, A., Arbeid, R., Caillabet, O., & Natusch, D. (2012). *The trade in southeast asian python skins*. International Trade Centre (ITC).

Khalsa, R. (2023). India's role in the global wildlife trade and the socio-economic drivers behind illegal wildlife trade. *Asian Journal of Environmental Studies, 41*(3), 223–240.

Kolipakam, V., Singh, S., Ray, S., Prasad, L., Roy, K., Wakid, A., et al. (2020). Evidence for the continued use of river dolphin oil for bait fishing and traditional medicine: implications for conservation. *Heliyon, 6*(8), e04690. https://doi.org/10.1016/j.heliyon.2020.e04690

Kumar, S. (2020). The growing threat of the illegal wildlife trade in India: An analytical study. *Journal of Environmental Protection, 8*(3), 22–33.

Kumar, V., & Dey, A. (2019). Economic impacts of pollinator decline in agricultural systems: A case study on bees and bats. *Journal of Agricultural Sustainability, 38*(5), 334–345.

Kumar, S., & Gupta, R. (2018). Wildlife trafficking in the Indian subcontinent: Transnational crime and global demand. *Asian Journal of Criminal Law, 4*(1), 105–120.

Kuswanda, A., Rahim, Z., & Singh, T. (2023). The nexus between corruption and illegal wildlife trade in India. *Asian Environmental Law Journal, 10*(2), 105–120.

Law, Y. H. (2019). New tarantula highlights illegal trade in spiders. *Science, 363*(6430), 914–915. https://doi.org/10.1126/science.363.6430.914

Lembo, T., Olival, K. J., & LeBreton, M. (2018). The role of middlemen in illegal wildlife trade: Investigating the smuggling networks. *Nature Sustainability, 1*(5), 173–179.

Liew, M., Jayasinghe, N., & Smit, D. (2021). Economic inequality and its role in international wildlife trafficking. *Global Trade Sustainability Development, 28*(5), 512–529.

Lim, P. H. C. (2017). Asian herbals and aphrodisiacs used for managing ED. *Translational Andrology and Urology, 6*(2), 167–175. https://doi.org/10.21037/tau.2017.04.04

Lyons, J. A., & Natusch, D. J. (2011). Wildlife laundering through breeding farms: Illegal harvest, population declines and a means of regulating the trade of green pythons (Morelia viridis) from Indonesia. *Biological Conservation, 144*, 3073–3081. https://doi.org/10.1016/j.biocon.2011.10.002

Maekawa, M., Lanjouw, A., Rutagarama, E., & Sharp, D. (2013). Mountain gorilla tourism generating wealth and peace in post-conflict Rwanda. *Natural Resources Forum, 37*, 127–137. https://doi.org/10.1111/1477-8947.12020

Magliolo, M., Prost, S., Orozco-terWengel, P., et al. (2021). Unlocking the potential of a validated single nucleotide polymorphism array for genomic monitoring of trade in cheetahs (Acinonyx jubatus). *Molecular Biology Reports, 48*, 171–181. https://doi.org/10.1007/s11033-020-06030-0

Mander, M., Ntuli, L., Diederichs, N., & Mavundla, K. (2007). Economics of the traditional medicine trade in South Africa care delivery. *South African Health Review, 2007*, 189–196.

Margulies, J. D., Moorman, F. R., Goettsch, B., et al. (2022). Prevalence and perspectives of illegal trade in cacti and succulent plants in the collector community. *Conservation Biology, 37*, e14030. https://doi.org/10.1111/cobi.14030

Marker, L. (2019). Loving a species to death. *Biodiversity, 20*, 50–55. https://doi.org/10.1080/14888386.2019.1591300

Mehta, K., Sharma, D., & Joshi, A. (2018). Ayurveda, Unani, and the wildlife trade: Traditional medicine systems and endangered species in India. *J Tradit Med Ethics., 14*(2), 112–120.

Mohanty, S., & Patra, D. (2021). Legal frameworks in wildlife conservation: The case of India. *Journal of Environmental Law Policy, 26*(3), 301–320.

Morton, D., et al. (2021). Impact of illegal wildlife trade on ecosystems and species: A case study from India. *Biodiversity and Conservation Journal, 32*(3), 135–144.

Musing, L., Harris, L., Williams, A., et al. (2019). *Corruption and wildlife crime: A focus on caviar trade*. Traffic.

Nalayini, M. (2024). The environmental and regulatory costs of illegal wildlife trade in India. *Journal of Biodiversity and Conservation, 19*(2), 117–132.

Nasi, R., & Fa, J. E. (2015). *The role of bushmeat in food security and nutrition*. XIV World Forestry Congress.

National Tiger Conservation Authority (NTCA). (2020). *Tiger poaching and illegal trade in India: Report 2018–2020*. Ministry of Environment, Forest and Climate Change. https://ntca.gov.in

Nellemann, C., Henriksen, R., Raxter, P., et al. (2014). *The environmental crime crisis: Threats to sustainable development from illegal exploitation and trade in wildlife and forest resources*. United Nations Environment Programme.

Nellemann, C., Henriksen, R., Kreilhuber, A., et al. (Eds.). (2016). *The rise of environmental crime: A growing threat to natural resources, peace, development and security*. UNEP.

Nijman, V., & Nekaris, K. A. I. (2017). The Harry Potter effect: The rise in trade of owls as pets in Java and Bali, Indonesia. *Global Ecology Conservation, 11*, 84–94. https://doi.org/10.1016/j.gecco.2017.04.004

Nittu, G., Shameer, T. T., Nishanthini, N. K., & Sanil, R. (2022). The tide of tiger poaching in India is rising! An investigation of the intertwined facts with a focus on conservation. *GeoJournal, 88*, 1–14. https://doi.org/10.1007/s10708-022-10633-4

Pande, S. (2021). Improving enforcement resources and tools for wildlife protection in India. *Indian Journal of Environmental Law, 34*(1), 23–35.

Phelps, J., & Webb, E. L. (2015). "Invisible" wildlife trades: Southeast Asia's undocumented illegal trade in wild ornamental plants. *Biological Conservation, 186*, 296–305. https://doi.org/10.1016/j.biocon.2015.03.030

Phelps, J., Piana, P., & Gallo, J. (2016). Coordinated approaches to wildlife conservation: Combining enforcement and community-based alternatives. *Environmental Conservation, 43*(3), 245–253.

Pradhan, K. (2021). Enhancing law enforcement in wildlife protection: The need for technological integration. *Indian Journal of Environmental Law, 34*(1), 23–35.

Prasad, A., Sharma, P., & Bhardwaj, R. (2022). The economics of illegal wildlife trade in India: An analysis of market drivers and policy responses. *Journal of Environmental Economics and Policy, 11*(4), 367–385.

Raja, V. (2024). The economic incentives driving India's illegal wildlife trade. *Conservation Development, 22*(1), 72–90.

Rajpoot, A., Kumar, V. P., Bahuguna, A., et al. (2018). Wildlife forensics in battle against veneration frauds in Uttarakhand, India: Identification of protected Indian monitor lizard in items available in the local market under the name of Hatha Jodi. *Mitochondrial DNA Part B Resources, 3*(2), 925–932. https://doi.org/10.1080/23802359.2018.1501284

Ramakrishnan, U. (2014). Indigenous communities and forest resource management in India: A study on the Khasi Tribe of Meghalaya. *Environmental Science & Policy, 41*, 79–88.

Roe, D., & Booker, F. (2019). Engaging local communities in tackling illegal wildlife trade: A synthesis of approaches and lessons for best practice. *Conservation Science and Practice, 1*, e26. https://doi.org/10.1111/csp2.26

Romero-Daza, N. (2002). Traditional medicine in Africa. *Annual American Academy of Political and Social Science, 583*, 173–176. https://doi.org/10.1177/000271620258300111

Ruenes, D., McEvoy, J., & Bell, K. (2023). Souvenir consumption and wildlife exploitation: A review of tourism's role in the trade of animal-derived products. *Tourism Management Perspectives, 45*, 101079.

Sarma, N. (2021). Rural poverty and the illegal wildlife trade: A case study from the Sundarbans. *Journal of Rural Development, 40*(2), 87–98.

Sethi, R. (2018). The role of intermediaries in India's illegal wildlife trade. *Asian Journal of Criminology, 11*(3), 178–190.

Shaffer, L. J., Khadka, K. K., Van Den Hoek, J., & Naithani, K. J. (2019). Human-elephant conflict: A review of current management strategies and future directions. *Frontiers in Ecology and Evolution, 6*, 235. https://doi.org/10.3389/fevo.2018.00235

Shairp, R., Veríssimo, D., Fraser, I., et al. (2016). Understanding urban demand for wild meat in Vietnam: Implications for conservation actions. *PLoS One, 11*, e0134787. https://doi.org/10.1371/journal.pone.0134787

Shairp, et al. (2025). The bushmeat trade and its implications on wildlife. *Global Environmental Impact Studies*, 1(1):1–15

Sharma, S. (2020). Human-wildlife conflict and the economic costs of wildlife depletion in India. *Conservation and Development Review, 21*(1), 98–115.

Sharma, R. (2021). Social media and its role in promoting illegal wildlife products. *J Environmental Ethics, 22*(1), 81–94.

Sharma, S., & Kumar, R. (2020). Leveraging economic alternatives to reduce illegal wildlife trade in India. *Biodiversity and Conservation, 29*(7), 1875–1886.

Sharma, C. P., Kumar, A., Vipin, V., et al. (2018). Online selling of wildlife part with spurious name: A serious challenge for wildlife crime enforcement. *International Journal of Legal Medicine, 133*, 65–69. https://doi.org/10.1007/s00414-018-1795-7

Shrestha, B., Bhatta, K., & Ghimire, S. (2016). Use of endangered species in traditional medicine in South Asia: Implications for wildlife conservation. *Biodiversity and Conservation, 25*(2), 365–378.

Singh, H. P., Batish, D. R., Dogra, K. S., et al. (2014). Negative effect of litter of invasive weed Lantana camara on structure and composition of vegetation in the lower Siwalik Hills, northern India. *Environmental Monitoring and Assessment, 186*(6), 3379–3388. https://doi.org/10.1007/s10661-014-3624-x

Singh, A., Ghosh, A., Dolker, S., et al. (2021). Species identification from seized animal oil: A case study of suspected Gangetic dolphin (*Platanista gangetica*). *International Journal of Legal Medicine, 135*(4), 1413–1416. https://doi.org/10.1007/s00414-021-02574-3

Singh, R., Sharma, A., & Pande, H. (2023). Illegal wildlife trade and its impact on biodiversity: India's role in global networks. *Journal of Conservation Science, 41*(2), 199–215.

Sinha, R., & Patil, V. (2020). The role of traditional medicine in India's illegal wildlife trade: Regional practices and policy gaps. *J Wildl Trade Policy., 8*(3), 201–210.

Sosnowski, M., & Petrossian, G. A. (2020). Luxury fashion wildlife contraband in the USA. *EcoHealth, 17*, 94–110. https://doi.org/10.1007/s10393-020-01467-y

Still, J. (2003). Use of animal products in traditional Chinese medicine: Environmental impact and health hazards. *Complementary Therapies in Medicine, 11*, 118–124. https://doi.org/10.1016/S0965-2299(03)00055-4

Symes, W. S., Edwards, D. P., Miettinen, J., Rheindt, F. E., & Carrasco, L. R. (2017). Combined impacts of deforestation and wildlife trade on tropical biodiversity are severely underestimated. *Nature Communications, 9*(1), 4052.

TRAFFIC. (2020). *The economic costs of wildlife trafficking: A global overview*. TRAFFIC International.

TRAFFIC. The role of digital platforms in facilitating illegal wildlife trade: A case study of Facebook. TRAFFIC Report; 2021.

Tricorache, P., Nowell, K., Wirth, G., et al. (2018). Pets and pelts: Understanding and combating poaching and trafficking in cheetahs. In *Cheetahs: Biology and conservation* (pp. 191–205). Elsevier.

van Uhm, D. P. (2016). Monkey business: The illegal trade in Barbary macaques. *Journal of Trafficking, Organized Crime and Security, 2*, 36–49.

UNODC. (2012). *Wildlife and forest crime analytic toolkit – revised edition*. United Nations Office on Drugs and Crime.

UNODC. (2016). *World wildlife crime report: Trafficking in protected species*. United Nations.

UNODC. (2025). *Global wildlife seizures and trafficking statistics*. UNODC Report.
UNODC. (2020). *World wildlife crime report 2020: Trafficking in protected species*. United Nations.
Uprety, N., Singh, B., & Bhattacharya, R. (2021). Corruption and weak enforcement mechanisms in India's wildlife protection system: A review. *Journal of Environmental Law and Policy, 24*(1), 34–48.
Veríssimo, D., MacMillan, D., & Tovar, D. (2020). Tackling the demand for wildlife products: A global approach to reducing illegal wildlife trade. *Conservative Sustainability Journal, 38*(2), 77–91.
Verma, S., & Sinha, A. (2019). Corruption and its impact on wildlife protection in India: A policy review. *Environmental Policy Journal, 29*(2), 67–82.
Vu, H. N. D., & Nielsen, M. R. (2018). Understanding utilitarian and hedonic values determining the demand for rhino horn in Vietnam. *Human Dimensions of Wildlife, 23*(5), 417–432. https://doi.org/10.1080/10871209.2018.1449038
Walpole, M. J., & Goodwin, H. J. (2001). Local attitudes towards conservation and tourism around Komodo National Park, Indonesia. *Environmental Conservation, 28*, 160–166. https://doi.org/10.1017/S0376892901000169
WCCB (Wildlife Crime Control Bureau). (2021). *Annual report on wildlife trafficking in India 2021*. Ministry of Environment, Forest and Climate Change. https://wccb.gov.in
WHO. (2013). *WHO traditional medicine strategy 2014–2023*. World Health Organization.
WHO. (2014). *Ebola virus outbreak: The link between wildlife trade and emerging zoonotic diseases*. WHO Report.
WHO (World Health Organization). (2019). *Traditional medicine and the wildlife trade: Impacts on biodiversity and human health*. World Health Organization.
World Bank. (2019). *Illegal logging, fishing, and wildlife trade: The costs and how to combat it*. World Bank.
World Bank. (2021). *Legal and institutional frameworks for combating wildlife crime*. World Bank Report.
WPSI. (2019). *Corruption and its impact on wildlife conservation efforts in India*. Wildlife Protection Society of India.
WWF. (2014). *More tigers in American backyards than in the wild*. https://www.worldwildlife.org/stories/more-tigers-in-american-backyards-than-in-the-wild
WWF India. (2018). *Hotspot analysis of illegal wildlife trade in Madhya Pradesh*. WWF India.
WWF India. (2021). *India's role in the smuggling and trafficking of elephant ivory: A case study of the North-East*. WWF India Wildlife Report.
WWF-India. (2020). *Understanding the link between poverty and illegal wildlife trade in India*. WWF-India.
Wyatt, T. (2018). Exploring the organization of Russia Far East's illegal wildlife trade: Two case studies of the illegal fur and illegal falcon trades. *Global Crime, 10*, 144–174. https://doi.org/10.1080/17440570902783947
Zhou, X. H., Wan, X. T., Jin, Y. H., & Zhang, W. (2016). Concept of scientific wildlife conservation and its dissemination. *Zoological Research, 37*(5), 270–274. https://doi.org/10.13918/j.issn.2095-8137.2016.5.270
Ziegler, S. (2010). Application of food balance sheets to assess the scale of the bushmeat trade in Central Africa. *Traffic Bull., 22*, 105–116.
Ziegler, S. (2025). *Ornamental plants and illegal logging*. Conservation Science and Practice.
Zimmerman, M. E. (2003). The black market for wildlife: combating transnational organized crime in the illegal wildlife trade. *Vanderbilt Journal of Transnational Law, 36*, 1657–1689.

Chapter 11
Sustainable Models for Wildlife Trade and Livelihood Alternatives

Rajinder Kaur and Abhinav Saxena

Contents

R. Kaur
Department of Zoology, Government Mohindra College, Patiala, Punjab, India

A. Saxena (✉)
Department of Zoology and Environmental Sciences, Punjabi University, Patiala, Punjab, India

A. Poonia et al. (eds.), *Wildlife Trade in India*,
https://doi.org/10.1007/978-981-95-3694-8_11

Abstract Wild animal and plant trade, whether legal or illegal, presents a huge challenge to financial security as well as to the conservation of biological diversity. The following chapter explores sustainable models that balance conservation efforts with livelihood opportunities for communities dependent on wildlife resources. With the application of advanced techniques such as orbital monitoring, distributed ledger-based monitoring, and scientific inspection, conservation initiatives can be improved. In addition to this, the enforcement of law and order with a strong mandate, public education campaigns, and stricter regulations must be implemented in order to mitigate illegal trade.

Community-based conservation, such as ecotourism, regulated harvesting of non-timber forest products, and alternative livelihoods, provide moral and economically viable alternatives. Indian experiences demonstrate successful models, such as the rhino conservation program at Kaziranga, wildlife stewardship by the Bishnoi community, and community forest rights in Maharashtra. The models acknowledge the significance of integrating local participation, policy enforcement, and sustainable economic activities.

An integrated approach-combining legal structures, technological solutions, and economic incentives-can make wildlife conservation sustainable in the long term and enrich livelihoods. Strengthening enforcement, promoting international cooperation, and promoting sustainable trade policy are essential in attaining ecological integrity and human well-being and eventually assuring a more ethical and resilient future for people and wildlife.

Keywords Wildlife conservation · Illegal wildlife trade · Sustainable livelihoods · Community-based conservation · Biodiversity protection · Eco-tourism · Forest management · Law enforcement · Technological innovations · Public awareness

11.1 Introduction

India is a nature that has many varieties of plants and animals. India is home to some of the most famous mammals, colorful birds, rare reptiles, and unique plants-some of which are found nowhere else in the world. The ecosystems which support wildlife and offer benefits like that of water purification, carbon storage, and the enrichment of soil. For millions, especially in rural and indigenous communities, these resources

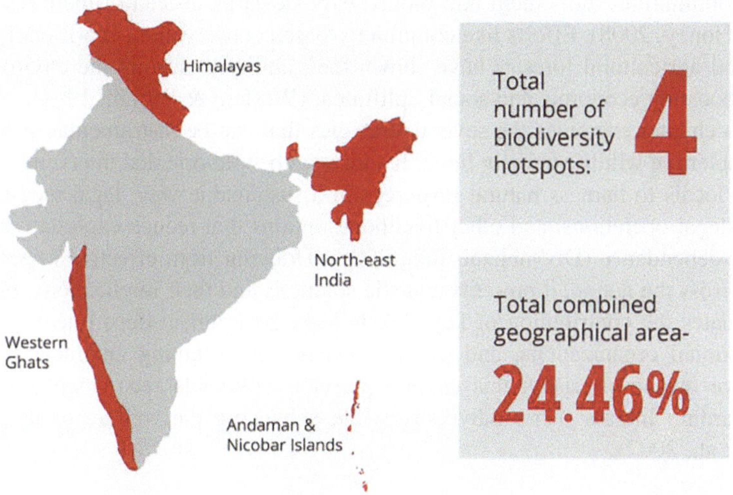

Fig. 11.1 Map of India's biodiversity hotspots highlighting key regions affected by wildlife trade

are essential to their life. They depend on forests for food, use traditional fishing and farming methods, and have a strong bond with nature which is essential for their well-being and existence (Myers et al., 2000; Rangarajan & Shahabuddin, 2006).

India is remarkably rich in biodiversity (Fig. 11.1). However, illegal wildlife trade is emerging as one of the most serious threats to it. This underground trade exploits the rich fauna and flora of the country for monetary gain, threatening the existence of many species and disturbing ecosystems. The pelts and bones of tigers, scales of pangolins and numerous medicinal plants are in high demand. Many of these are being unsustainably harvested. This is posing serious ecological and economic problems in the country. The disappearance of a key species can trigger a chain reaction that alters the entire ecosystem, while communities dependent on these ecosystems for their income face lack of resources and increasing hardships (Divyabhanusinh., 2005; Karanth, 2006).

Confronting this crisis needs something more than the strict application of wildlife protection laws, which are still vital. Laws, such as the Wildlife Protection Act, 1972 or international agreements like CITES, are quite useful but problems with enforcement, such as lack of resources or lack of coordination between agencies, reduce their efficacy (Berkes, 2004). To be sustainable, punishment alone is insufficient; the people that live in and near wildlife areas should be involved. With their unique knowledge, these communities can contribute greatly if their views and livelihoods are brought into the fold of long-term solutions for poaching and habitat destruction. They are the first line of defense against such threats.

To achieve success in long-term conservation definite models are needed, which should include environmental protection along with human well-being. Providing alternative means of livelihood, these measures address the economic incentive behind illegal wildlife trade. Providing incentives and decision-making power to local communities turns them into biodiversity stewards instead of mere resources users (Honey, 2008). Efforts like community-based conservation, eco-friendly tourism, and agricultural forestry have shown their ability to sustain the environment while boosting economic and social upliftment (Western & Wright, 1994).

This chapter examines the several strategies that can be instrumental in solving the problem of wildlife trade in India. It addresses people-oriented interventions that enable locals to harness natural resources in a sustainable way, legal measures to deter illegal operations, and other livelihood options that reduce exploitative operations' dependence (Divyabhanusinh., 2005). Drawing from effective experiences from across the nation, it presents realistic solutions and their implications. Further, it evaluates the contribution of key stakeholders-government departments, NGOs, international organizations, and private agencies-in designing an integrated and effective response. Lastly, this chapter presents actionable recommendations for safeguarding India's rich biodiversity while enhancing the welfare of its people (Roe et al., 2002).

11.2 The Challenges of Wildlife Trade

Wildlife trade, i.e., global trade in plant and animal species, is a highly complicated and contentious phenomenon. It carries profound economic, environmental, and ethical implications. Unregulated and regulated markets, and illegal trade, place tremendous pressures on world security, international public health, and biodiversity. Its complication is something we should appreciate if we have to formulate realistic plans of reducing its impacts.

11.2.1 Ecological Impacts

Among the serious issues that trouble wildlife trade is its adverse effect on biodiversity. Over-exploitation of species for trade can cause severe depletion in their numbers and, in worst case scenarios, leads to extinction. For instance, illegal ivory trade has been a leading cause of decreasing the number of elephants in Africa, and pangolins in India (Challender et al., 2015) (Fig. 11.2) have been similarly threatened due to high demand. Likewise, overexploitation of certain plant species for medicine has pushed them to the brink of extinction, disrupting the balance of the ecosystem and the services they render (Roe et al., 2020).

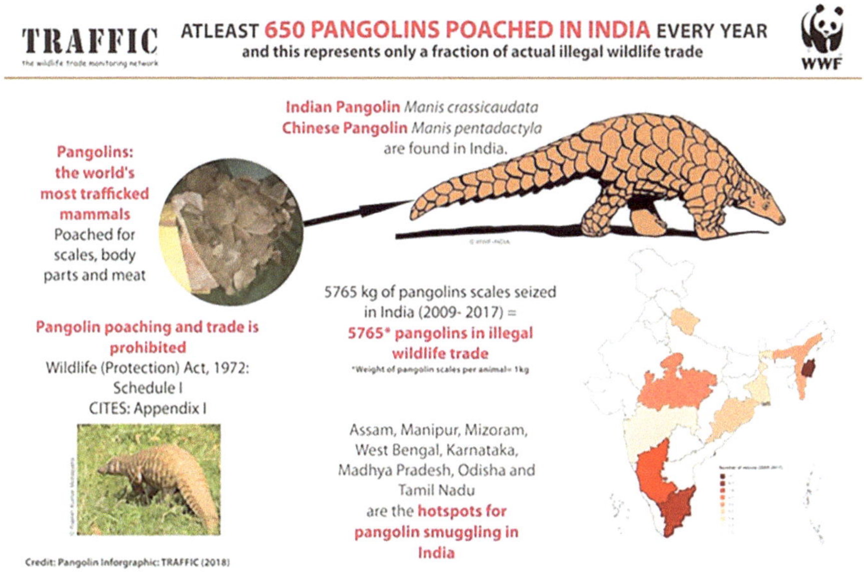

Fig. 11.2 Endangered species most affected by wildlife trade. (Source: TRAFFIC, 2018)

11.2.2 Economic and Social Impacts

The economic impacts of wildlife trade have far-reaching, both positive and negative, impacts. On the one hand, legalized wildlife trade benefits the economy through sustainable livelihood activities. On the other hand, illegal trade avoids taxation and the regulatory system and therefore causes huge economic losses. Additionally, it targets poor communities, compelling them to engage in illegal activities (Challender et al., 2019). Besides the economic impacts, the social impact is equally poignant, particularly on the health front. The zoonotic disease outbreaks, as seen with the COVID-19 pandemic, have been attributed to wildlife trade, emphasizing the danger of illegal trade (Karesh et al., 2020).

11.2.3 Regulatory and Enforcement Problems

Wildlife trade is difficult to regulate due to its complexity and international nature. Global regulations such as CITES seek to regulate and oversee trade (Table 11.1). It is still difficult to enforce, due to inadequate resources, corrupt mechanisms, and government failure (Esmail et al., 2020). Furthermore, economic gains from illegal wildlife trade are lucrative, inviting criminal syndicates to search for and exploit loopholes in existing control systems, rendering it even harder to eliminate the illegal trade.

Table 11.1 Comparison of legal vs. illegal wildlife trade. (Esmail et al., 2020)

Aspect	Legal wildlife trade	Illegal wildlife trade
Regulation	Governed by national and international laws (e.g., CITES).	Operates outside legal frameworks, avoiding regulations.
Economic impact	Generates revenue for legal businesses and communities.	Supports black markets, leading to economic instability.
Ecological consequences	Managed harvesting ensures sustainability.	Leads to biodiversity loss and ecosystem disruption.
Social impact	Creates jobs and promotes ethical practices.	Exploits vulnerable communities and endangers human health.

11.2.4 Ethical Considerations

Wildlife trade is of very serious moral concern, particularly as it concerns animal welfare subjected to capture, transport, and captivity. Most animals suffer tremendous stress and agony, and there has been broad disagreement over animal welfare. With growing public concern over these issues, so too has the demand for stricter controls and the promotion of more sustainable, humane options (Rosen & Smith, 2010).

11.2.5 The Path Forward

Successfully solving wildlife trade problems requires a master plan. Enabling global cooperation, restoring law enforcement, and community-based conservation are all indispensable steps. Enhancing public education can be supplemented by awareness programs and alternative livelihoods offered to wildlife trade-reliant communities, which can also help in reducing demand and towards sustainability (Roe et al., 2020).

11.3 Principles of Sustainable Wildlife Trade

A properly managed wildlife trade aims to balance environmental protection, economic benefit, and social equity. It is based on sustainable use of resources, with a focus on safeguarding wildlife populations and habitats for future generations. A few of the major features of sustainable wildlife trade are the safeguarding of ecological balance, imposing legal protection, and involvement of local communities in conservation (Fig. 11.3).

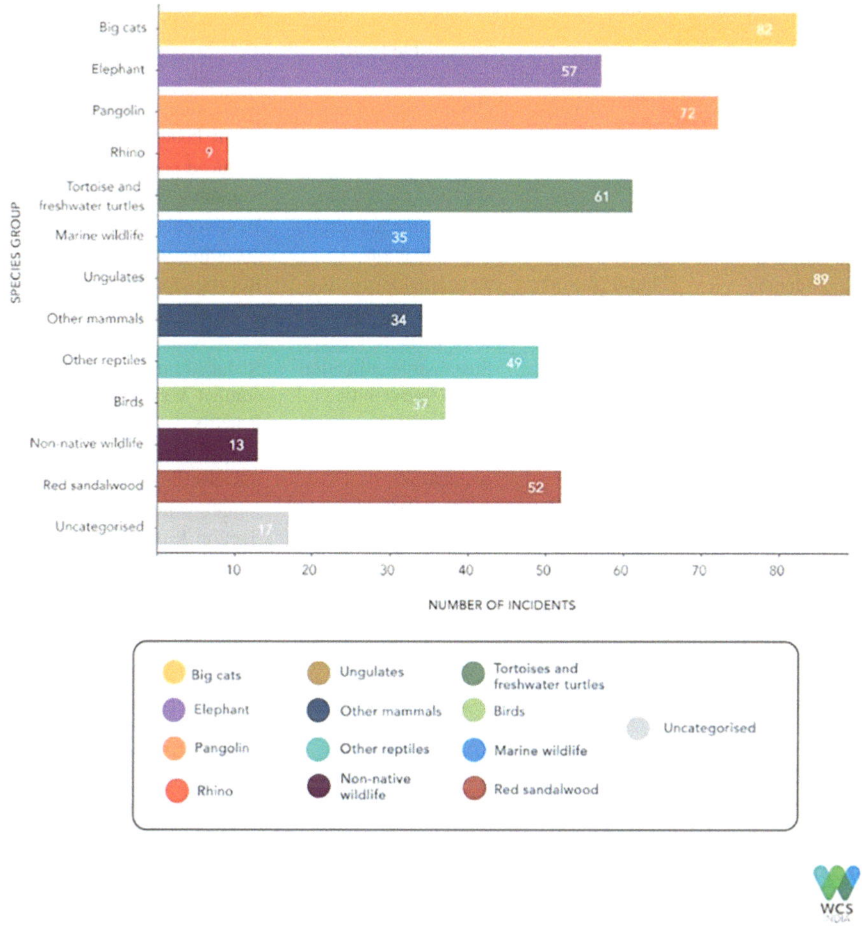

Fig. 11.3 Comparison between regions suffering from poaching vs. areas with successful conservation efforts. (Source: WCS, India, 2020)

11.3.1 Ecological Sustainability

The ecological sustainability principle promotes the wise control of wildlife populations to ensure that exploitation does not exceed the natural renewing capacity of the species. This balance demands careful scientifically driven analysis to define sustainable harvesting and prevent resource degradation. A good example is controlled trade in some plants and animals between nations such as South America, Australasia, and India, which has demonstrated the confluence of conservation and economic benefits (Roe et al., 2020) (Table 11.2).

Table 11.2 Sustainable wildlife trade models. (Roe et al., 2020)

Species	Country/Region	Sustainable trade strategy	Outcome
Vicuña (Camelid)	South America	Community-managed fiber harvesting	Population recovery & economic benefits.
Crocodiles	Australia	Regulated farming for skin and meat trade	Reduced poaching, economic gain.
Medicinal plants	India	Sustainable harvesting & certification	Biodiversity conservation & economic upliftment.
Red Sanders (*Pterocarpus santalinus*)	India (Andhra Pradesh)	Regulated harvesting under strict quotas	Revenue for local communities & conservation.
Honey production	India (Sundarbans)	Sustainable collection of mangrove honey	Reduced human-wildlife conflict & economic benefits.
Agarwood (*Aquilaria* species)	India (Assam, northeast)	Community-managed plantations	Supports livelihoods & reduces illegal harvesting.
Snake venom extraction	India	Ethical and regulated venom farming	Supports anti-venom production & conservation.
Sandalwood farming	India (Karnataka, Tamil Nadu)	Legal cultivation with government oversight	Prevents illegal logging & provides revenue.

11.3.2 Legal and Regulatory Frameworks

A strong legal system is the essence of making wildlife trade sustainable. International treaties like CITES are essential tools for controlling trade and curbing illegal activities. But the success of such measures is dependent on collaborative action from governmental, non-governmental, and local agencies. Implementing policies more strictly, closing loopholes in law, and curbing corruption are essential measures for strengthening such initiatives (Esmail et al., 2020).

11.3.3 Community Involvement

Active involvement of the local communities is essential to sustainable wildlife trade's long-term viability. The giving of ownership to such communities as well as issuing them monetary benefits can encourage the communities to engage more in wildlife conservation activities. Various models of community-based conservation, such as managed trophy hunting programs in some regions of Africa, have proved to be of potential in serving biodiversity conservation while fostering economic growth (Challender et al., 2019).

11.3.4 Ethical Considerations

Ethical values are an intrinsic component of sustainable wildlife trade, involving humane treatment of the animals at the time of capture, transport, and commercial sale. Public awareness through targeted public education can influence consumer demand and a need for ethically and sustainably sourced wildlife products. It can be an effective tool to put an end to the market demand for illegally traded species (Rosen & Smith, 2010).

11.4 Livelihood Alternatives to Unsustainable Wildlife Trade

Providing sustainable livelihood options to unsustainable wildlife trade groups is important in promoting a decrease in dependence on illicit activity. Poverty can be minimized, dependence on wildlife exploitation can be reduced, and sustainable conservation can be facilitated through economic options.

11.4.1 Promoting Sustainable Agriculture and Ecotourism

Building ecologically sustainable farming practices and nature-based ecotourism has been successful methods of income diversification in communities. Through growing high-value crops or through ecotourism activities with the aim of conservation, communities can be economically sustainable without wildlife exploitation. Kenya's community-based ecotourism has been a successful model, as it has promoted the conservation of threatened species like elephants and rhinos by associating conservation with economic incentives (Honey, 2008).

11.4.2 Handicrafts and Non-Timber Forest Products (NTFPs)

Promotion of handicraft production and promotion of sustainable harvesting of non-timber forest products (NTFPs) provide economically sustainable opportunities for local communities while conserving biodiversity. Sustainably harvested products such as honey, medicinal herbs, and hand-carved wood products provide alternative sources of income that do not destroy natural ecosystems. By combining traditional knowledge and sustainable harvesting, these programs provide a means of ensuring that local resources are sustainably harvested. Furthermore, exposure to wider markets, including environmentally aware consumers and fair-trade networks, enhances the economic sustainability of these sustainable activities. For instance, programs across various regions have proved that well-managed NTFP harvesting can yield economic stability and conservation dividends (Roe et al., 2020).

11.4.3 Capacity Building and Education

Capacity development programs and education initiatives are a core investment aimed at facilitating communities to transition to sustainable livelihoods. By facilitating the exchange of practical skills and technical know-how, the programs ensure access to long-term economic autonomy with diminished dependence on environmentally destructive methods. Training schemes have to prioritize sustainable use of resources, entrepreneurship, and market-access mechanisms for empowering local communities to participate meaningfully in alternative economic activities.

For example, organized workshops in agroforestry, green craftsmanship, and responsible tourism management have empowered communities to establish sustainable businesses. Besides, bridging with cooperatives and fair-trade networks provides access to markets for sustainably produced commodities. Where unsustainable wildlife trade has been a main source of income in some areas, such intervention has been beneficial in promoting economic resilience and environmental sustainability (Elliott et al., 2014).

11.4.4 Policy Support and Market Access

The role of government policies and market intervention is crucial in triggering the transition towards sustainable livelihoods, particularly for those communities that have traditionally depended on wildlife trade. By applying well-designed subsidies and financial incentives, governments can trigger sustainable activities that guarantee economic security as well as conservation. For example, direct financial assistance to sustainable agriculture, sustainable forestry, and sustainable tourism can create sustainable income alternatives that reduce pressure on wildlife populations.

Infrastructure development complements these interventions by improving access to basic services, transport, and market access. Investment in better roads, digital connectivity, and storage infrastructure improves the ability of local producers to transport sustainably produced products at competitive prices. In addition, regulatory systems that improve transparency and fair trade agreements ensure communities receive fair remuneration for their labor.

A policy-backed, coordinated intervention that includes market-based interventions has been effective in many regions. For example, community-based conservation programs backed by financial incentives have decreased poaching and habitat destruction as well as encouraged local economies (Challender et al., 2019).

Prioritizing policies that are sustainability-oriented allows governments to create long-term economic opportunities that balance conservation with economic development.

11.4.5 Long-Term Benefits

The creation of alternative livelihoods is a critical strategy in reducing human dependence on wildlife exploitation, encouraging environmental conservation as well as economic resilience among communities. When well-designed, these programs reduce direct pressure on wildlife populations by offering sustainable sources of income that are compatible with conservation. The success of such programs relies on their ability to connect ecological sustainability with socio-economic development, creating long-term benefits for people and nature.

One of the significant aspects of alternative livelihood programs is that they provide poverty reduction and conservation simultaneously. Most wildlife-trading communities undertake the activity for necessity and not for preference on economic grounds. Alternatives such as sustainable agriculture, ecotourism, and NTFP harvesting provide tangible economic incentives that reduce dependency on unsustainable use of wildlife. In places where ecotourism has been promoted, local people have attained economic benefits from tourism employment, guiding, and handicraft manufacture, and the resultant decrease in economic incentives for poaching and habitat destruction.

Second, economic security through alternative livelihoods raises resilience at the community level. By diversifying the sources of income, individuals and households become more resilient to economic shocks such as market volatility or environmental degradation. Conservation training, financial capability, and business development incorporated in community-based activities further enhance the sustainability of alternatives in the long term. Where similar initiatives have been successful in places, there has been spectacular decline in illegal wildlife trade and an increase in local participation in conservation programs (Roe et al., 2020).

The integration of conservation and poverty reduction creates a win-win situation. By associating economic incentives with environmental conservation, these programs affect attitudes toward wildlife stewardship and provide access to stable and ethical sources of income. Governments, NGOs, and international agencies play an important role in such transitions through policy support, finance, and technical assistance. Generally, an adequately designed strategy toward alternative livelihoods not only saves biodiversity but also empowers people with the resources and opportunities for sustainable development.

11.5 Case Studies of Successful Models from India

India has innovated various highly effective models of wildlife protection and sustainable resource exploitation, and they are considered worldwide benchmarks in conservation of biodiversity. These projects reflect the importance of combining people's participation, sound legal framework, and innovative approaches to conserving natural ecosystems alongside local livelihood.

Amongst the most notable instances is the Rajasthan and Madhya Pradesh's community-led conservation program. Communities which were exclusively dependent on forest ecosystems have been induced to assume guardianship over hotspot biodiversity patches through participatory regimes enabling stewardship rights to such communities. By associating biodiversity conservation with earnings, for example, through the prospects of ecotourism and sustainable returns from non-wood forest products, these projects have been able to effectively address human-conflict with wild species and illegal picking and harvesting of floral and faunal products.

India's legal tradition, particularly the Wildlife Protection Act of 1972 and subsequent amendments to it, has been instrumental in facilitating conservation (Government of India, 1972). Sanctions against wildlife-related crime, along with initiatives like Project Tiger and Project Elephant, have restored many species to a considerable extent. Tiger density in large areas under protection, for instance, rose directly as a result of judicious law enforcement operations and measures focused on habitat restoration.

11.5.1 The Bishnoi Community: Wildlife Guardians

The Bishnoi people of Rajasthan have been a shining example of unyielding commitment to wildlife conservation for centuries, embedded in their religious and cultural beliefs. Founded on a fifteenth-century philosophy established by Guru Jambheshwar, the Bishnoi way of life is founded on non-violence, biodiversity conservation, and harmonious coexistence with nature. Their worth as conservationists goes beyond mere advocacy, speaking through action on the behalf of wildlife, particularly the blackbuck (*Antilope cervicapra*), a species once devastated by severe population declines due to poaching and habitat loss.

The Bishnois' reverence for life has turned into a successful grass-roots conservation campaign. Local people have traditionally risked their own lives to safeguard blackbucks and other wildlife from poachers, even standing up to hunters in defiance of exploitation. This commitment is evident in the historical 1730 Khejarli Massacre, where over 350 Bishnois sacrificed their lives to protect the felling of holy Khejri trees-a testament to their environmental dedication. In recent times, the community has played a pivotal role in court cases against celebrity poaching cases, further reinforcing the impact of indigenous conservation efforts.

The success of Bishnoi-led conservation model demonstrates the robustness of traditional ecological knowledge in modern biodiversity conservation. Its approach demonstrates the robustness of combining spiritual beliefs with conservation of the environment as a model for community-based conservation worldwide. By living together with wildlife, the Bishnois not only saved the blackbuck from extinction but also provided a role model for sustainable conservation deeply rooted in cultural identity (Kothari, 2009).

11.5.2 Kaziranga National Park: A Model for Rhino Conservation

Kaziranga National Park in Assam, which has been awarded the status of UNESCO World Heritage Site, is among the most effective wildlife conservation efforts in India. The park is famous for its phenomenal work towards conserving the Indian rhinoceros (Rhinoceros unicornis), an animal species that had previously threatened to be extinct because of constant poaching and destruction of habitats. Through the decades, conservation policies, stringent anti-poaching, and habitat management have helped maintain a steady increase in the number of rhinos, consolidating Kaziranga as a conservation success story.

One of the most significant factors in Kaziranga's success is that it has engaged local people in conservation. By providing sustainable livelihood alternatives through ecotourism, park authorities have secured people's cooperation, reducing conflict between people and wildlife and enhancing local involvement in conservation activities. Many locals are employed as park guides, forest guards, and eco-tourism operators, which acts as an economic incentive for conservation. Additionally, robust anti-poaching efforts, supported by sophisticated surveillance equipment and task forces, have greatly helped to stem illegal hunting. As a result, cases of poaching have declined significantly, testifying to the efficacy of comprehensive and strongly enforced conservation actions (Talukdar et al., 2018).

11.5.3 Vulture Conservation and the Ban on Diclofenac

India's vulture populations, once ubiquitous across the subcontinent, suffered a devastating population decline in the second half of the twentieth century after widespread veterinary use of diclofenac, a nonsteroidal anti-inflammatory drug (NSAID) commonly administered to treat livestock. When vultures fed on the carcasses of diclofenac-poisoned animals, they suffered acute kidney failure, leading to a record population reduction of over 90% for species such as the Oriental white-backed vulture (*Gyps bengalensis*), the long-billed vulture (*Gyps indicus*), and the slender-billed vulture (*Gyps tenuirostris*). This crisis had profound ecological impacts, since vultures play a critical role in maintaining ecosystems' balance via scavenging and disease suppression.

A true understanding of the situation's significance led India's government to enforce a nation-wide prohibition on veterinary diclofenac in the year 2006, which is a breakthrough policy intervention by India in wildlife conservation. Sustained with intense awareness creation campaigns, lobby efforts, and the establishment of vulture breeding and conservation units all over India, this ban was made absolute. Alternative, safer medications such as meloxicam were promoted as substitutes for diclofenac so that treatment of livestock did not have to come at the expense of the vulture populations.

Conservation efforts have yielded promising results, with steady stabilization of populations in some of the vulture species. Programme success indicates the contribution of evidence-based policy intervention towards averting risk to wildlife and is a testament to the efficacy of inter-institutional conservation policies among government, scientists, and environmental agencies (Prakash et al., 2012).

11.5.4 Community Forest Rights in Maharashtra: A Model for Sustainable Resource Management

Implementing Community Forest Rights (CFRs) under India's Forest Rights Act (2006) has made a revolutionary difference among indigenous and forest-living communities in Maharashtra. By acknowledging their traditional rights over forests through the Act, local residents have been provided legal powers to take active roles towards sustainable usage of forests. The historic policy has, in addition to boosting the autonomy of the indigenous peoples, contributed towards initiating the conservation of the forest, better living standards, as well as checked illegal deforestation.

Several villages in Maharashtra have been successful in harvesting their CFRs, using forest land in ways that foster ecological integrity alongside revenue generation through sustainable forestry of minor forest products such as bamboo, tendu leaves, and medicinal plants. In Gadchiroli district, for instance, a leader in CFR implementation, communities have established cooperative forest management systems that offer a balance between conservation and economic benefits. The revenues from sustainable forest-based enterprises have stimulated local development, funding education, healthcare, and infrastructure development.

Beyond economic benefits, the CFR model has led to minimized encroachment and illegal timber harvesting, since community-based enforcement and monitoring frameworks have proved superior to conventional state-led models of conservation. The program illustrates how participatory governance and traditional ecological knowledge can advance biodiversity conservation and promote social justice and economic resilience (Agarwal & Saxena, 2018).

11.5.5 Sundarbans Mangroves: Balancing Conservation and Livelihoods

The Sundarbans, the world's largest single mangrove forest, is situated across India and Bangladesh, which sustains the vital habitat of a variety of species like the Royal Bengal Tiger (*Panthera tigris tigris*), estuarine crocodiles, and other migratory birds. However, danger to this fragile ecosystem comes from the rising sea levels, encroachment by humans, and habitat loss.

Acknowledging the need for sustainable management, community-based conservation has played an important role in keeping the Sundarbans highly biodiverse and ensuring economic viability for the residents. Schemes promoting sustainable fishing habits, crab culture, and beekeeping have provided alternative income sources that alleviate the pressure on the ecosystem. For instance, the practice of adopting non-destructive methods of fishing-like using wider mesh nets in order not to catch juvenile fish-has been able to maintain fish populations and provide livelihood for thousands of fishers (Western & Wright, 1994).

Honey harvesting, an old Sundarbans way of living, has been restructured via sustainable harvesting and responsible beekeeping to ensure that honey gatherers do not displace the very fragile mangrove ecosystem or move into tiger sanctuaries. Organic Sundarbans honey marketing and branding created a premium sector, increasing income while compensating conservation.

In addition, afforestation and rehabilitation of mangrove forests, with support from the government initiatives and NGOs, has increased the cyclone and sea-level rise resilience of coasts. Apart from preserving biodiversity, these programs have also secured people's habitations by acting as a natural protection against storm surges and tidal erosion.

The Sundarbans case study shows the effectiveness of community conservation in reconciling economic development and biodiversity conservation, proving that sustainable, broad-based solutions are crucial for maintaining livelihoods as well as nature in ecologically sensitive regions (Ghosh et al., 2015).

11.6 Policy Recommendations and Future Directions

The challenges of wildlife trade (Table 11.3), coupled with successful case studies, underscore the need for forward-looking policies and innovative strategies to promote sustainable practices.

Here are some policy recommendations and directions for future action:

Table 11.3 Challenges in enforcing wildlife trade regulations. (Roe et al., 2020; Challender et al., 2019)

Challenge	India-specific issues	Global challenges
Lack of resources	Insufficient funding, outdated technology, and staff shortages make it difficult to monitor vast forests and trade routes.	Many developing nations lack financial and technical capacity for effective enforcement, leading to weak surveillance and poaching hotspots.
Corruption	Bribery at multiple levels weakens enforcement, allowing illegal traders to evade penalties and continue operations.	Corruption in customs, law enforcement, and government agencies enables trafficking networks to operate globally.
International trafficking	India serves as both a source and transit hub for smuggled wildlife products like ivory, pangolin scales, and exotic species.	Smugglers use fraudulent permits, online black markets, and hidden routes to traffic endangered species worldwide.
Consumer demand	Domestic demand for wildlife in traditional medicine, pet trade, and luxury goods fuels illegal activities.	Global markets for exotic pets, fashion, and traditional medicines drive poaching and unsustainable harvesting.
Weak legal frameworks	Despite strong laws like the Wildlife Protection Act (1972), inconsistent enforcement and lenient penalties reduce effectiveness.	Inconsistent wildlife laws across countries create loopholes that traffickers exploit, undermining global regulations.
Limited community participation	Rural communities engaged in illegal wildlife trade lack viable alternative livelihoods, making conservation difficult.	Many indigenous communities worldwide depend on wildlife trade due to economic necessity, requiring sustainable livelihood programs.
Technological gaps	Limited use of AI-powered surveillance, satellite tracking, and forensic tools hampers enforcement and monitoring.	Many countries lack access to advanced conservation technologies like DNA barcoding and real-time tracking systems.
Inadequate cross-border cooperation	Weak coordination with Nepal, Myanmar, and China allows illegal trade routes to thrive.	Weak enforcement of international treaties like CITES leads to ineffective global action against wildlife crime.
Human-wildlife conflict	Encroachment into forests and habitat destruction increase conflicts between communities and wildlife, leading to retaliatory killings.	Growing human populations worldwide are putting pressure on wildlife habitats, escalating conflicts and poaching.
Climate change	Rising temperatures and erratic monsoons affect species distribution, making them more vulnerable to poaching and trade.	Global climate shifts are altering ecosystems, forcing species migration and increasing exploitation risks.
Weak judicial prosecution	Low conviction rates and prolonged legal cases in India fail to deter wildlife criminals effectively.	In many countries, wildlife crime is not treated as a serious offense, leading to weak legal deterrence.
Illegal online trade	India is witnessing an increase in wildlife trafficking through social media platforms and dark web transactions.	Online marketplaces and encrypted communication apps allow global traffickers to evade detection.

(continued)

Table 11.3 (continued)

Challenge	India-specific issues	Global challenges
Lack of public awareness	Many Indian consumers unknowingly purchase illegal wildlife products due to limited awareness campaigns.	Global consumers often remain unaware of the environmental impact of their purchasing decisions.
Biosecurity risks & zoonotic diseases	Illegal wildlife markets in India pose risks of zoonotic disease outbreaks, similar to global concerns seen in the COVID-19 pandemic.	Unregulated wildlife trade worldwide increases the chances of new pandemics emerging from animal-human interactions.

11.6.1 Improving Wildlife Conservation in India

India's biodiversity is under threat from illegal wildlife trade, habitat loss, and poaching (Esmail et al., 2020). Enhancing national and international cooperation, such as compliance with conventions like the Convention on International Trade in Endangered Species of Wild Fauna and Flora (CITES), is critical for conservation (Challender et al., 2019).

11.6.2 Areas of Cooperation

- **Enhanced Enforcement**—Consolidating the Wildlife Protection Act, 1972 (Wildlife Protection Act, 1972), and enhancing coordination between customs, forest departments, and law enforcement agencies can check illegal trade (Talukdar et al., 2018).
- **Enhanced Intelligence Sharing**—Real-time information sharing, augmented by AI-powered analysis, can trace poaching networks and intercept smuggled animals like tigers, pangolins, and elephants (Challender et al., 2015).
- **Coordinated Anti-Poaching Activities**—Programs such as the Wildlife Crime Control Bureau (WCCB) and Interpol cooperation have been successful in confronting organized wildlife crime (Rosen & Smith, 2010). Cross-border initiatives have helped reduce trafficking routes and dismantle smuggling networks (Karesh et al., 2020).

Conservation, the Indian government, and local communities must unite in order to protect India's wildlife heritage.

11.6.3 Creating Sustainable Livelihoods

Breaking reliance on wildlife exploitation needs substitute economic opportunities (Roe et al., 2020). Conservation groups and government policies need to promote community-based sustainable livelihood programs (Elliott et al., 2014).

11.6.4 Strategies for Sustainable Livelihoods

- **Eco-Friendly Farming**—Organic farming, agroforestry, and climate-resilient agriculture can enhance income while maintaining natural habitats (Berkes, 2004; Agarwal & Saxena, 2018).
- **Community-Based Ecotourism**—National parks like Jim Corbett National Park and Ranthambore National Park have successfully blended tourism with conservation, and benefited local communities (Honey, 2008).
- **Sustainable Use of Forest Resources**—Ethical harvesting of non-timber forest products (NTFPs) such as honey, medicinal plants, and bamboo offers sustainable alternatives to poaching and logging (Roe et al., 2002; Ghosh et al., 2015).

11.6.5 Empowering Communities for Conservation

Empowering local involvement in conservation is central to conserving biodiversity. Skills development programs, cooperative business enterprises, and financial inclusion can enable communities to participate in sustainable conservation activities (Elliott et al., 2014).

11.6.6 Strengthening Legal Protections

There are effective wildlife protection laws in India, yet loopholes in enforcement and in law continue to be utilized (Esmail et al., 2020). It is important to strengthen legal frameworks to stop illegal trade and improve conservation (Challender et al., 2019).

11.6.7 Key Legal Reforms

- **Greater Penalties for Poaching and Trafficking**—Enhanced fines and jail terms in the Wildlife Protection Act can be an effective deterrent against wildlife crime (Challender et al., 2015).
- **Linkages with National Development Plans**—Introducing conservation policies into India's economic and rural development plans can ensure sustainable growth (Roe et al., 2002; Agarwal & Saxena, 2018).
- **Improved Training for Law Enforcement**—Offering training for forest officials, police, and judicial staff in specialized aspects of wildlife crimes can enhance detection and prosecution of such crimes (Karesh et al., 2020).

Legislation needs to be updated periodically to respond to new threats such as cyber-enabled wildlife trafficking and high-tech smuggling techniques (Roe et al., 2002).

11.6.8 Raising Public Awareness

Demand reduction for illegal wildlife products depends on widespread awareness (Rosen & Smith, 2010). Consumers are often not aware of the impact of their actions on biodiversity. Education and media campaigns can bring about changep (Esmail et al., 2020).

11.6.9 Effective Awareness Strategies

- **Social Media and Digital Outreach Campaigns**—Social media sites such as Instagram, YouTube, and Twitter can be used to emphasize the significance of wildlife conservation and the effects of illegal trade (Challender et al., 2015; Adams & Hutton, 2007).
- **Influencer and Celebrity Support**—Celebrities who support wildlife protection can help raise awareness and promote ethical decisions (Honey, 2008).
- **School and University Education Programs**—Incorporating conservation into school curricula promotes long-term dedication to wildlife protection (Elliott et al., 2014).

India's achievement of reducing demand for ivory and tiger parts with continuous awareness campaigns demonstrates that public education can be an influential conservation weapon (Wildlife Protection Act, 1972).

11.6.10 Promoting Research and Development

Scientific study and technological development are essential to monitor and protect India's biodiversity (Roe et al., 2020). Investment in state-of-the-art tools can make enforcement and conservation more effective (Esmail et al., 2020).

11.6.11 Research and Development Priorities

- **Wildlife Monitoring Technologies**—Drones, satellite monitoring, and camera traps can be used to monitor endangered animals like tigers, rhinos, and elephants in real time (Karesh et al., 2020).

- **Forensic Science in Wildlife Crime Investigation**—Isotopic tracking and DNA analysis assist in tracing the source of trafficked wildlife products to facilitate prosecution (Rosen & Smith, 2010).
- **Blockchain for Transparency in Wildlife Trade**—Blockchain technology can be utilized to provide proof of the origin of legally traded wildlife products and avoid fraud (Elliott et al., 2014).

There is a need for cooperation among government agencies, research institutions, and conservation organizations to effectively apply these innovations and protect India's wildlife for generations to come (Berkes, 2004).

Collaboration between government agencies, research institutions, and conservation organizations is necessary to implement these innovations effectively and safeguard India's wildlife for future generations.

References

Adams, W. M., & Hutton, J. (2007). People, parks, and poverty: Political ecology and biodiversity conservation. *Conservation & Society, 5*(2), 147–183.

Agarwal, A., & Saxena, N. C. (2018). Community forest rights in India: Promoting sustainable forest management and livelihoods. *For Political Economy., 92*, 102–110. https://doi.org/10.1016/j.forpol.2018.03.014

Berkes, F. (2004). Rethinking community-based conservation. *Conservation Biology, 18*(3), 621–630. https://doi.org/10.1111/j.1523-1739.2004.00077.x

Challender, D. W. S., Harrop, S. R., & MacMillan, D. C. (2015). Understanding markets to conserve trade-threatened species in CITES. *Biological Conservation, 187*, 249–259. https://doi.org/10.1016/j.biocon.2015.04.015

Divyabhanusinh. (2005). *The story of Asia's last lions*. Marg Publications.

Elliott, J., Schlaepfer, R., & Pechacek, P. (2014). Community-based conservation programs as a tool for poverty alleviation. *Ecological Economics, 107*, 493–504. https://doi.org/10.1016/j.ecolecon.2014.09.019

Esmail, N., Wintle, B. C., Sas-Rolfes, M. 't., Athanas, A., Beale, C. M., Bending, Z., Dai, R., Fabinyi, M., Gluszek, S., Haenlein, C., Harrington, L. A., Hinsley, A., Kariuki, K., Lam, J., Markus, M., Paudel, K., Shukhova, S., Sutherland, W. J., Verissimo, D., Wang, Y., Waugh, J., Wetton, J. H., Wright, J., & Milner-Gulland, E. J. (2020). Emerging illegal wildlife trade issues: A global horizon scan. *Conservation Letters, 13*(4), e12715. https://doi.org/10.1111/conl.12715

Ghosh, S., Hazra, S., & Mukhopadhyay, A. (2015). Community-based conservation in the Sundarbans: Balancing livelihoods and mangrove forest sustainability. *The Journal of Coastal Conservation, 19*(6), 517–526. https://doi.org/10.1007/s11852-015-0408-9

Government of India. (1972). *The Wildlife (Protection) Act, 1972*. Ministry of Law and Justice.

Honey, M. (2008). *Ecotourism and sustainable development: Who owns paradise?* Island Press.

Karanth, K. U. (2006). *A view from the Machan: How science can save the fragile predator*. Permanent Black.

Karesh, W. B., Dobson, A. P., Lloyd-Smith, J. O., Lubroth, J., Dixon, M. A., Bennett, M., Aldrich, S., & Daszak, P. (2020). Ecology of zoonoses: Natural and unnatural histories. *Lancet., 380*(9857), 1936–1945. https://doi.org/10.1016/S0140-6736(12)61678-X

Kothari, A. (2009). The Bishnois: The world's first environmentalists. *Economic and Political Weekly., 44*(52), 38–44.

Myers, N., Mittermeier, R. A., Mittermeier, C. G., Da Fonseca, G. A., & Kent, J. (2000). Biodiversity hotspots for conservation priorities. *Nature., 403*(6772), 853–858. https://doi.org/10.1038/35002501

Prakash, V., Green, R. E., Rahmani, A. R., Pain, D. J., & Virani, M. Z. (2012). Saving Asia's vultures from diclofenac poisoning: The success of advocacy and policy interventions. *Bird Conservation International, 22*(3), 389–403. https://doi.org/10.1017/S0959270912000269

Rangarajan, M., & Shahabuddin, G. (2006). Displacement and relocation from protected areas: Towards a biological and historical synthesis. *Conservation and Society, 4*(3), 359–378.

Roe, D., Mulliken, T., Milledge, S., Mremi, J., Mosha, S., & Grieg-Gran, M. (2002). *Making a killing or making a living? Wildlife trade, trade controls, and rural livelihoods* (Biodiversity Livelihoods Issues No. 6). International Institute for Environment and Development (IIED).

Roe, D., Booker, F., Day, M., Zhou, W., & Shepherd, G. (2020). Sustainable wildlife management: Wild meat, food security, and conservation. *Frontiers in Ecology and Evolution, 8*, 566800. https://doi.org/10.3389/fevo.2020.566800

Rosen, G. E., & Smith, K. F. (2010). Summarizing the evidence on the international trade in illegal wildlife. *EcoHealth., 7*(1), 24–32. https://doi.org/10.1007/s10393-010-0317-y

Talukdar, B. K., Barua, M., & Singh, R. (2018). Kaziranga National Park: Community engagement and conservation success. *Parks Wildlife., 24*(2), 18–26. https://doi.org/10.1111/pw.2018.24.2

TRAFFIC. (2018). *Pangolin Infographic: Illegal Wildlife Trade in India*. TRAFFIC – The Wildlife Trade Monitoring Network & WWF India.

't Sas-Rolfes, M., Challender, D. W. S., Hinsley, A., Veríssimo, D., & Milner-Gulland, E. J. (2019). Illegal wildlife trade: Scale, processes, and governance. *The Annual Review of Environment and Resources, 44*, 201–228. https://doi.org/10.1146/annurev-environ-101718-033253

WCS India. (2020). *Species group-wise frequency of wildlife poaching and illegal trade incidents recorded during 2020*. Wildlife Conservation Society (WCS) India.

Western, D., & Wright, R. M. (1994). *Natural connections: Perspectives in community-based conservation*. Yale University Press.

Chapter 12
Spectrum of Species in Wildlife Trafficking: The Rising Pet Trade and Its Socio-economic Impact

Sharda Kalra, Asha Poonia, Kapish Samota, and Preeti Devi

Contents

S. Kalra · K. Samota
Department of Zoology, Chaudhary Bansi Lal University, Bhiwani, Haryana, India

A. Poonia (✉)
Faculty of Life Sciences, Chaudhary Bansi Lal University, Bhiwani, Haryana, India
e-mail: asha.poonia@cblu.ac.in

P. Devi
Faculty of Commerce and Management, Chaudhary Bansi Lal University, Bhiwani, Haryana, India
e-mail: preeti.commerce@cblu.ac.in

A. Poonia et al. (eds.), *Wildlife Trade in India*,
https://doi.org/10.1007/978-981-95-3694-8_12

Abstract Wildlife trafficking remains a critical threat to biodiversity, driven by diverse demands, including the rising interest in exotic pets. The illegal and unsustainable trade of species for the pet market has accelerated biodiversity loss and poses significant socio-economic challenges. This chapter examines the scope of species affected, the drivers of the trade, and its socio-economic implications, with a focus on the need for robust policy interventions and international collaboration. The trafficking of wildlife for the pet trade has emerged as a substantial driver of global biodiversity loss. Unlike subsistence-driven exploitation, the pet trade caters to a consumer base seeking exotic species for private ownership. This demand has expanded with advancements in technology and globalization, leading to widespread illegal trade. The increasing removal of species from their natural habitats disrupts ecosystems, threatens vulnerable populations, and creates socio-economic pressures on source and transit countries. This chapter delves into the spectrum of species involved, examines the socio-economic impact, and underscores the urgency of addressing this burgeoning issue.

Keywords Demand drivers · Wildlife trafficking · Pet trade · Socio-economic · Exotic species

12.1 Introduction

Wildlife trafficking along with habitat loss and climate change (Caro et al., 2022) is the major threat to decline the species diversity for economic gain all over the globe (Fukushima et al., 2021; Oyanedel et al., 2021). Unsustainable extraction of wildlife for the pet trade accelerates population declines, disrupts ecosystems, and increases the extinction risk for many species (Cardoso et al., 2021) Vertebrates, in particular, are disproportionately affected due to their slow reproductive rates, which hinder population recovery and contribute to ecological imbalances (Hughes et al., 2023). Increasing human population has led to local pressure on wildlife and drive the demand for international trade and in response biodiversity hotspots have negative impacts (Gallego-Zamorano et al., 2020; Di Minin et al., 2019). Wildlife trafficking is a multifaceted global challenge driven by various market demands, including luxury goods, traditional medicine, economic stability, livelihood and the pet trade (Sorenson, 2011). Among these, the exotic pet trade has witnessed unprecedented growth, particularly in developing countries such as India, where increasing affluence and urbanization have fueled demand for rare and non-native (exotic) species (Singh, n.d.; Dutta, 2023; Sorenson, 2011). The pet trade include a broad spectrum of wildlife species, ranging from highly charismatic vertebrates to lesser-known invertebrates, with significant ecological and socio-economic consequences (Singh, n.d.; Dutta, 2023; Sorenson, 2011). This chapter examines the diverse range of species trafficked for the pet trade, the socio-economic drivers of demand, and the broader implications of this industry on biodiversity conservation and public health.

12.2 Laws for Exotic Species

The Wildlife Protection Act 1972 prohibits the trade of any animals in India specified in the schedules (Wildlife Protection Act, 1972). Prior to 2022 amendment, there is no provision for trade of exotic species. But this amendment enlisted CITES listed species in Schedule IV. EXIM (Export –Import) policy of Directorate General of Foreign trade restrict the trade of any wildlife and their parts through its guidelines (Wildlife Protection Act, 1972). However, it presented a gap in monitoring and regulating the trade of CITES listed species beyond the trade point (Ghanghas & Poonia, 2021). The ministry of Environment, Forest, and Climate Change (MoEFCC), the Government of India issued an advisory to regulate the import of exotic species in india and declaration of stock of exotic species. This advisory have significant role in regulation of exotic pet ownership and provide an estimate of exotic species in India. Only 32,645 individuals of reptile, birds and mammals were declared through "Voluntary Disclosure Scheme" of advisory. Some of the species which are difficult to imagine as pets such as Kangaroos and Lemurs were also recorded (Ghanghas & Poonia, 2021; Poonia et al., 2022). The incorporation of these species into national legislation facilitates the enforcement of regulatory measures against violations of CITES provisions. This amendment will strengthen the implementation of CITES objectives, ensuring that the trade in wild animal and plant specimens does not threatened the survival of the species (Wildlife Protection Act, 1972).

Many people seeking to keep wildlife as exotic pets often lack awareness of the species' conservation and legal status, the potential risks of zoonotic disease transmission, and the inhumane conditions these animals endure before reaching them (Manral et al., 2016). Public awareness campaigns focused on disseminating this critical information could play a pivotal role in discouraging the purchase of threatened wildlife and reducing violations of national and international wildlife protection laws.

12.3 India's Exotic Pet Trade

The illegal trade in exotic pets has a significant rise in India, as evidenced by the increasing number of wildlife seizures conducted by Indian Customs in recent years. This upward trend suggests a growing demand for non-native wildlife species, particularly among affluent urban residents, where the ownership of exotic pets is perceived as a status symbol. A major factor facilitating this illicit trade is the disparity in legal protections, while the commercial trade of indigenous species is strictly prohibited under Indian law, many exotic species lack comparable regulatory framework. According to the Directorate of Revenue Intelligence (DRI), traffickers have exploited these legal gaps, shifting their focus towards non-native wildlife species. This article examines key smuggling routes into India, the primary retail markets for exotic pets, and the species most frequently trafficked (Kalra et al., 2024; Ghanghas & Poonia, 2021; Poonia et al., 2022). Furthermore, it explores the legislative shortcomings that have contributed to the persistence and expansion of the illegal exotic pet trade (https://news.mongabay.com/2025/01/increase-in-gibbon-trafficking-into-india-has-conservationists-worried/).

Northeastern India is gaining recognition as a key area for illegal wildlife trafficking, largely because its states share a land border with Bangladesh. Conservationists have pinpointed this region as a source for the trafficking of many species into India and Myanmar, and it serves as a central hub for wildlife trafficking activities in the region. The data analysis from the online media reports showed the network of trade route followed by wildlife smugglers (Fig. 12.1). Uttar Pradesh and West Bengal have porous border with Nepal and Bangladesh respectively. Apart from these porous Channels, air channels were also actively critical importance in

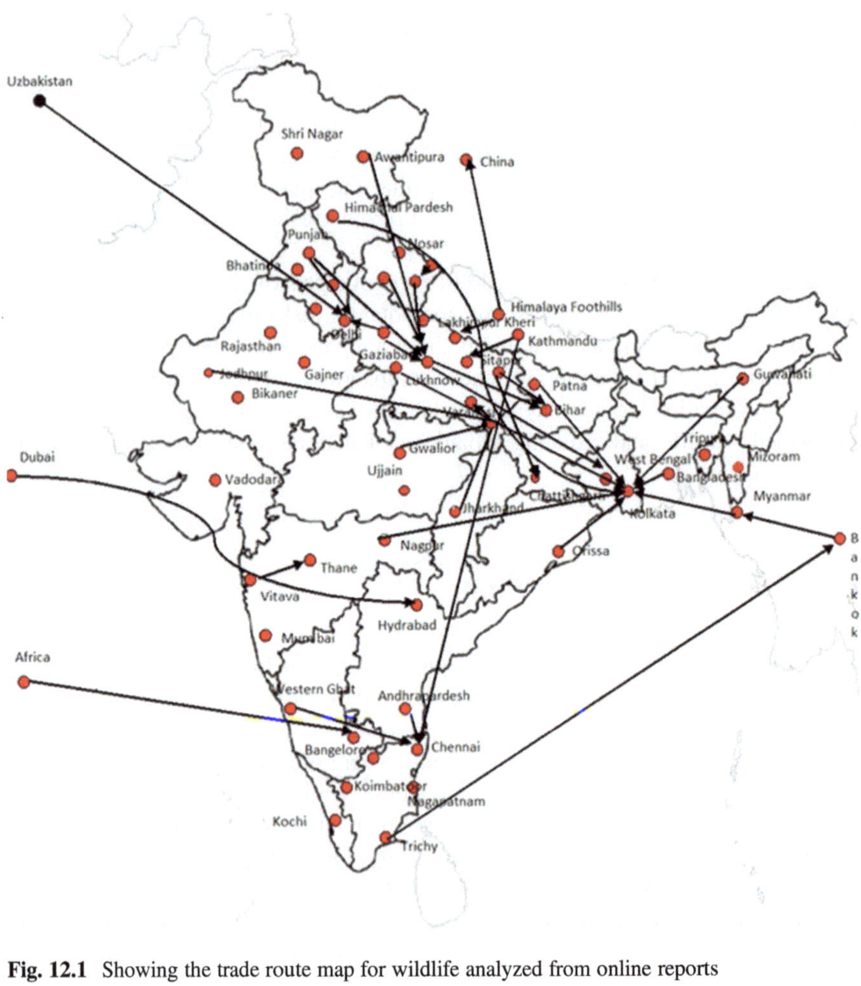

Fig. 12.1 Showing the trade route map for wildlife analyzed from online reports

international smuggling network. Many wildlife seizure were also takes place at national airports such Chennai, Delhi, Mumbai etc. during exports and imports. Many transactions were seized at these airports destined from Bankok, Uzbakistan, Dubai, Africa, Malasiya, Singapore etc. (Kalra et al., 2024) Red Circle denotes the seizure recorded in online reports. Arrow soes the direction of transaction in the map Various species of birds, reptiles and mammals are illegally traded to meet the demand of international black markets.

In India, animals are transported by road or rail to major cities such as Kolkata, Chennai, Hyderabad, Ahmedabad, Lucknow, Delhi, Bangalore, Mumbai, and Cochin. The long-distance movement of animals across national borders under highly restrictive conditions results in considerable physiological and psychological distress, confined in small cages and containers to hide their presence. Many do not survive this journey, with mortality rates for certain species, especially birds, estimated to be between 30% and as high as 90% in the illegal trade. Exotic animals can be found for sale in markets, physical stores, and online platforms (https://www.ecojust.eu/indias-exotic-pet-trade/; https://wildlifesos.org/exotic-pet-trade-threatens-indias-endangered-species/).

Most of the illegal trade thorough social media like Whatsapp, Facebook, Instagram and Telegram. A study conducted in January 2022 found no detectable evidence of exotic wildlife trade on e- commerce Platform like OLX.com and Quicker.com. However, it is possible that these sites implemented stricter policies to prohibit wildlife advertisements following a complaint lodged by the Animal Welfare Board of India with the Ministry of Electronics and Information Technology in July 2020 (https://wildlifesos.org/exotic-pet-trade-threatens-indias-endangered-species/).

12.4 Socio-economic Impact of Pet Trade on Local Communities

The pet trade has significant socio-economic implications for local communities, offering both economic benefits and ecological challenges. Economically, it provides direct employment opportunities for trappers, breeders, traders, and retailers, particularly in rural areas where wildlife collection serves as a primary or supplementary source of income. The increasing demand for exotic pets has expanded markets, fostering businesses such as specialty pet stores, online platforms, and veterinary services, especially in urban centers. However, unregulated extraction for the pet trade contributes to the overexploitation of wildlife, leading to population declines and the degradation of ecosystems, which, in turn, diminishes long-term economic benefits from sustainable resource use, such as ecotourism (Wildlife Protection Act, 1972, 1972; Manral et al., 2016; Kalra et al., 2023). Moreover, overharvesting can deplete natural resources, adversely affecting alternative livelihoods like fishing and non-timber forest product collection. Legal consequences also pose economic risks, as individuals engaged in illicit wildlife trade face fines and imprisonment, leading to financial instability within communities. Beyond economic concerns, the pet trade alters human-wildlife relationships. Additionally, it increases public health risks by facilitating zoonotic disease transmission (Hughes et al., 2023; Gallego-Zamorano et al., 2020), posing threats to both traders and consumers while potentially triggering economic disruptions due to disease outbreaks and stricter regulatory measures. The involvement of organized criminal networks in illegal wildlife trade can further result in resource conflicts, exploitation of vulnerable populations, and human rights violations. Addressing these challenges requires the implementation of community-based conservation initiatives, promoting alternative livelihoods such as sustainable handicrafts and agro forestry, and strengthening law enforcement mechanisms. Raising awareness about the ecological and ethical consequences of the pet trade is crucial for mitigating its negative impacts while ensuring the economic stability of affected communities.

Many tribes like Nyishi tribe, Baheliyas, Mirshikar, Phasiya/Shikaris, Bhatiyara, Pathami or Jabjalies, Hakkipukki, Kuruvikkara/Amblakaran/Narikurava, Kalandars, Vagri, Pardis/Pashe-pardi, Lodies and Harries, Bawarias, Yanadi, Lambadas and Yerukula, Kalbaleyas/Jogis or Badiya, Karori or Kurmi, Damora, Bhil and Munda, Garo, Karbi and other north-eastern tribes (TRAFFIC,2010) were involved (https://www.wwfindia.org/news_facts/feature_stories/trafficking_of_exotic_wildlife_species_in_india/). These are the major communities from different regions which act as sub dealers for supply the wildlife animal in clandestine manner and contributed to the illicit network of bird trade (https://www.wwfindia.org/news_facts/feature_stories/trafficking_of_exotic_wildlife_species_in_india/).

India is most bio diverse region in south Asia includes four out of 34 global hotspots region. Apart from these, there are many protected areas such as National Parks, Wildlife sanctuaries, Conservation reserve and community reserve were created under the wildlife Protection Laws,1972 to preserve the biodiversity. These protected areas are not only protecting the wildlife animals but, it also provide the source of livelihood to local community. The Local community can be engaged in conservation projects and as guide for ecotourism in the buffer zone of protected areas. The map showing approx. 574 wildlife sanctuaries in India, Play significant role in protecting the species diversity (Fig. 12.2).

12.5 Spatial Pattern of Pet Trade Network

The pet trade commodity chain is complex and multistage process composed of origination source transport to next node along the commodity chain until the destination were achieved. These commodity chain includes source, collector, Aggregator, exporter, port of export, re-export, Port of import, importer, Whole seller, Retailer and consumer (Reino et al., 2017; Vall-llosera & Cassey, 2017; Sinclair et al., 2021). The literature provided the general structure of pet trade by

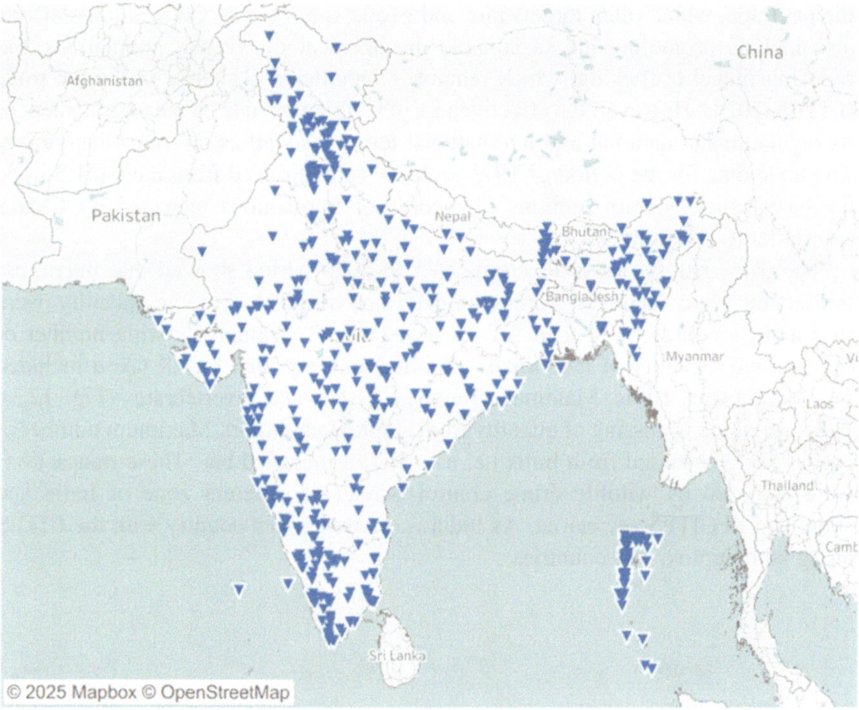

Fig. 12.2 Wildlife sanctuaries in India

identifies these 11 actors participate actively in international wildlife trade network. By using this representation of commodity chain, one can better analyze the spatial and temporal pattern of pet trade and can reveal the key targets to intercept the chain (Reino et al., 2017; Vall-llosera & Cassey, 2017; Sinclair et al., 2021).

12.6 Spectrum of Species in Trade

The CITES trade data has been analyzed to know the spectrum of species in wildlife trade. CITES (Convention on international trade on endangered species) is multinational agreement that regulate the trade at international level and protect the species from the threat of extinction (CITES, 2024). CITES aims to ensure that international trade in wild animals and plants does not threaten their survival in the wild (CITES, 2024). Through its system of permits and monitoring, CITES regulates the trade of species listed in its appendices, which include both commonly traded animals and highly endangered species (CITES, 2024). By establishing legal frameworks for the trade of wildlife, CITES works to mitigate the risk of over-harvesting, particularly in the pet trade, which often targets rare and exotic species. The Convention's efforts also include promoting the sustainable management of wildlife populations and reducing illegal trafficking, which remains a significant challenge in the pet trade (CITES, 2024). However, the effectiveness of CITES depends on the enforcement of its regulations at national and international levels, as well as on the collaboration. The trade data for the period of 1975 to 2020 were retrieved from the CITES trade database which contain millions of records of transactions managed by CITES secretariat (CITES, 2024).

The data analysis for India from 1975 to 2020 period showed that maximum transaction were recorded from mammals and Reptiles, but the quantity were maximum for birds more than 2 Lacs in exports. The graph showing number of Species and Quantity of individuals exported and imported for all taxon included i.e. Amphibians, Birds, Mammals, Reptiles, Fishes and invertebrates (Fig. 12.3). However, there is missing of quantity from amphibians export. Maximum number of species were recorded from birds i.e. n = 157 in imported bar. These transactions were recorded by wildlife crime control bureau of different zone of India and submitted to CITES secretariat. As India is the participant country with the CITES along with approx. 194 countries.

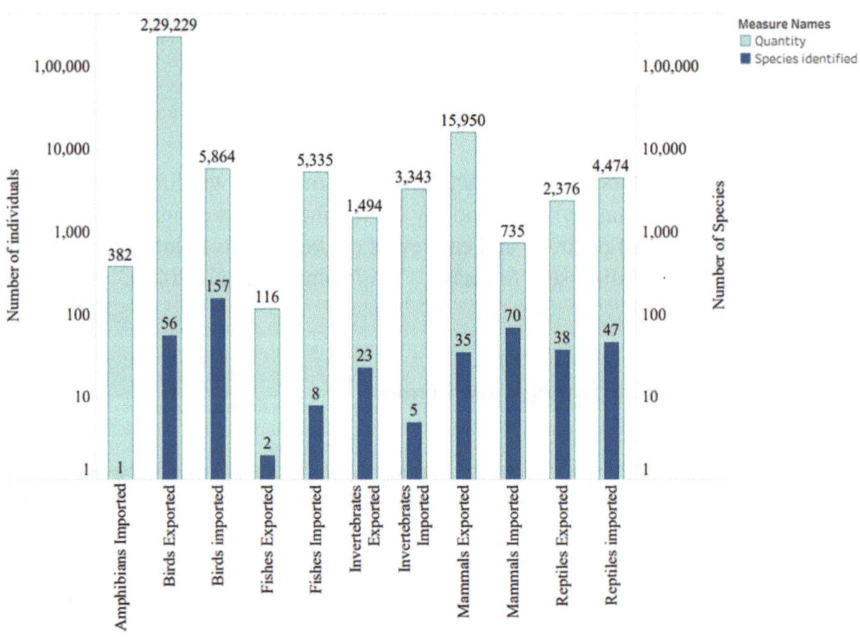

Fig. 12.3 Live quantity of species and their individual recorded in CITES trade database from India

12.6.1 Hotspots and Black Market for Illegal Wildlife Trade

There are many blackmarkets including weekly pet markets in Chennai (Tamilnadu); Crawford Market in Mumbai (Maharashtra); Russell Market in Bangalore (Karnataka); Jama Masjid market and Chakla area in Dehli; Gallif street market in Kolkata (West Bengal); Mulki Bazaar and Mehboob Chowk bird market in Hyderabad (Telangana); Mishrikar toll market inPatna(Bihar), Bahelia Toll market in Varanasi (Uttar Pradesh); and Nakhas market in Lucknow (Uttar Pradesh) (Fig. 12.4). The sources for the wildlife animals traded in these markets were primarily local inhabitants who possessed extensive knowledge of the natural habitats of various species (https://www.wwfindia.org/news_facts/feature_stories/trafficking_of_exotic_wild life_species_in_india/; Kalra et al., 2023). These individuals had acquired, often

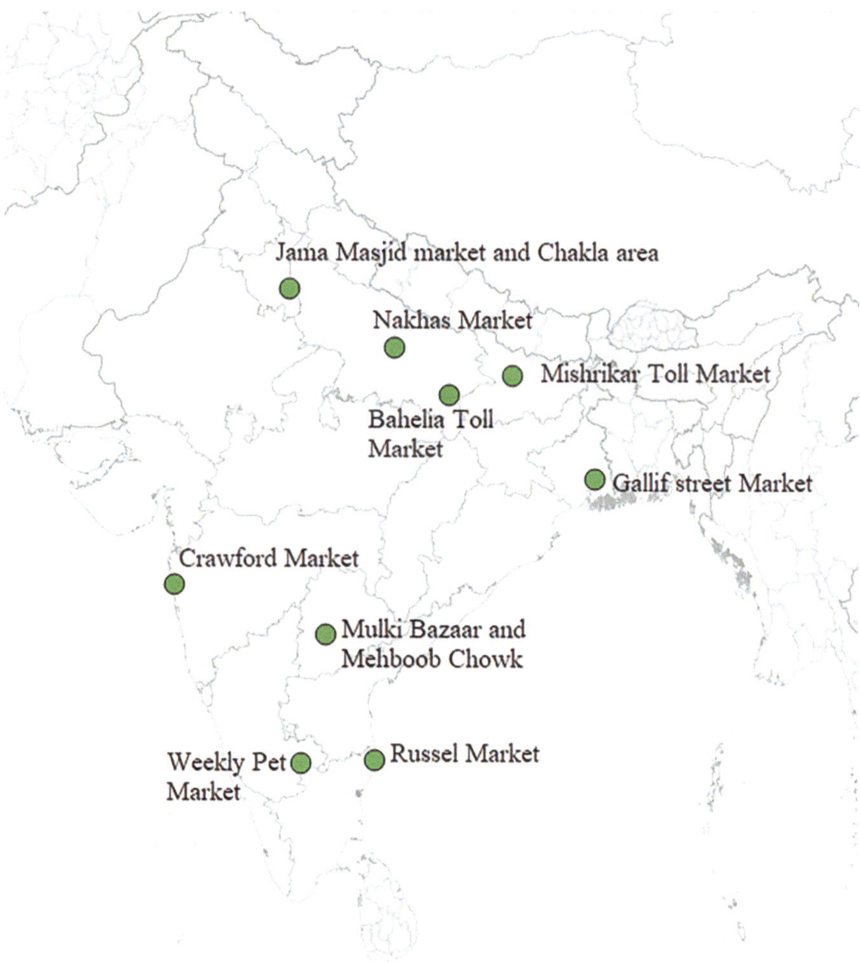

Fig. 12.4 Black market for wildlife trade in India

through generational experience, specialized techniques for capturing wildlife. Their intimate understanding of the environment enabled them to locate these species with high precision. Furthermore, these local individuals were well-versed in methods to evade the surveillance efforts of wildlife authorities, allowing them to continue their activities with a degree of secrecy (https://www.wwfindia.org/news_facts/feature_stories/trafficking_of_exotic_wildlife_species_in_india/; Kalra et al., 2023). This knowledge, passed down through cultural practices, provided a significant advantage in the illegal wildlife trade, as it allowed for the systematic exploitation of biodiversity while minimizing detection by regulatory bodies and law enforcement agencies.

12.7 Spectrum of Species in the Wildlife Pet Trade

Wildlife trafficking for the pet trade includes species from various taxonomic groups, each targeted for their unique aesthetic appeal, rarity, or symbolic value.

12.7.1 Avian Species in the Pet Trade

A large number of bird species are involved in the illegal and unregulated pet trade primarily driven by their colored appearance, mimicry ability and perceived rarity (Ribeiro et al., 2019). Parrots, cockatoos, and various songbirds remain highly sought after by pet owners and collectors. For example Alexandrine parakeet (Psittacula alexandrine), Rose ringed parrot (Psittacula krameri), plum headed parrot (*Psittacula cyanocephala*), African grey parrots (*Psittacus erithacus*), macaws (*Ara* spp.), and cockatoos (*Cacatua* spp.), are extensively traded due to their cognitive abilities, capacity for vocal mimicry, and striking plumage (Kalra et al., 2024; Harris et al., 2017). The African grey parrot is recognized for its exceptional intelligence and ability to replicate human speech, making it one of the most remarkable species in the pet trade. Despite being listed under Appendix I of the Convention on International Trade in Endangered Species of Wild Fauna and Flora (CITES), illicit trafficking continues to supply global markets, including India (Chan et al., 2021; Martin et al., 2018). Macaws and cockatoos which are predominantly native to South America and Australia, are also targeted, with many individuals sourced from illegal breeding operations or smuggled into the country (Chan et al., 2021; Poole & Shepherd, 2017).

Native Indian passerine birds, including munias (*Lonchura* spp.), mynas (*Acridotheres* spp.), and parakeets (*Psittacula* spp.), remain highly susceptible to illegal capture and trade (Kalra et al., 2024; Fink et al., 2021). Despite being legally protected under the Wildlife Protection Act (WPA), 1972, these species are frequently trafficked due to their melodious calls, vibrant plumage, and cultural significance. The continued demand for these birds is exacerbated by traditional bird-keeping practices and religious customs, where certain species are believed to bring good fortune (Pandey et al., 2024). Poaching and unregulated trade pose significant threats to wild populations, often leading to habitat depletion and declining numbers in their native ecosystems. Strengthening enforcement mechanisms, public awareness, and regulatory frameworks is imperative to mitigate the unsustainable exploitation of these avian species (Dutta, 2023).

12.7.2 Reptiles in the Pet Trade

Reptiles are extensively reported in exotic pet trade due to their unique morphological characteristics, perceived rarity, and symbol of status value (Pragatheesh et al., 2021). These are particularly vulnerable due to their slow reproductive rates, habitat-specific requirements, and high mortality rates during transit and captivity (Valdez, 2021). The trade is increased by the rising popularity in social media such as facebook, Instagram, Whatsapp, Telegram etc. and the perception of reptiles as low-maintenance pets (Pragatheesh et al., 2021). Snake, Lizard and turtle are heavily traded reptiles advertised through online platform (Pragatheesh et al., 2021; Valdez, 2021). Many species including Indian Star Tortoise (*Geochelone elegans*), Monitor Lizards (*Varanus* spp.), Leopard Gecko (*Eublepharis macularius*), Tokay Gecko (*Gekko gecko*), Bengal monitor (*Varanus bengalensis*), yellow monitor (*Varanus flavescens*), Ball Pythons (*Python regius*), Green Iguanas (*Iguana iguana*) and Red-eared Slider Turtles (*Trachemys scripta elegans*) are the most heavily trafficked reptile species in India which are highly exploiated for the exotic pet trade, religious rituals, and traditional medicine (Sengottuvel et al., 2024; Valdez, 2021). The trade poses a serious threat to wild populations, leading to population declines and loss of genetic diversity (Valdez, 2021). Many individuals are smuggled into India via illegal shipments or fraudulent captive breeding claims. Green Iguanas (*Iguana iguana*) are frequently kept as pets despite requiring specialized husbandry and dietary needs. Many die due to poor care, stress, and inadequate housing conditions. Iguana iguana has been recognized as a potential source of Salmonella infections in humans (Warwick et al., 2001). Red-eared Slider Turtles (*Trachemys scripta elegans*) are generally kept in aquarium but invasive in nature and pose a significant threat to native turtle species in India when released into the wild. Despite the legal protection granted to native species under the Wildlife Protection Act (WPA), 1972, these species have contributed to the illegal trade. However, some website has already banned the online reptile's trade such as Facebook (Valdez, 2021).

12.7.3 Amphibians in the Pet Trade

The CITES trade recorded for 45 years showed that there is negligible quantity of amphibians were observed. However, it is recorded that 5424 individuals of Rana were exported to United States of America from India illegally in 2015 (CITES,

2024). it is observed that two species of amphibians were green pond frog *Euphlyctis Hexadactylus* and the Indian bullfrog *Hoplobatrachus tigerinus* highly exploited for frog's leg (Altherr et al., 2011). Approximately, 9000 tonne of frogs were removed from the wild for the frog's leg trade in 1985, as a result, CITES put these species in Appendix II and banning of the exports of frog's leg in 1987 in India (Altherr et al., 2011). As per the IUCN red list "Small numbers of Rhacophorus reinwardtii appear in the pet trade, but probably not at a level to constitute a major threat to the species." (https://iucn.org/our-union/commissions/group/iucn-ssc-amphibian-specialist-group).

12.7.4 Mammals in the Pet Trade

The network of internet and social media has increased demand for mammal pets, which led to serious conservation issues (Pragatheesh et al., 2021). A large number of mammals species are exploited for pet trade in India driven by rarity and their products in fashion, traditional medicines etc. (Singh Parmar, A., 2024). Despite being protected under the Wildlife Protection Act (WPA), 1972, these species continue to be illegally captured and sold in local markets and private dealers (Singh, n.d.; Sethi et al., n.d.; Valdez, 2021; Azad & Valson, 2020). Several mammalian species particularly primates, large civet and small carnivores are trafficked for their cultural significance and popularity (Sethi et al., n. d.). However tiger, elephants, Rhinoceros, Pangoline etc. are highly poached animals for their skin, teeth, nails, tusks, horns, scale etc. (Sethi et al., n.d.).

The Bengal slow loris (Nycticebus bengalensis) (Swapna et al., 2010; Musing et al., 2015), a nocturnal primate, Indian Palm Civet (Paradoxurus hermaphrodites) (Kannan, 2024), Otters (Gomez & Bouhuys, 2018), Indian Flying Squirrel (Petaurista philippensis), Red Kangaroo (*Osphranter rufus),* capuchin monkeys (*Cebus* spp) is frequently trafficked for the exotic pet trade (Swapna et al., 2010; Musing et al., 2015). The demand for this species is driven popularity among exotic pets. However, slow lorises suffer extreme distress in captivity, often having their teeth forcibly removed to prevent biting, and leading to high mortality rates (Musing et al., 2015). Indian Palm Civet (Paradoxurus hermaphrodites) are illegally captured and kept as pets due to their semi-domesticated nature and distinctive appearance (Kannan, 2024). The Indian palm civet is also exploited for its role in producing kopi luwak, an expensive coffee made from partially digested coffee beans excreted by civets (Kannan, 2024).

A large number of primate including Moor Macaque (*Macaca Maura),* grey monkey (*Semnopithecus* spp.), Siamang (*Symphalangus syndactylus),* pygmy marmoset (*Cebuella* spp.), marmoset, Spot-nosed Monkey (*Cercopithecus petaurista),* potto (*Perodicticus* spp.), Myanmar Snub-nosed Monkey (*Rhinopithecus strykeri),* Red-chested Mustached Tamarin (*Saguinus labiatus),* spider monkey (*Ateles* spp.), Panamanian White-throated Capuchin (*Cebus imitator),* Black-capped Capuchin (*Sapajus paella),* orangutan (*Pongo* spp.), guereza (*Colobus* spp.), Golden-headed Lion Tamarin (*Leontopithecus chrysomelas),* tamarin monkey (*Saguinus* spp.),

Dusky Langur (*Trachypithecus obscures)*, De Brazza's Monkey (*Cercopithecus neglectus)*, Chimpanzee (*Pan troglodytes)*, and Indri (*Indri indri*) were reported in 22 seizure in 2022 in india (https://www.wwfindia.org/news_facts/feature_stories/trafficking_of_exotic_wildlife_species_in_india/).

12.8 Demand Drivers for Wildlife Trade

Wild animal and their products are traded mainly for sustenance, economic gain, financial stability, cultural practice, symbol of status and traditional medicine etc. These drivers enhancing the wildlife trade and making the wildlife vulnerable to extinction.

12.8.1 For Sustenance or Livelihood

Many communities and local people dependent on wildlife And thrir products for their survival. Among these, hunting of wild animal for food is most common and important practice. For instance, many people in central Africa dependent on bushmeat for their survival as well as financial benefits (Ziegler et al., 2016). Not only in africa, hunting of wild animals for food and their products is usual such as pangolin trade in asian countries were remarkable. A total of four species of pangolin are prevalent in trade for their meat and scale (Challender et al., 2014; Mohapatra et al., 2015). It has been studied that the destination point of major portion of trade were china. Apart from these, India also acts as the source country for many wild animals and birds smuggled in international black markets (Zhang et al., 2017; Mohapatra et al., 2015; Kalra et al., 2024). However, western countries rely on domestic animal for protein source but those animal were further need to feed protein source like soya when grow ultimately lead to deforestation and habitat destruction (Levitt, 2020). To maintain the viability of wild animals along with the sustainable development of local community, there is need of balanced solution through implementing strong legislative framework.

12.8.2 For Financial Stability

Wildlife animals and their products have great demand in international market. To accomplish the demand, high value were provided to the suppliers. The supplier may be poor people of the local community which have not any alternate source of their livelihood. The local community provides the wild animals and their products to the intermediate, which ultimately supply them to the international markets. To prevent this, many community based conservation program has been initiated to provide alternate source of livelihood such as employing them in legal land use near the wildlife habitat, incentive based scheme, revenue sharing scheme and involving them in decision making of cases related wildlife crime (Kalra et al., 2023).

12.8.3 For Traditional Medicine

The traditional practice of using wild animal and their products for medicine in India were from ancient time. India has signed a agreement with World health organization to create a global center for traditional medicine with the investment of USD 250 million. This center seeks to use contemporary science and technology to harness the potential of traditional medicine from around the world to enhance human and environmental health. Approximate 80% people use traditional medicines. Among all member countries of World health organization, 170 have acknowledged using traditional medicine and their governments have asked for WHO's support in creating a body of reliable evidence and data on traditional medicine practices and products (World Health Organization, 2022). Currently, traditional medicine which are part of Ayurveda, Siddha, Unani etc. under the scheme "Ayush" represent a significant contribution of USD 18.1 billion in Indian economy. The growing economy is the result of high demand of herbal products all over the globe (RIS(Research and information System for developing countries), 2021). Musk deer gland, snake venom, pangoline scale, tiger or leopard bones were used to make medicine.

12.8.4 For Religious and Spirituals Belief

Many wildlife animals and their body parts are used in religious rituals, black magic and tantric practices (Singh, n.d.; Sethi et al., n.d.; Azad & Valson, 2020). For instances, owls were used on Diwali festival for wealth, monitor lizard's hemipenis also called as 'Hatha Jodi" is used as tantric objects and Turtle shells are used in religious ceremonies. As a result, there is a continuous illegal demand for these animals, fueling wildlife trafficking networks (Pragatheesh et al., 2021). Despite strict legal protections under the Wildlife Protection Act, 1972 (WPA) and CITES regulations, the underground market continues to thrive due to illegal smuggling, black-market networks, and online sales (Pragatheesh et al., 2021; Singh, n.d.; Azad & Valson, 2020).

12.8.5 For Pet and Symbol of Status

The illegal pet trade in India is largely driven by wealthy people and influencers who want to display wild animals as a symbol of status and luxury (Dutta, 2023; Sethi et al., n.d.). The price of animals increases with their rarity, which fuels an increase in illicit commerce. Many birds are trafficked from Southeast Asia, Africa, and South America, entering India through Nepal, Myanmar, and Bangladesh (Kalra et al., 2024). Not only birds, many other taxon also traded illegally for pet such as tiger, turtle, marmosets, ball pythons, slow loris, alexandrine parrots, macaws, munias etc. (Pragatheesh et al., 2021; Mendiratta et al., 2017). The lack of stringent regulations governing the trade of non-native (exotic) species in India has further contributed to the expansion of this illicit industry, allowing traders to exploit legal loopholes and circumvent enforcement mechanisms. Expanding legal protections to include CITES-listed exotic species would eliminate legal loopholes that allow traders to exploit non-native wildlife (Singh, n.d.; Fink et al., 2021). The Collaboration with social media companies and cyber security agencies to detect and disrupting online wildlife trade networks may help in improving wildlife conservation. Through Implementing AI-driven monitoring systems to track suspicious online transactions related to exotic pets.

12.8.6 Luxury Goods, Fashion, and Decoration

The demand for wildlife-derived products in the luxury goods, fashion, and home décor industries continues to be a significant driver of illegal wildlife trade in India and globally (Suvra & Ontar, 2024). High-end fashion brands, jewelry designers, and interior decorators often seek rare animal-derived materials such as fur, skin, teeth, leather products, elephant ivory, rhino horns, deer antlers which are perceived as symbols of wealth, exclusivity, and status (Sorenson, 2011; Altherr et al., 2011). Products crafted from python and crocodile leather are highly valued in the luxury handbag and footwear industries, while shawls made from shahtoosh wool (derived from the endangered Tibetan antelope) remain in demand despite strict legal prohibitions (Van Uhm, 2016; Suvra & Ontar, 2024). The increasing role of social media and online platforms in marketing and distributing these products has further exacerbated the problem, making enforcement efforts more challenging. Addressing the exploitation of wildlife for luxury goods requires stricter regulatory frameworks, enhanced monitoring of online markets, and consumer awareness campaigns to reduce demand and promote ethical alternatives (Suvra & Ontar, 2024).

12.9 Conclusion

The rising demand for exotic pets has fueled a global wildlife trafficking crisis, impacting a diverse spectrum of species, including mammals, reptiles, birds, and ornamental fish. As the exotic pet trade continues to expand, it exacerbates the illegal wildlife trade, leading to the overexploitation of wild populations and increased risks of zoonotic disease transmission. The pet trade is driven by many factors such as social media, increasing affluence, and ease of access through online marketplaces posing significant threats to biodiversity, public health, and socio-economic stability extend far beyond individual ownership. The socio-economic ramifications include the proliferation of organized crime networks, the displacement of traditional livelihoods, and Increasing difficulties for law enforcement agencies involved in curbing of wildlife trade. Addressing these challenges requires a multifaceted approach, including stronger international collaboration, stricter enforcement of wildlife trade regulations, and widespread public awareness campaigns. Educating potential buyers on the ethical, legal, and ecological implications of the pet trade can play a crucial role in reducing demand. Furthermore, enhancing conservation efforts, promoting sustainable alternatives, and integrating stricter monitoring mechanisms can help mitigate the adverse impacts of wildlife trafficking. Ultimately, ensuring the survival of species threatened by the pet trade demands a global commitment to ethical wildlife management, conservation, and the enforcement of policies aligned with CITES and national legislation. Without urgent and coordinated action, the rising exotic pet trade will continue to drive biodiversity loss and compromise ecological balance on a global scale.

References

Altherr, S., Goyenechea, A., & Schubert, D. J. (2011). *The international trade in FROGS'LEGS and its ecological impact*. Pro Wildlife Munich (DE).

Azad, K. R., & Valson, M. C. (2020). Trade on wild animals and animal products: Effectiveness of law in India a critical evaluation.

Cardoso, P., Amponsah-Mensah, K., Barreiros, J. P., Bouhuys, J., Cheung, H., Davies, A., Kumschick, S., Longhorn, S. J., Martínez-Munoz, C. A., Morcatty, T. Q., & Peters, G. (2021). Scientists' warning to humanity on illegal or unsustainable wildlife trade. *Biological Conservation, 263*, 109341.

Caro, T., Rowe, Z., Berger, J., Wholey, P., & Dobson, A. (2022 May). An inconvenient misconception: Climate change is not the principal driver of biodiversity loss. *Conservation Letters., 15*(3), e12868.

Challender, D., Baillie, J., Ades, G., Kaspal, P., Chan, B., Khatiwada, A., Xu, L., Chin, S., KC, R., Nash, H., Hsieh, H.. (2014). Manis pentadactyla. The IUCN Red list of threatened species 2014: e. T12764A45222544 [Internet].

Chan, D. T., Poon, E. S., Wong, A. T., & Sin, S. Y. (2021). Global trade in parrots–Influential factors of trade and implications for conservation. *Global Ecology and Conservation, 30*, e01784.

CITES. (2024). https://trade.cites.org/

Di Minin, E., Brooks, T. M., Toivonen, T., Butchart, S. H., Heikinheimo, V., Watson, J. E., Burgess, N. D., Challender, D. W., Goettsch, B., Jenkins, R., & Moilanen, A. (2019). Identifying global centers of unsustainable commercial harvesting of species. *Science Advances., 5*(4), eaau2879.

Dutta, H. (2023). Illegal avian and reptilian pets: Global perspectives and challenges. *Cuadernos de Biodiversidad., 65*, 4–22.

Fink, C., Toivonen, T., Correia, R. A., & Di Minin, E. (2021). Mapping the online songbird trade in Indonesia. *Applied Geography, 134*, 102505.

Fukushima, C. S., Tricorache, P., Toomes, A., Stringham, O. C., Rivera-Téllez, E., Ripple, W. J., Peters, G., Orenstein, R. I., Morcatty, T. Q., Longhorn, S. J., & Lee, C. (2021). Challenges and perspectives on tackling illegal or unsustainable wildlife trade. *Biological Conservation, 263*, 109342.

Gallego-Zamorano, J., Benitez Lopez, A., Santini, L., Huijbregts, M. A., & Schipper, A. M. (2020). Combined effects of land use and hunting on distributions of tropical mammals. *Conservation Biology, 34*, 1271–1280.

Ghanghas, A., & Poonia, A. (2021). Advisory for dealing with import of exotic live species in India and declaration of stock: A boom or bane for illegal wildlife trade in India. *Environmental Sciences and Ecology: Current Research, 2*, 1026.

Gomez, L., & Bouhuys, J. (2018). *Illegal otter trade in Southeast Asia* (p. 37). TRAFFIC, Petaling Jaya.

Harris, J. B., Tingley, M. W., Hua, F., Yong, D. L., Adeney, J. M., Lee, T. M., Marthy, W., Prawiradilaga, D. M., Sekercioglu, C. H., Suyadi, N., & Winarni, N. (2017 Apr). Measuring the impact of the pet trade on Indonesian birds. *Conservation Biology., 31*(2), 394–405.

https://cites.org/sites/default/files/common/docs/meeting_info/amphibians/International%20Trade%20in%20Amphibians%20DRAFT%2015%20Nov%202023.pdf

https://iucn.org/our-union/commissions/group/iucn-ssc-amphibian-specialist-group

https://news.mongabay.com/2025/01/increase-in-gibbon-trafficking-into-india-has-conservationists-worried/

https://wildlifesos.org/exotic-pet-trade-threatens-indias-endangered-species/

https://www.ecojust.eu/indias-exotic-pet-trade/

https://www.wwfindia.org/news_facts/feature_stories/trafficking_of_exotic_wildlife_species_in_india/

Hughes, L. J., Morton, O., Scheffers, B. R., & Edwards, D. P. (2023 Jun). The ecological drivers and consequences of wildlife trade. *Biological Reviews., 98*(3), 775–791.

Kalra, S., Poonia, A., Sharma, R., Kaur, R., Sharma, P., Kumar, A., & Gulia, N. (2023). Wildlife conservation through local community engagement in India. *Uttar Pradesh Journal of Zoology., 44*(2), 13–24.

Kalra, S., Davies, A., Martin, R. O., & Poonia, A. (2024 Jan). Insights from the media into the bird trade in India: An analysis of reported seizures. *Oryx, 58*(1), 69–77.

Kannan, V. (2024). A note on hunting common palm civet (Paradoxurus Hermaphroditus) in the Tropical Dry Evergreen Forests of Uluru Coastal Village, Kancheepuram district, Tamil Nadu, India. *Acta Scientific Veterinary Sciences (ISSN: 2582-3183), 6*(7).

Levitt, T. (2020 November 25). How can we stop using soya linked to deforestation?. Guardian..

Manral, U., Sengupta, S., Hussain, S. A., Rana, S. A., & Badola, R. U. (2016). Human wildlife conflict in India: A review of economic implication of loss and preventive measures. *The Indian Forester, 142*(10), 928–940.

Martin, R. O., Senni, C., & D'Cruze, N. C. (2018). Trade in wild-sourced African grey parrots: Insights via social media. *Global Ecology and Conservation., 15*, e00429.

Mendiratta, U., Sheel, V., & Singh, S. (2017). Enforcement seizures reveal large-scale illegal trade in India's tortoises and freshwater turtles. *Biological Conservation, 207*, 100–105.

Mohapatra, R. K., Panda, S., Acharjyo, L. N., Nair, M. V., & Challender, D. W. (2015 Apr). A note on the illegal trade and use of pangolin body parts in India. *Traffic Bulletin., 27*(1), 33–40.

Musing, L., Suzuki, K., & Nekaris, K. A. (2015). Crossing international borders: The trade of slow lorises, Nycticebus spp., as pets in Japan. *Asian Primates Journal., 5*(1), 12–24.

Oyanedel, R., Gelcich, S., & Milner-Gulland, E. J. (2021). A framework for assessing and intervening in markets driving unsustainable wildlife use. *Science of The Total Environment., 792*, 148328.

Pandey, S., Samal, A., & Kar, N. B. (2024). Enforcement and capacity building to combat illegal trade of alexandrine parakeets Psittacula eupatria (Linnaeus 1766) in Odisha, India. *Sustainability and Biodiversity Conservation., 3*(3), 1–4.

Poole, C. M., & Shepherd, C. R. (2017 Jul). Shades of grey: The legal trade in CITES-listed birds in Singapore, notably the globally threatened African grey parrot Psittacus erithacus. *Oryx, 51*(3), 411–417.

Poonia, A., Ghanghas, A., Sharma, S., Jangra, M., & Sindhu, V. (2022). Trends in CITES listed bird's trade in south Asian countries in view of evolution of Indian laws during last four decades. *International Journal of Ecological Economics and Statistics., 43*(9), 115.

Pragatheesh, A., Deepak, V., Girisha, H. V., & Tomar, M. S. (2021). A looming exotic reptile pet trade in India: Patterns and knowledge gaps. *Journal of Threatened Taxa., 13*(6), 18518–18531.

Reino, L., Figueira, R., Beja, P., Araújo, M. B., Capinha, C., & Strubbe, D. (2017). Networks of global bird invasion altered by regional trade ban. *Science Advances., 3*(11), e1700783.

Ribeiro, J., Reino, L., Schindler, S., Strubbe, D., Vall-llosera, M., Araújo, M. B., Capinha, C., Carrete, M., Mazzoni, S., Monteiro, M., & Moreira, F. (2019 Oct). Trends in legal and illegal trade of wild birds: A global assessment based on expert knowledge. *Biodiversity and Conservation, 28*, 3343–3369.

RIS(Research and information System for developing countries). (2021). Ayush sector in India: Prospects and challenges. https://www.ris.org.in/sites/default/files/Publication/Ayush%20Study-Public%20Policy%20and%20Economic-FINAL%20REPORT%2028%20OCT_0.pdf#:~:text=With%20the%20current%20turnover%20of%20US$%2018.1,grown%20by%2017%20per%20cent%20during%202014%2D2020.&text=The%20present%20study%20has%20estimated%20the%20global,to%20touch%20US$%20746.9%20billion%20by%202022.

Sengottuvel, R. R., Mendis, A., Sultan, N., Shukla, S., Chaudhuri, A., & Mendiratta, U. (2024 Jan). From pets to plates: Network analysis of trafficking in tortoises and freshwater turtles representing different types of demand. *Oryx, 58*(1), 78–89.

Sethi, S., Goyal, S. P., & Choudhary, A. N. Poaching, illegal wildlife trade, and bushmeat hunting in India and South Asia. In *International wildlife management: Conservation challenges in a changing world* (Vol. 12, p. 157).

Sinclair, J. S., Stringham, O. C., Udell, B., Mandrak, N. E., Leung, B., Romagosa, C. M., & Lockwood, J. L. (2021 Sep). The international vertebrate pet trade network and insights from US imports of exotic pets. *Bioscience, 71*(9), 977–990.

Singh P. A.. Exotic pet ownership in India: Exploring motivations, legal influences, and informal exchanges (Doctoral dissertation, Carleton University).

Sorenson, J. (2011). Ethical fashion and the exploitation of nonhuman animals. *Critical Studies In Fashion & Beauty, 2*(1–2), 139–164.

Suvra, A. N., & Ontar, M. H. (2024). Illegal wildlife trade in Bangladesh and India: A study of comparative Laws. *Uti Possidetis: Journal of International Law., 5*(3), 488–547.

Swapna, N., Radhakrishna, S., Gupta, A. K., & Kumar, A. (2010 Feb). Exudativory in the Bengal slow Loris (Nycticebus bengalensis) in Trishna Wildlife Sanctuary, Tripura, Northeast India. *American Journal of Primatology: Official Journal of the American Society of Primatologists, 72*(2), 113–121.

Valdez, J. W. (2021). Using Google trends to determine current, past, and future trends in the reptile pet trade. *Animals, 11*(3), 676.

Vall-llosera, M., & Cassey, P. (2017 Mar). 'Do you come from a land down under?'characteristics of the international trade in Australian endemic parrots. *Biological Conservation, 207*(1), 38–46.

Van Uhm, D. P. (2016 Nov 15). *The illegal wildlife trade: Inside the world of poachers, smugglers and traders*. Springer.

Warwick, C., Lambiris, A. J., Westwood, D., & Steedman, C. (2001). Reptile-related salmonellosis. *Journal of the Royal Society of Medicine., 94*(3), 124–126.

Wildlife Protection Act, 1972, No. 53, Acts of Parliament, 1972 (India).

World Health Organization. (2022). *WHO establishes the global Centre for Traditional Medicine in India*. World Health Organization. https://www.who.int/news/item/25-03-2022-who-establishes-the-global-centre-for-traditional-medicine-in-india#:~:text=This%20global%20knowledge%20centre%20for,traditional%20medicine%20practices%20and%20products

Zhang, M., Gouveia, A., Qin, T., Quan, R., & Nijman, V. (2017). Illegal pangolin trade in northernmost Myanmar and its links to India and China. *Global Ecology and Conservation., 10*, 23–31.

Ziegler, S., Fa, J. E., Wohlfart, C., Streit, B., Jacob, S., & Wegmann, M. (2016 May). Mapping bushmeat hunting pressure in Central Africa. *Biotropica, 48*(3), 405–412.

Part V
Technological Interventions in Wildlife Conservation and Trade Regulation

Chapter 13
The Digital Era: E-Commerce and Illegal Wildlife Trade

Kumudani Bala Gautam, Diksha Kumari, and Akhilesh Kukreti

Contents

Abstract The illegal Wildlife Trade (IWT) is a lucrative industry. Similar to illegal drugs and weapons, wildlife and animal parts are trafficked by deadly cross-border networks. It is nearly impossible to get accurate estimates of the trade's worth due to its elusive nature. An estimated billions of dollars are involved. It seriously jeopardizes public and global security as well as biodiversity. With the advent of digital age, the traditional wildlife trade is quickly embracing new technology as a medium. The chapter addresses how e-commerce presents significant difficulties for law enforcement since it facilitates purchases and sales of criminals. Wildlife products are being traded over social networking sites and e-commerce sites like Facebook, Instagram, Amazon, OLX and eBay, where maintaining anonymity is easier. Further, scale, scope and available statistics of illegal wildlife trade have been discussed across countries.

The chapter also presents the factors supporting the emergence of India as a major hotspot of illegal wildlife trading activities. Factors encompassing heightened demand include traditional medicines, luxury goods and exotic pets; such utilization of wildlife products in cultural or religious rituals and facilitated transportation.

K. B. Gautam (✉)
Department of Biosciences, Graphic Era (Deemed to be University), Dehradun, Uttarakhand, India

D. Kumari
Defence Institute of Physiology & Allied Sciences (DIPAS), New Delhi, India

A. Kukreti
Government PG College Karanprayag, Chamoli, Uttarakhand, India

A. Poonia et al. (eds.), *Wildlife Trade in India*,
https://doi.org/10.1007/978-981-95-3694-8_13

Authors also discuss how the technology that have helped criminals in the illegal activities, can also be used to identify and curb wildlife crime in collaboration with enforcement agencies with social networking sites. Innovative tools for surveillance and intervention can be developed using social media mining and machine learning. Incorporating societal consciousness, technical innovation, adaptability, policies, and international cooperation are all necessary to ensure the protection of wildlife.

Keywords Cross-border · E-commerce · Hotspot · Wildlife products

13.1 Introduction

The trade of wildlife is antique and archaic, as human history traces back to the early Egyptian and Greek civilizations with commercial transactions continuing uninterrupted through the ages (Hughes, 2003; Kalof, 2007). The Roman Empire was known to engage in extensive, unsustainable commercial wildlife utilization which is now internalized in all the societies (Whatmore & Thorne, 1998; Milner-Gulland, 2018). Wild flora and fauna is harvested for various reasons such as food, medicine, ornaments, fuel, building materials and other related recreational and cultural utility. Illegal wildlife trade covers the illicit purchase, storage and sales of plants, animals, animal body parts or goods made from them; such species are also protected and characterized as endangered through national and international regulations (UNODC, 2016). Since the longest time, the illegal wildlife trade (IWT) continues to be the most widespread and adversely affecting types of organized crime, endangering public health, regional economies and biodiversity worldwide. Recently, the trade has increased for number of species including keystone species and umbrella species as well as lower taxa such as rhinoceros, elephant, tiger, whale, minke, small cats, pangolin, hornbills, amphibians, reptiles, turtle, tortoises and numerous marine species (Warchol, 2004; Nijman, 2010; Beastall et al., 2016; Gautam et al., 2023; Sahoo et al., 2024). To combat the illegal wildlife trafficking, stricter laws, community involvement, technical monitoring tools and improved international collaboration are necessary components to formulate some multi-pronged strategies. Offering alternative livelihood options to pertinent stakeholders of the locality and supporting vital financial backing for region based conservation initiatives aid in appropriately managed wildlife trade which eventually help in reducing associated threats to biodiversity such as habitat loss due to land conversion or fragmentation from urbanisation. The escalated trafficking of wildlife poses a substantial risk to biodiversity as the intense collection of species from their natural habitat is pushing them to edge of extinction. Additionally, the rapid loss of habitats, fragmentation and anthropogenic activities are further contributing to the annihilation for various species. The advent in the field of internet technology is one factor that enables poachers to circumvent enforcement agencies and market is evolving quickly.

The twenty first century world is living in the digital era. The technologies and networking provided by the internet is increasing the productivity and decreasing the costs which facilitates the new market opportunities. Among various facets of information and communication technology (ICT), one is E-Commerce. E-Commerce is also known as E-Business which refers to the transaction of goods and services through the electronic communications. Hence, any type of business transaction in which the parties communicate electronically instead of face-to face or through physical exchanges. The functioning of E-Commerce has been around for over 30 years. However, general public became familiar to this only in last 10 years. There are three basic types of E-Commerce: namely, Business to Business (B2B), Business to Consumer (B2C) and Consumer to Consumer (C2C). The B2B types businesses deal with the interaction between two or more companies (e.g. Intel, Boeing). The B2C type business usually sells their product or services to the customer (e.g. Netflix, Coca-Cola). On the contrary, C2C type business majorly is an interaction between two consumers (e.g. eBAY, OLX). Subsequently, E-Commerce represents the modernization of key business processes through the implementation of internet technology, where buying and selling transactions takes place in a digital environment. The communication regarding product information, payments, and other related activities occur digitally, through telephonic lines or networks rather than through physical interactions and confirmation. Business process is also automated and the entire operation is conducted within a technological framework. Recently, the services provided by the E-Commerce sector have gained appreciation compared to traditional physical marketing for their convenience, speed and consumer centric approach, often surpassing the reach and efficiency of traditional systems. The improvement in the quality, availability and delivery of goods/products/services have enriched user trust and engagement in digital marketplaces. A general pathway of products processing through E-Commerce is illustrated in Fig. 13.1. Commonly a user/consumer places an order through an online platform followed by a successful digital payment which upon confirmation triggers message to merchant who is responsible for forwarding the order details to the warehouse for dispatch. Afterwards, the product is then packed and sent for the delivery eventually reaching the user's doorstep. The merchant's anonymity is a noteworthy aspect of this procedure. The consumer remains oblivious of the merchant's identity throughout the process and interacts digitally. On the contrary, merchant gains access to the user's information such as delivery address, contact information etc. through the order receipt, which is necessary for successful delivery but not reciprocated in terms of identity transparency. The smooth functionality of e-commerce system is beneficial for the consumer in terms of privacy and operational efficiency but it also introduces a critical risk and often misused by delinquent individuals. The well-maintained anonymity of the merchant facilitates the clandestine trade of prohibited or illicit items including wildlife derivatives. The digital transactions combined with the encrypted communication, drop shipping, inconspicuousness of merchant and minimal regulatory oversight has catered a ground for illegal wildlife trade to flourish with minimum detection globally. Traditional poachers and traffickers working as middlemen are

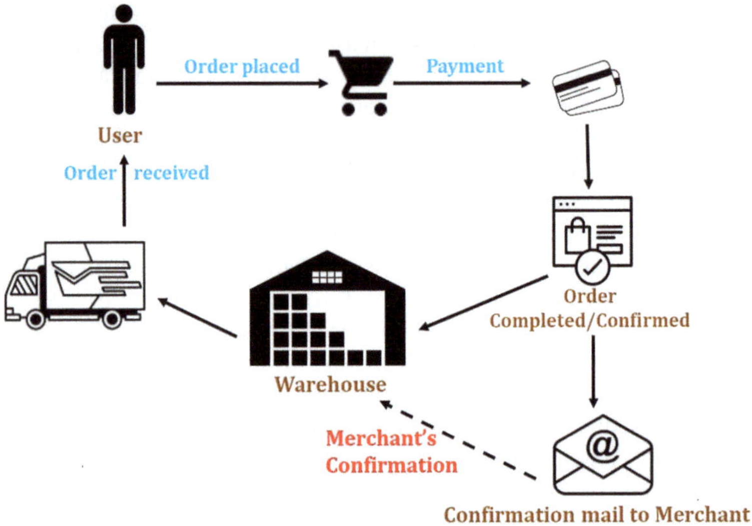

Fig. 13.1 A typical flow of product processing

increasingly adapting the procedure of digital market. Currently, rather than relying on the risky, open market, cash transactions or physical trade routes, which are easier to monitor by enforcement agencies, traffickers are now operating as anonymous digital merchant. This paradigm shift has drastically reduced their chances of being apprehended by enforcement agencies, thus enabling them to bypass the traditional law enforcement mechanisms. The rise of digital shopping has reduced the visibility and physical footprint of the illegal wildlife trade and consecutively allowed traffickers to reach global buyers under the guise of legitimate products and services. E-Commerce platforms that are inadequately monitored for illicit content or that allow third-party vendors without stringent verification protocols contribute further to expansion of illicit trade worldwide. Therefore, while E-commerce brings undeniable comfort and benefits to legal trade and consumer access, it also poses serious conservation problem and enforcement challenges.

13.2 Scale and Scope of Illegal Wildlife Trade

The wildlife trade is the fourth most lucrative criminal enterprise worldwide after the smuggling of firearms, illegal drug sales and human trafficking (Doody et al., 2021). In addition to drug, weapon and human trafficking, the illegal wildlife trade represents one of the largest economies valued at billions of dollars annually (UNODC, 2016). This illicit market is fueled by the high demand for luxury items made from animal parts such as ivory, tiger bones, rhino horn, pangolin scales, turtle calipee or shell along with traditional medicine, aphrodisiac item and exotic pets ('t Sas-Rolfes

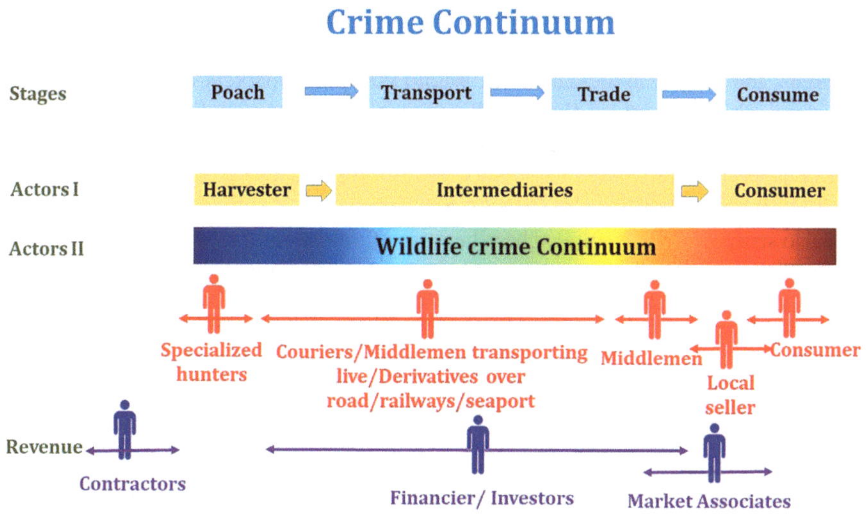

Fig. 13.2 Components of wildlife crime continuum

et al., 2019). India is recognized as a biodiversity hotspot and acts as both transit and supply hub for trafficking of wildlife (Mendiratta et al., 2017; Sahoo et al., 2024). Such illegal trade in wildlife is leading to the extensive economic, ecological and socio-political consequences, hence recognized as one of the most significant forms of multi-national organized crime (Cardoso et al., 2021; Morton et al., 2021). The number of threatened, vulnerable and endangered species has increased due to the rising demand globally (Butchart et al., 2010). The heightened global demand is facilitated by the trans-landscape trade which unknowingly induce the pathogens risk to a new area leading to the general health concerns and spread of infectious diseases among human population (Daszak et al., 2000). The wildlife trade affects every region of the world with Asia and Africa being hotspots for sourcing and demand has an approximate value between $19–$23 billion annually (Osorio, 2024). High valued wildlife trade enterprise has been characterized as an invisible market. Attempts to gather precise and thorough statistics on its scope and impact are hampered by the lack of transparency in the trade. To grasp how anonymous merchant operates and how digital devices are used, it is essential to comprehend the various elements of the trade continuum that function together seamlessly (Fig. 13.2).

Wildlife trade consists of four phases: poaching, transportation, trading and consumer's possession. The actors involved effectively carry out and finalize the stages of this continuum. Harvesters are often skilled poachers or individuals who illegally capture or hunt wildlife. In the intermediary category, many individuals participate in various tasks including transportation handling, transactions, retail marketing as well as brokers and traders who negotiate deals and facilitate smooth trading. The final actor is consumer who purchases and utilizes the wildlife products or their derivatives. Conversely, revenue actors generally operate behind the scenes and have less physical involvement in the continuum. Contractors recruit hunters while investors provide funding for the poaching operations and trade logistics whereas, the market associates work within consumer markets and profiting from the demands. To further obscure the trade's visibility, wildlife traffickers also employ sophisticated strategies and convert techniques to escape law enforcement and conservation agencies. Traffickers use corruption and weak border controls to their advantage, enabling goods and wildlife to cross undetected. The illegal trade has also spread into the digital sphere because of online platforms where transactions are carried out anonymously thus making it more difficult to monitor and control (Esmail et al., 2020).

13.3 Trends of Illegal Wildlife Trade in India

Continuous poaching due to increasing demand poses a key threat to biodiversity (Scheffers et al., 2019). Widespread trade of wild animals is happening as live pets, bush meat, ornaments, medicine and magic occults (Gray et al., 2018; Nijman, 2010; Sharma et al., 2019). Mass slaughter of majestic species like elephants and rhino has drawn them to verge of extinction during twentieth century which led to determinant rehabilitation missions (Milliken & Shaw, 2012). But majority of species in trade are not high priority flagships and are vastly traded both internationally and domestically (UNODC, 2016). Approximately, 152 million reptiles were traded globally over the span of four decades (Harfoot et al., 2018).

According to Wildlife Crime Control Bureau (WCCB) and other enforcement agencies working in India, between 2012–2018, approximately 9253 poachers were apprehended in various incidents, yet only 2% of them were convicted. Over the past 8 years, National Crime Records Bureau (NCRB) documented that Tamil Nadu, Karnataka, Maharashtra, Madhya Pradesh, West Bengal, Assam, Rajasthan and Uttar Pradesh are major states of India involved in the cases related to wildlife crime. Ease of transportation such as railway connectivity, roadways, airways and presence of busiest port, is the major factor for becoming a hub for illegal trade market or transit routes. TRAFFIC-India documented that approximately 70,000 native and exotic species weighing around 4000 Kg were trafficked between 2011–2020. The list of trafficked exotic species includes red-eared slider turtle (*Trachemys scripta elegans*), One-toed hedgehog (*Atelerix albiventris*), Iguana (*Iguana iguana*), Marmosat (*Callithrix jacchus*), Tamarin (*Saguinus* spp.), Tricolor squirrels (*Callosciurus prevostii*), Kangaroos (*Macropus* spp.), Hornbills and various exotic birds (Pers. Comm.). Furthermore, numerous plant species such as agarwood (*Aquilaria malaccensis*), red sanders (*Pterocarpus santalinus*), sandalwood (*Santalum album*) and parts of medicinal and aromatic plants like Kuth (*Saussurea costus*) roots also have been seized at different airports in India (Rana & Kumar, 2023). The primary factors driving illegal wildlife trade (IWT) include the demand for traditional medicines, increased affluence that creates new markets for luxury items and exotic pets and the use of wildlife products in cultural or religious ceremonies. Globalisation is also playing a vital factor for increased exotic pet trade in India, the flow of K-pop related music, movies, serials, cosmetic, toys etc. is evident in India along with 'Kawaii' which means "Cute" (Crossley, 2020).

13.4 Role of E-Commerce in IWT

The commerce in wildlife has always taken place in the actual or physical marketplaces. Its scope and character have drastically evolved in recent years with the emergence of internet a significant space and market place for numerous wildlife become accessible (Lavorgna, 2014). The IWT seems to have shifted to other channels, especially social media, despite law enforcement's relative success in curbing it on major E-Commerce sites (Yu & Jia, 2015). The rapid advancement of information and communication technology has made virtual markets on the Internet platforms a smooth channel for the dissemination of availability, trafficking, selling and distribution of animal products (Herrel & Van der Meijden, 2014). Recently, social media platforms have grown in popularity as a conduit for the legal and illicit wildlife trade due to the rise in Internet access (South & Wyatt, 2011). There are many advantages of e-commerce when compared to traditional commerce as shown in Table 13.1. Due to the ease of doing business without hassle, users are now relying more upon the digital marketplace. The online market has minimized the time spent travelling to the physical stores, is more cost-effective, reduces reliance on the cash transactions and offers consumers the ability to compare different sellers for the same item within a similar timeframe, therefore these are all key reasons customers prefer digital shopping over traditional methods. Additionally, the IWT is benefiting from the advancements in social media and e-commerce platforms, hence traffickers can easily lure potential buyers while evading the law enforcement and selling illegal flora and fauna and their products.

Online marketing and sales are possible for items derived from endangered species, such as apparel or jewellery fashioned from animal scales, tusks, or skins. Such goods may nonetheless find receptive consumers in other nations even if they are prohibited in one. Legal or illegal, animal e-commerce has expanded rapidly on social media platforms because of its user-friendliness and growing user base. In the majority of the world's areas, e-commerce is free, unfettered and mostly unregulated which makes it possible for criminal activities to occur through straightforward anonymous sales and purchases. For example, with over 730 million members, China has the largest online community in the world. Their e-commerce platforms have been the primary means of selling animals, with half of that being related to the ivory trafficking (Komosny & Mehic, 2018). Online platforms provide a vast marketplace where traffickers can advertise and sell illegal wildlife products to a global audience. These platforms range from major e-commerce websites to smaller niche platforms and social media groups. E-commerce bypasses geographical barriers, enabling transactions across countries without physical trade hubs. For luring potential buyer, wildlife traders share images and details about their products on

Table 13.1 Key elemental differences between Traditional commerce and E-commerce

Key elements	E-Commerce	Traditional Commerce
Value creation	Information	Product or service
Ease of doing business	Yes	No
Buying/selling	From home/warehouse	Needs a designated shop/physical marketplace
Pricing flexibility	Efficient	Fixed/non-flexible
Coverage	Wide (globally)	Restricted to surrounding of marketplace
Payment	Online transaction/cash	Predominantly cash/online transaction
Operational strategy	Responsive to customer choice data and trends	Primarily follows traditional business norms and routines
Competitive strength	Agile, speedy and innovative	Reliable on quality and price control
Pressure on market	Highly influenced by the customer and low entry barriers	Influenced by suppliers and product substitution threats
Resource orientation	Concentrated on understanding and meeting the customer needs	Mostly emphasizes management of supply chains and inventory
Customer engagement	Digital interactions occur via interfaces	Face-to-face communication with the salespersons
Mode of communication	Depend on digital media, AI bots and virtual support	Relies on personal conversations and direct assistance
Approachability	Round the clock; 24 × 7	Limited to specific store timings
Customer experience	Self-navigated with miniscule staff involvement	Seller guides and influences purchase decisions
Marketing approach	Personalized targeting user data analytics	Targeting mass audiences follows one-size fits all philosophy
Promotional tactics	Compelled by online reviews, influencers and sharing	Depends on in-store promotions and physical displays
Product type	Common for standardized or digital products	Better suited for tangible or perishables goods, requires inspection

social media and networks with people who have found their products through word-of-mouth. Hence it is essential to monitor such advertisement, news, information, shared posts on different social media by enforcement agencies. In this sense, machine learning presents an abundance of new possibilities, assisting in the creation of equipment that facilitate the analysis of enormous amount of raw data extracted from social media sites in order to track trade of wildlife more rapidly and effectively. In particular, by automatically analysing photos, words, and videos shared online, machine learning models and algorithms have improved our ability to recognize the existence of various species present in the dataset of different social media sites and retrieved from all networks. In recent times, social media has become an accessible platform with billions and millions of user base (Xiao et al., 2017). According to the Global Social Media Statistics by DataReportal, the most used social media is Facebook followed by YouTube, WhatsApp, Instagram, TikTok and many more with monthly active users of 3.07 billion, 2.54 billion, 2 billion, 2 billion and 1.84 billion respectively. These social networking sites have multi lingual characteristics allowing user or organisation to easily form different profiles, share material/information and connect via message without hindrance. Advertisements for wildlife trading have been regularly sorted or seen on such sites through latest investigations and news reports (CNBC, 2018).

To evade detection and make financial tracking more difficult, some unlawful traders employ anonymous payment methods, such as cryptocurrencies, making it more challenging for law authorities to investigate illicit transactions (Dupuis & Gleason, 2020). Cross-border e-commerce transactions frequently provide jurisdictional challenges for law enforcement. It might be difficult to implement laws that are prohibited in one country but legal in another. Illegal wildlife trade on dark web has not evolved much, probably, because it requires high order technical knowledge and skills to access the dark web and locate illegal wildlife products there.

13.5 Utilization of E-Commerce to Combat IWT

Cybercrimes, such as ransomware, identity theft, computer and network intrusions, and illicit online trafficking of goods and services, have become worldwide phenomena with increasing scope and severity due to the internet's rapid development and spread. Fighting illicit wildlife transactions online is a fight against time with criminals. Regular activity monitoring to identify transactions where animal products are being purchased and sold illegally is made difficult by the speed and anonymity of online commerce. Any effort to keep an eye on online activity must be flexible enough to quickly adjust to new situations as they emerge. Rapid detection of cybercrime involving wildlife is made possible by big data and sophisticated data mining techniques, especially when utilizing computational techniques like deep algorithm based learning, machine learning and natural language processing that are influenced by thorough manual annotation to describe buyer-seller interactions and related evidences (Xu et al., 2019). More thorough surveillance in multiple languages, including the use of regional keywords and wildlife trade-specific vocabulary, should be the first step towards comprehensive monitoring of wildlife trafficking via SMS-based services and various platforms on internet. These efforts ought to be flexible enough to adjust to changes in language and selling tactics. Policy coherence is extremely necessary while formulating national, regional or international policies since they are specific to the technical challenges of online wildlife trafficking as opposed to those needed in the non-cyber field. Initiatives like CITES (Convention on International Trade in Endangered Species of Wild Fauna and Flora) set international standards that e-commerce platforms can follow to ensure compliance with global efforts to combat IWT. Future policies should explicitly necessitate and positively compel social networking sites, e-commerce sites, domain registration companies, internet service providers and search engines to remove content that violates their own terms of use or obligations under national law or conservation treaties. This is in addition to sharing information on illicit traffickers with law enforcement and conservation authorities to support offline investigations. To support the efforts of conservation organizations, researchers must also keep coming up with new detection and categorization techniques. For example, previous studies have proposed a framework that utilises machine learning to mine, filter and select pertinent data in order to examine the illegal wildlife trade on social media platforms (Di Minin et al., 2015; Di Minin et al., 2019).

To reduce the IWT, there is a necessity to educate local people who eventually become buyers of traded articles, public need to be more educated and vigilant about the catastrophic effects of poaching/hunting of wild fauna and flora on the environment. Infusion of advertisement, news and games might educate customers regarding the effects of the illegal wildlife trade and its repercussions by leveraging their global presence. They can promote ethical alternatives, provide awareness messages, and instruct consumers on how to spot illicit goods. Consumer education can increase demand for sustainable, lawful wildlife-related products and deter involvement in illicit commerce. Recent development in artificial intelligence like machine learning can be of great use in tracking online illegal trading. In order to successfully analyse the content of any digital dataset such as digital advertisement, different social media content, web pages etc., the machine learning technique can apply what it has learned from earlier assignments to new datasets. One crucial step in identifying potential solution for this emerging issue is the mining of social media dataset and the creation of machine learning models and algorithms that can help in identification of species targeted for trading. AI, machine learning, and other technologies can be used by e-commerce businesses to identify and highlight questionable wildlife-related products that are being offered online. These tools can automatically notify platform administrators and law enforcement when they detect keywords, photos, or descriptions associated with prohibited animals or goods. User-generated text, photos, and videos, along with related metadata like the location and time of the content's uploading and user connections, can all be accessed using the application programming interface offered by many social media sites. It is wasteful and time-consuming to process such material manually, but expert machine-learning algorithms can identify relevant information by filtering the whole database. It is possible to train these algorithms to recognize different kinds of wildlife through the images or videos.

Deep-learning algorithms are very effective in processing natural language and categorizing images according to the content and identifying specific objects publi cized within those images. These techniques can be applied to social media databases to study and analyse the human behaviour on a scale that was not possible before the age of machine learning. However, the drivers of the biodiversity catastrophe are still infrequently addressed using these methods and data sources (Di Minin et al., 2015). Logistics and courier service companies must take a vigorous role in the fights against the illicit wildlife trade on the internet by regularly and thoroughly inspecting their cargoes to make sure they adhere to current laws.

The companies owing different social media platforms have the technical capacity to monitor and search for key code words. These companies could play a significant and active role in combatting issue of IWT along with the relevant law enforcement agencies of the particular country. Given the example of E-commerce platforms like Amazon, Snapdeal, eBay, Alibaba and others have been stepping up efforts to combat illegal wildlife trade by implementing stricter policies and monitoring mechanism (Sharma et al., 2019). The alliance among government bodies, industries and non- governmental organization could act foundational in combating the IWT in digital era. The mandate of such collaborations is data exchange, cross-promotional efforts with cooperative operations leading to curtail illicit transactions. Therefore, there is an urgent necessity for integrated monitoring system with improved digital forensics and enhanced policy frameworks that can detect and deter the online trafficking of wildlife and their derivatives. Thereby protecting the biodiversity from escalated exploitation in this digital era.

References

Beastall, C., Shepherd, C. R., Hadiprakarsa, Y., & Martyr, D. (2016). Trade in the Helmeted Hornbill Rhinoplax vigil: The 'ivory hornbill'. *Bird Conservation International, 26*, 137–146.

Butchart, S. H. M., Walpole, M., Collen, B., Van Strien, A., Scharlemann, J. P. W., Almond, R. E. A., et al. (2010). Global biodiversity: Indicators of recent declines. *Science, 328*(5982), 1164–1168.

Cardoso, P., Amponsah-Mensah, K., Barreiros, J. P., Bouhuys, J., Cheung, H., Davies, A., et al. (2021). Scientists' warning to humanity on illegal or unsustainable wildlife trade. *Biological Conservation, 263*, 109341.

CNBC. (2018). Facebook puts ads on pages illegally selling animal parts [www document]. https://www.cnbc.com/2018/04/10/facebook-puts-ads-onpages-illegally-selling-animal-parts.html.

Crossley, É. (2020). Japanese animal tourism and the kawaii (cute) aesthetic. *Scope (Art & Design), 19*, 65–71.

Daszak, P., Cunningham, A. A., & Hyatt, A. D. (2000). Emerging infectious diseases of wildlife—Threats to biodiversity and human health. *Science, 287*(5452), 443–449.

Di Minin, E., Tenkanen, H., & Toivonen, T. (2015). Prospects and challenges for social media data in conservation science. *Frontiers in Environmental Science, 3*, 63.

Di Minin, E., Fink, C., Hiippala, T., & Tenkanen, H. (2019). A framework for investigating illegal wildlife trade on social media with machine learning. *Conservation Biology, 33*(1), 210–213.

Doody, J. S., Reid, J. A., Bilali, K., Diaz, J., & Mattheus, N. (2021). In the post-COVID-19 era, is the illegal wildlife trade the most serious form of trafficking? *Crime Science, 10*, 19.

Dupuis, D., & Gleason, K. (2020). Money laundering with cryptocurrency: Open doors and the regulatory dialectic. *Journal of Financial Crime, 28*(1), 60–74.

Esmail, N., Wintle, B. C., 't Sas-Rolfes, M., Athanas, A., Beale, C. M., Bending, Z., et al. (2020). Emerging illegal wildlife trade issues: A global horizon scan. *Conservation Letters, 13*, e12715.

Gautam, K. B., Kumar, A., Das, A., & Gupta, S. K. (2023). Himalayan upliftment and Shiwalik succession act as a cradle for divergence in Bengal monitor lizard Varanus bengalensis (Reptilia: Varanidae) in India. *Cladistics, 39*(5), 382–397.

Gray, T. N., Hughes, A. C., Laurance, W. F., Long, B., Lynam, A. J., O'Kelly, H., et al. (2018). The wildlife snaring crisis: An insidious and pervasive threat to biodiversity in Southeast Asia. *Biodiversity and Conservation, 27*, 1031–1037.

Harfoot, M., Glaser, S. A., Tittensor, D. P., Britten, G. L., McLardy, C., Malsch, K., & Burgess, N. D. (2018). Unveiling the patterns and trends in 40 years of global trade in CITES-listed wildlife. *Biological Conservation, 223*, 47–57.

Herrel, A., & Van der Meijden, A. (2014). An analysis of the live reptile and amphibian trade in the USA compared to the global trade in endangered species. *Herpetological Journal, 24*(2), 103–110.

Hughes, J. D. (2003). Europe as consumer of exotic biodiversity: Greek and Roman times. *Landscape Research, 28*, 21–31.

Kalof, L. (2007). *Looking at animals in human history*. Reaktion Books.

Komosny, D., & Mehic, M. (2018). The value of geographic locations submitted by internet users. *IEEE Access, 6*, 62699–62706.

Lavorgna, A. (2014). Wildlife trafficking in the internet age. *Crime Science, 3*, 5.

Mendiratta, U., Sheel, V., & Singh, S. (2017). Enforcement seizures reveal large-scale illegal trade in India's tortoises and freshwater turtles. *Biological Conservation, 207*, 100–105.

Milliken, T., & Shaw, J. (2012). *The South Africa–Viet Nam rhino horn nexus*. TRAFFIC.

Milner-Gulland, E. J. (2018). Documenting and tackling the illegal wildlife trade: Change and continuity over 40 years. *Oryx, 52*(4), 597–598.

Morton, O., Scheffers, B. R., Haugaasen, T., & Edwards, D. P. (2021). Impacts of wildlife trade on terrestrial biodiversity. *Nature Ecology & Evolution, 5*(4), 540–548.

Nijman, V. (2010). An overview of international wildlife trade from Southeast Asia. *Biodiversity and Conservation, 19*(4), 1101–1114.

Osorio, C. P. (2024). Battling the illegal wildlife trade through regulatory finance: The southeast Asian context. *Journal of Academics Stand Against Poverty, 5*(1), 44–74.

Rana, A. K., & Kumar, N. (2023). Current wildlife crime (Indian scenario): Major challenges and prevention approaches. *Biodiversity and Conservation, 32*(5), 1473–1491.

Sahoo, S., Kumar, A., Rai, J., & Gupta, S. K. (2024). Unlocking the genetic enigma: A paradigm shift towards evidence-based conservation of Indian star tortoise (Geochelone elegans). *Biological Conservation, 290*, 110421.

Scheffers, B. R., Oliveira, B. F., Lamb, I., & Edwards, D. P. (2019). Global wildlife trade across the tree of life. *Science, 366*(6461), 71–76.

Sharma, C. P., Kumar, A., Vipin, Sharma, V., Singh, B., Kumar, G. C., & Gupta, S. K. (2019). Online selling of wildlife parts with spurious names: A serious challenge for wildlife crime enforcement. *International Journal of Legal Medicine, 133*, 65–69.

South, N., & Wyatt, T. (2011). Comparing illicit trades in wildlife and drugs: An exploratory study. *Deviant Behavior, 32*(6), 538–561.

't Sas-Rolfes, M., Challender, D. W., Hinsley, A., Veríssimo, D., & Milner-Gulland, E. J. (2019). Illegal wildlife trade: Scale, processes, and governance. *Annual Review of Environment and Resources, 44*, 201–228.

United Nations Office on Drugs and Crime. (2016). World Drug Report 2016 (United Nations publication, Sales No. E.16).

Warchol, G. L. (2004). The transnational illegal wildlife trade. *Criminal Justice Studies, 17*, 57–73.

Whatmore, S., & Thorne, L. (1998). Wilderness: Reconfiguring the geographies of wildlife. *Transactions of the Institute of British Geographers, 23*(4), 435–454.

Xiao, Y., Guan, J., & Xu, L. (2017). *Wildlife cybercrime in China*. Traffic.

Xu, Q., Li, J., Cai, M., & Mackey, T. K. (2019). Use of machine learning to detect wildlife product promotion and sales on Twitter. *Frontiers in Big Data, 2*, 28.

Yu, X., & Jia, W. (2015). Moving targets: Tracking online sales of illegal wildlife products in China. *TRAFFIC Briefing, 1009649*, 1–10.

Chapter 14
Conservation Technologies: AI, Drones and Beyond in Wildlife Conservation and Mitigating Illegal Wildlife Trade

Vikram Delu, Yomto Mayi, and Shantabala Devi Gurumayum

Contents

Abstract The accelerating loss of biodiversity and the persistent threat of illegal wildlife trade have spurred the development and adoption of innovative digital technologies for wildlife conservation. Artificial intelligence (AI), drones, satellite remote sensing, genomic forensics, Internet of Things (IoT) devices, blockchain and autonomous systems are among the emerging tools revolutionising conservation practice across terrestrial and marine ecosystems. These technologies enable real-time monitoring, improve decision-making and support enforcement agencies in detecting and preventing wildlife crime. This chapter reviews the current applications of conservation technologies, highlighting recent advances in AI and automated image analysis, UAV-based surveillance, habitat mapping, forensic genetics

V. Delu
Scientist -B, Zoological Survey of India, Arunachal Pradesh Regional Centre, Itanagar, Arunachal Pradesh, India

Y. Mayi (✉) · S. D. Gurumayum
Zoological Survey of India, Government of India, Arunachal Pradesh Regional Centre, Itanagar, India

A. Poonia et al. (eds.), *Wildlife Trade in India*,
https://doi.org/10.1007/978-981-95-3694-8_14

and digital traceability systems. It also discusses the integration of big data and decision support platforms, and outlines challenges related to data interoperability, technical capacity and ethical considerations. Finally, the chapter presents future prospects and recommendations for scaling technology-driven conservation in an inclusive and sustainable manner.

Keywords Artificial intelligence · Drones · Remote sensing · Wildlife forensics · Internet of things · Blockchain · Illegal wildlife trade mitigation · Big data integration · Conservation technology · Biodiversity monitoring

14.1 Introduction

Over the past few decades, the unprecedented decline in global biodiversity has intensified efforts to identify and implement innovative conservation strategies. Human-induced land use changes, habitat degradation, and illegal wildlife trade remain among the most pervasive threats to wildlife populations globally (IPBES, 2019). Conventional conservation approaches, although significant, often face limitations in scale, precision and response time, particularly in remote and complex ecosystems. In response to these challenges, technological interventions have emerged as transformative tools to support evidence-based wildlife conservation and management (Pimm et al., 2015).

Advancements in artificial intelligence (AI), unmanned aerial vehicles (UAVs; commonly known as drones), remote sensing, and environmental forensics are reshaping how species are monitored and protected in real-time. AI-driven algorithms have enabled rapid, automated processing of large and heterogeneous datasets, including camera trap images, acoustic recordings, and satellite imagery, thereby enhancing species detection, modelling of spatial dynamics, and predictive conservation planning (Christin et al., 2019). Likewise, drones provide cost-effective and flexible solutions for monitoring wildlife in inaccessible terrains, reducing the risks associated with field patrols and improving the detection of illegal activities such as poaching and habitat encroachment (Koh & Wich, 2012).

The illegal wildlife trade (IWT) remains one of the primary drivers of biodiversity decline, with an estimated annual economic value of USD 7–23 billion (UNEP, 2020). Emerging conservation technologies, including forensic genomics and blockchain-based traceability systems, are increasingly strengthening enforcement mechanisms by enabling precise species identification and transparent supply chains (Ogden et al., 2020; CITES, 2021). Consequently, a convergence of digital technologies and conservation science is fostering a paradigm shift towards proactive and adaptive management of wildlife populations.

This review aims to provide a comprehensive overview of the emerging technologies—with emphasis on AI, drones and allied tools—that are revolutionizing wildlife conservation practices and mitigating the illegal trade in wildlife and wildlife-derived products. In doing so, we highlight current applications, case studies, existing challenges, and future prospects for integrating these technologies into global conservation planning.

14.2 Artificial Intelligence (AI) in Wildlife Conservation

Artificial intelligence has increasingly become a central component of modern conservation practice, providing powerful analytical capabilities that enhance both the efficiency and accuracy of wildlife monitoring and ecological forecasting. One of the most widely adopted applications is the use of AI-driven image recognition to automate the processing of large volumes of camera trap and drone images (Tabak et al., 2019). Convolutional neural networks (CNNs) can classify species with high accuracy and substantially reduce the time and labour requirements traditionally associated with manual image classification (Norouzzadeh et al., 2018). These AI-based systems also facilitate rapid feedback for rangers and conservation managers by detecting the presence of endangered or invasive species in near real time.

Machine learning algorithms have also demonstrated great promise in analysing multi-source datasets to predict spatiotemporal patterns of wildlife movements and poaching events (Kalan et al., 2019). For example, integration of historical poaching incidents, topographic variables and patrol effort data allows predictive modelling that can guide the allocation of anti-poaching patrols in highly threatened areas. Combining AI with acoustic monitoring technologies has further enabled the automatic detection of species-specific vocalisations, supporting non-invasive monitoring in dense or inaccessible habitats (Joppa, 2020).

Beyond monitoring, AI has been applied to understand ecosystem-level processes through the analysis of remote sensing data for habitat suitability assessments and corridor mapping (Reddy et al., 2025). Emerging AI-powered decision support platforms provide conservation planners with data-driven insights and scenario-based outputs that assist in prioritizing conservation interventions under rapidly changing environmental conditions.

While the potential of AI is considerable, its effective deployment requires high-quality training datasets, robust ground-truthing, and interdisciplinary collaboration between ecologists, computer scientists and local stakeholders. In the coming years, successful conservation strategies will increasingly rely on the integration of AI systems with other digital technologies to support adaptive conservation planning and effective protection of wildlife populations.

14.3 Drone Applications in Wildlife Conservation

Unmanned aerial vehicles (UAVs), commonly referred to as drones, have rapidly transformed field-based conservation practice, offering a flexible and cost-effective alternative to conventional ground and aerial surveys. Their ability to capture high-resolution imagery across large and remote areas enables efficient assessment of wildlife populations, habitat conditions and anthropogenic disturbances. Drones equipped with thermal or multispectral sensors can detect animals even in dense vegetation or during low-light conditions, thus improving the accuracy of wildlife counts and minimizing observer bias (Chrétien et al., 2016). In addition, rapid data acquisition and processing allow real-time monitoring of population dynamics and seasonal movement patterns, facilitating timely interventions by conservation managers.

In the context of anti-poaching efforts, drones have proven particularly valuable due to their capacity for discreet surveillance of protected areas. Real-time video streaming and geo-tagged imagery allow rangers to identify suspicious activities and direct patrol units to high-risk sites before illegal incidents occur (Mulero-Paules et al., 2022). This proactive approach increases the likelihood of deterrence and reduces the operational risks faced by frontline staff. Furthermore, the integration of drones with AI-powered detection systems enhances their effectiveness, enabling automated recognition of poachers and illegally harvested wildlife products.

Beyond wildlife monitoring, drones serve as vital tools for landscape-level assessments of habitat degradation, deforestation and the expansion of agricultural frontiers. Detailed spatial data generated by UAVs contribute to habitat suitability models and assist in the design of wildlife corridors for fragmented populations (van Andel et al., 2015). Such high spatial resolution data can also verify remotely-sensed satellite information and generate base maps for ecological restoration projects.

Although the use of drones continues to grow, challenges remain related to regulatory frameworks, local acceptance, and technical capacity in resource-limited regions. Moreover, ethical considerations relating to disturbance of wildlife and privacy of local communities must be addressed through appropriate guidelines and stakeholder participation. Nevertheless, the expanding technological capabilities and decreasing operational costs of drones underscore their increasing importance in the global conservation toolbox.

14.4 Camera Traps and Automated Image Analysis

Camera traps have become an essential non-invasive tool for wildlife monitoring, enabling the collection of continuous data on species presence, behaviour and activity patterns across remote landscapes. These devices can operate for long periods under varying environmental conditions and generate large image datasets that facilitate the assessment of population trends, even for cryptic and nocturnal species. When combined with automated image analysis, the efficiency of camera trap monitoring is greatly increased. Recent advances in machine learning and convolutional neural networks enable the rapid classification of species and the identification of individual animals from large image libraries. Global citizen-science initiatives such as **Snapshot Serengeti** and **eMammal** have demonstrated the potential of integrating public participation with AI-assisted classification, producing high-quality biodiversity datasets for conservation analyses (Swanson et al., 2016). Automated pipelines also allow for the detection of poaching activities and the identification of illegal human presence in protected areas, providing early warnings to law enforcement agencies. Moreover, integration of camera trap data with GIS and spatial modelling platforms improves the detection of hot-spots of biodiversity and informs management decisions in priority conservation areas. While challenges related to data storage, model training and species misclassification remain, continuous refinement of algorithms and the growing availability of open-access training datasets are progressively expanding the utility of automated image analysis for wildlife conservation.

14.5 Satellite Remote Sensing and GIS

Satellite remote sensing has become indispensable in wildlife conservation, providing continuous and large-scale information on habitat structure, land use change and anthropogenic pressures. Multispectral and hyperspectral satellite imagery from programs such as Landsat, Sentinel and MODIS are widely used to detect deforestation, monitor vegetation health, and track changes in habitat quality, thereby supporting long-term conservation planning. When combined with Geographic Information Systems (GIS), remotely sensed data can be translated into spatial models that identify core habitats and potential movement corridors, which is particularly valuable for species living in fragmented landscapes (Pettorelli et al., 2014). For example, time-series data derived from the Normalized Difference Vegetation Index (NDVI) have been successfully used to predict ungulate migration patterns in East Africa by linking vegetation productivity to seasonal resource availability (Boone et al., 2006). Likewise, spatial risk models generated from satellite datasets have allowed conservation managers to allocate anti-poaching efforts more efficiently by identifying high-risk areas based on accessibility, vegetation cover and historical poaching incidents (Kariuki et al., 2021). Remote sensing is also increasingly applied to detect environmental threats such as forest fires, droughts and floods, enabling rapid responses and minimizing negative impacts on wildlife populations. Despite challenges related to cloud cover and the need for on-ground validation, the integration of remote sensing with GIS continues to improve the accuracy and responsiveness of conservation management across multiple spatial and temporal scales (Turner et al., 2003).

14.6 Genomic and Forensic Technologies

The rapid development of molecular techniques has revolutionised wildlife conservation and enforcement by enabling accurate species identification, population structure assessments and traceability of wildlife products. DNA barcoding, which relies on standardized genetic markers such as mitochondrial cytochrome c oxidase I (COI), is now widely used to identify species from biological remains and confiscated animal products (Hebert et al., 2003). In the context of illegal wildlife trade, genetic reference libraries allow enforcement agencies to determine the origin and taxonomic identity of seized samples, thereby improving the likelihood of successful prosecution. Whole-genome sequencing and microsatellite analysis are also increasingly employed to assess genetic diversity and detect population bottlenecks, which are critical for managing small and fragmented wildlife populations (Allendorf et al., 2010).

Forensic genetics has been successfully applied to high-profile cases of ivory, rhino horn and big cat trafficking. For example, Wasser et al. (2015) used mitochondrial and nuclear DNA markers to pinpoint the geographic origin of seized ivory, linking it to specific poaching hotspots in Africa and informng targeted anti-poaching operations. Similarly, genetic assignment methods have been applied to illegally traded tiger skins to trace them back to source populations, contributing to improved transboundary enforcement and conservation planning. As genomic sequencing becomes faster and more affordable, the integration of forensic tools with real-time surveillance technologies and law enforcement databases is expected to significantly strengthen efforts to combat wildlife crime worldwide.

14.7 Internet of Things (IoT) and Smart Sensors

The integration of Internet of Things (IoT) technologies into wildlife conservation has enabled real-time, automated acquisition of ecological data through interconnected devices and smart sensors. GPS collars, satellite transmitters and bio-loggers are widely used for tracking the spatial movements and behavioural patterns of wildlife, providing precise information that can be used to identify critical habitats, migration routes and human–wildlife conflict zones (Kays et al., 2015). Modern collars are increasingly equipped with accelerometers, heart-rate monitors and temperature sensors, allowing detailed monitoring of physiological stress and energy expenditure in free-ranging animals.

Smart environmental sensors deployed across protected areas continuously record abiotic conditions such as temperature, humidity and water quality, enabling the detection of sudden changes that could negatively impact local biodiversity (Porter et al., 2009). The advancement of wireless sensor networks and low-power data transmission technologies (e.g. LoRaWAN) makes it possible to relay data from remote field sites to central servers in near real time. In anti-poaching applications, motion sensors and acoustic monitors linked to IoT platforms are used to detect unauthorized human activities and send automated alerts to enforcement personnel (Stokes, 2021). The combination of IoT devices with AI-driven analytics provides a powerful early-warning system for proactive intervention. While challenges related to battery life, network coverage and device maintenance remain, improvements in energy efficiency and miniaturization are expected to further expand IoT applications in conservation and wildlife management.

14.8 Blockchain and Digital Certification in Wildlife Trade Mitigation

Blockchain technology has recently emerged as a promising tool for improving transparency and traceability in the wildlife trade supply chain. By creating immutable, time-stamped records of transactions, blockchain-based systems enable real-time verification of product origin and movement, making it more difficult for illegally sourced wildlife products to enter legitimate markets (Gonçalves & Costa, 2021). Wildlife certification programs can integrate blockchain platforms to link physical products—such as ivory carvings, reptile skins or pashmina shawls—to secure digital identities, thereby providing end-to-end traceability from source to consumer. These systems are particularly valuable in complex supply chains involving multiple intermediaries, where traditional paper-based documentation is vulnerable to falsification.

Digital certification approaches have already been piloted for sustainably sourced timber and fisheries products, and similar methods are being explored for high-risk wildlife commodities. For example, decentralised ledgers combined with QR-coded provenance tags have been deployed by the World Wildlife Fund (WWF) to track the trade of tuna and ornamental fish, providing open-access verification for buyers and enforcement officials (WWF, 2020). When integrated with genetic databases and customs enforcement systems, blockchains have the potential to substantially increase detection of illicit products and reduce laundering of illegal wildlife into legal supply chains. However, widespread adoption will require standardised protocols, cross-border cooperation and careful consideration of data security and governance.

14.9 Robotics and Autonomous Systems

Robotics and autonomous systems are increasingly being integrated into field-based conservation activities to perform tasks that are difficult, dangerous or labour-intensive for humans. Ground-based robots equipped with cameras and environmental sensors are used to monitor hard-to-access areas such as cliff nesting sites or deep forest interiors, enabling the collection of wildlife data with minimal disturbance (van der Wal et al., 2020). In marine environments, autonomous underwater vehicles (AUVs) have been employed for coral reef mapping, sea turtle monitoring and detecting illegal fishing activities. Their ability to operate for extended periods and collect high-resolution data from complex underwater habitats provides valuable insights into the health and dynamics of marine ecosystems (Leonard et al., 2019).

Autonomous aerial systems, including fixed-wing and multi-rotor drones programmed to follow predefined routes, are being used to conduct regular surveillance flights over protected areas, reducing the need for human-operated patrols and ensuring consistent data collection. Recent advances in robotics also include the development of bio-inspired robots that mimic animal movement, making it possible to study natural behaviors in situ without disturbing wildlife. The integration of robotics with AI and remote sensing technologies enhances the detection and monitoring capabilities of these systems and supports timely decision-making in conservation management. Although high initial costs, technical complexity and maintenance constraints limit widescale deployment in some regions, collaborative efforts among researchers, conservation practitioners and engineers are steadily improving the accessibility and robustness of autonomous approaches in wildlife conservation.

14.10 Data Integration and Decision Support Systems

As conservation technologies generate increasingly large and heterogeneous datasets, the integration of information from multiple sources has become essential for informed decision-making. Data integration platforms combine outputs from field sensors, camera traps, drones, remote sensing, and genetic analyses into unified databases, facilitating holistic assessment of ecosystem status and threats (Hampton et al., 2017). Decision support systems (DSS) built on these integrated datasets employ sophisticated algorithms and spatial modelling tools to evaluate management scenarios, forecast outcomes, and prioritise conservation actions in real time. For example, the Spatial Monitoring and Reporting Tool (SMART) integrates patrol data, threat records, and geographic information to identify poaching hotspots and optimise ranger deployment across protected areas.

Multi-criteria decision analysis and Bayesian network models are increasingly incorporated into DSS to account for uncertainty and the complex trade-offs associated with conservation planning (Knight et al., 2023). By synthesising ecological, socio-economic, and logistical information, these systems enable managers to explore alternative intervention strategies and select context-appropriate solutions. Cloud-based platforms further enhance collaboration among researchers and state agencies by allowing secure data sharing and joint interpretation of results. Although challenges remain in terms of interoperability, data standardisation and access inequity, decision support systems represent a critical link between rapidly expanding digital datasets and effective, evidence-based wildlife conservation.

14.11 Challenges and Limitations

Despite the significant progress in conservation technologies, a number of technical, operational and ethical challenges limit their full-scale implementation. One of the most persistent constraints is the lack of financial and infrastructural capacity in many biodiversity-rich regions, which hampers the acquisition, maintenance and long-term deployment of advanced systems such as drones, smart sensors and genomic laboratories (Turnbull et al., 2023). In addition, the successful use of AI and data integration platforms depends heavily on the availability of high-quality training datasets and skilled personnel capable of managing and interpreting complex outputs. Data interoperability and standardisation remain major hurdles, as conservation data are often collected using disparate methodologies and stored in non-compatible formats.

Operational challenges also include regulatory restrictions on the use of UAVs in protected areas, potential disturbance to wildlife caused by drones or robotic devices, and data privacy concerns within local communities. Ethical considerations arise when automated monitoring is perceived as intrusive or when technological solutions displace traditional ecological knowledge and community-based monitoring practices (Sandbrook et al., 2021). Furthermore, overreliance on digital tools may create a false perception of precision and lead to inadequate consideration of socio-political factors during decision-making.

Mitigating these limitations requires sustained investment in capacity building, inclusive technology design, and the development of interdisciplinary partnerships that bridge engineering, ecology and social sciences. Establishing clear ethical and regulatory guidelines will be essential to ensure that conservation technologies enhance, rather than undermine, equitable and effective wildlife governance.

14.12 Future Prospects and Recommendations

The rapid evolution of digital and autonomous technologies presents unprecedented opportunities to enhance the effectiveness and responsiveness of wildlife conservation efforts. In the coming decade, integration across technological domains—such as combining AI-driven analytics with real-time IoT monitoring and blockchain-enabled traceability—will facilitate adaptive, data-informed conservation strategies at local, national and global scales. Increased adoption of open-source platforms and interoperable data standards will improve collaboration among conservationists, policymakers and technology developers, while also fostering greater transparency and public engagement.

Future efforts should prioritise the development of cost-effective, modular systems that can be adapted to the socio-economic contexts of biodiversity-rich countries. Capacity building and training programs are urgently needed to empower local practitioners and communities to use, maintain and co-design technological solutions. Incorporating equity and ethics into the early stages of technological innovation will help ensure that conservation tools support rather than marginalise traditional knowledge systems and local governance structures. Investment in interdisciplinary research, particularly at the interface of ecology, computer science and social sciences, will be critical to unlock the full potential of emerging technologies. Finally, mainstreaming technology into national conservation policies and facilitating cross-border data sharing will strengthen global efforts to combat illegal wildlife trade and sustain threatened populations in rapidly changing environments.

14.13 Conclusion

Technological innovation is reshaping the field of wildlife conservation, offering powerful new tools to monitor species, assess habitat change and combat the illegal wildlife trade. The integration of artificial intelligence, drone surveillance, genomic forensics, remote sensing, IoT networks and blockchain-based traceability systems is enhancing the accuracy, speed and scale of conservation actions across terrestrial and marine ecosystems. These technologies are shifting conservation from reactive to proactive approaches by enabling real-time data collection and predictive modelling that inform timely interventions. Importantly, the value of these tools lies not only in their technical capabilities, but in their ability to complement and strengthen existing ecological expertise and on-the-ground conservation efforts.

To fully realise the potential of these technologies, it is essential to address current constraints related to capacity, interoperability, ethics and governance. Equitable access to technological resources and participatory approaches that include local stakeholders in design and implementation will ensure socially just and effective conservation outcomes. As biodiversity continues to decline in the face of land-use change, climate impacts and illicit trade, the strategic deployment of conservation technologies—grounded in sound science and guided by ethical principles—will play a critical role in safeguarding the world's wildlife for future generations.

References

Allendorf, F. W., Luikart, G., & Aitken, S. N. (2010). *Conservation and the genetics of populations* (2nd ed.). Wiley-Blackwell.

Boone, R. B., Thirgood, S. J., & Hopcraft, J. G. (2006). Serengeti wildebeest migratory patterns modelled from rainfall and new vegetation growth. *Ecology, 87*(8), 1987–1994. https://doi.org/10.1890/0012-9658(2006)87[1987:SWMPMF]2.0.CO;2

Chrétien, L. P., Théau, J., & Ménard, P. (2016). Wildlife multispecies remote sensing using very low altitude UAV imagery. *Remote Sensing, 8*(11), 1064. https://doi.org/10.3390/rs8111064

Christin, S., Hervet, É., & Lecomte, N. (2019). Applications for deep learning in ecology. *Methods in Ecology and Evolution, 10*(10), 1632–1644. https://doi.org/10.1111/2041-210X.13256

Cites (2021) https://cites.org/eng/prog/eCITES

Gonçalves, A., & Costa, A. P. (2021). Blockchain technology as a tool to tackle the illegal wildlife trade. *Journal for Nature Conservation, 63*, 126071. https://doi.org/10.1016/j.jnc.2021.126071

Hampton, S. E., Jones, M. B., Wasser, L. A., et al. (2017). Skills and knowledge for data-intensive environmental research. *Bioscience, 67*(6), 546–557. https://doi.org/10.1093/biosci/bix025

Hebert, P. D., Cywinska, A., Ball, S. L., & deWaard, J. R. (2003). Biological identifications through DNA barcodes. *Proceedings of the Royal Society B: Biological Sciences, 270*(1512), 313–321. https://doi.org/10.1098/rspb.2002.2218

IPBES (2019). *Global Assessment Report on Biodiversity and Ecosystem Services*. https://ipbes.net/global-assessment

Joppa, L. (2020). AI for earth: Automating conservation monitoring and data analysis. *Nature Ecology & Evolution, 4*(9), 1196–1197. https://doi.org/10.1038/s41559-020-1273-4

Kalan, A. K., et al. (2019). Armed conflict and delimiting tropical wildlife refuges. *Nature, 569*(7755), 108–111. https://doi.org/10.1038/s41586-019-1134-2

Kariuki, M., Mukeka, J., Ogutu, J. O., Kanga, E., & Riginos, C. (2021). Spatial and temporal patterns of wildlife poaching in Kenya. *Biological Conservation, 266*, 109402. https://doi.org/10.1016/j.biocon.2021.109402

Kays, R., Crofoot, M. C., Jetz, W., & Wikelski, M. (2015). Terrestrial animal tracking as an eye on life and planet. *Science, 348*(6240), aaa2478. https://doi.org/10.1126/science.aaa2478

Knight, A. T., Cook, C. N., Redford, K. H., & Biggs, D. (2023). Improving the evidence base for conservation policy and practice through systems thinking and modelling. *Conservation Biology, 37*(1), e13998. https://doi.org/10.1111/cobi.13998

Koh, L. P., & Wich, S. A. (2012). Dawn of drone ecology: Low-cost autonomous aerial vehicles for conservation. *Tropical Conservation Science, 5*(2), 121–132.

Leonard, J. J., Bahr, A., Fallon, M., & Hollinger, G. (2019). Autonomous underwater vehicle navigation and mapping for marine environments. *Annual Review of Marine Science, 12*, 405–432. https://doi.org/10.1146/annurev-marine-010318-095209

Mulero-Paules, E., et al. (2022). UAV-based surveillance for anti-poaching operations: A review of current applications and technological advances. *Conservation Science and Practice, 4*(12), e12897. https://doi.org/10.1111/csp2.12897

Norouzzadeh, M. S., et al. (2018). Automatically identifying, counting, and describing wild animals in camera-trap images with deep learning. *PNAS, 115*(25), E5716–E5725. https://doi.org/10.1073/pnas.1719367115

Ogden, R., Linacre, A., & Foran, D. (2020). Wildlife forensic science and the rise of 'forensic conservation'. *Trends in Ecology & Evolution, 35*(6), 501–513. https://doi.org/10.1016/j.tree.2020.02.009

Pettorelli, N., Ryan, S., Mueller, T., Bunnefeld, N., Jędrzejewska, B., Lima, M., & Kausrud, K. (2014). The normalized difference vegetation index (NDVI): Unforeseen successes in animal ecology. *Climate Research, 46*(1), 15–27. https://doi.org/10.3354/cr00936

Pimm, S. L., Alibhai, S., Beresford, A. E., et al. (2015). Emerging technologies to conserve biodiversity. *Trends in Ecology & Evolution, 30*(11), 685–696. https://doi.org/10.1016/j.tree.2015.08.008

Porter, J., Hanson, P., Lin, C., et al. (2009). Wireless sensor networks for ecology. *Bioscience, 59*(7), 561–572. https://doi.org/10.1525/bio.2009.59.7.7

Reddy, C. S., Athira, K., & Mudalkar, S. (2025). Evolution of global biodiversity monitoring: Innovations, modernapproaches and integration with essential variables. *Journal of the Indian Society of Remote Sensing*, 1-14

Sandbrook, C., Luque-Lora, R., & Adams, W. M. (2021). Human bycatch: Conservation surveillance and the social implications of capture. *Conservation and Society, 19*(2), 95–104. https://doi.org/10.4103/cs.cs_17_136

Stokes, E. J. (2021). Smart sensors and the future of wildlife protection. *Frontiers in Conservation Science, 2*, 671171. https://doi.org/10.3389/fcosc.2021.671171

Swanson, A., Kosmala, M., Lintott, C., et al. (2016). Snapshot Serengeti, high-frequency annotated camera trap images of 40 mammalian species in an African savanna. *Scientific Data, 3*, 150026. https://doi.org/10.1038/sdata.2015.26

Turnbull, J., Searle, A., Hartman Davies, O., Dodsworth, J., Chasseray-Peraldi, P., von Essen, E., & Anderson-Elliott, H. (2023). Digital ecologies: Materialities, encounters, governance. *Progress in Environmental Geography,2*(1-2), 3-32

Tabak, M. A., Norouzzadeh, M. S., Wolfson, D. W., Sweeney, S. J., VerCauteren, K. C., Snow, N. P., ... & Miller, R.S. (2019). Machine learning to classify animal species in camera trap images: Applications in ecology. *Methods inEcology and Evolution, 10*(4), 585-590

Turner, W., Spector, S., Gardiner, N., Fladeland, M., Sterling, E., & Steininger, M. (2003). Remote sensing for biodiversity science and conservation. *Trends in Ecology & Evolution, 18*(6), 306–314. https://doi.org/10.1016/S0169-5347(03)00070-3

UNEP (2020) https://www.thegef.org/what-we-do/topics/illegal-wildlife-trade

van Andel, A. C., Wich, S. A., Boesch, C., et al. (2015). Locating chimpanzee nests and identifying fruiting trees with UAVs in a tropical rainforest. *PLoS One, 10*(6), e0129758. https://doi.org/10.1371/journal.pone.0129758

van der Wal, R., Siddharthan, A., Lamb, A., et al. (2020). Autonomous robotics in geography and ecology: A new frontier in field research. *Journal of Applied Ecology, 57*(11), 2081–2089. https://doi.org/10.1111/1365-2664.13790

Wasser, S. K., Brown, L., Mailand, C., Mondol, S., Clark, W., Laurie, C., & Weir, B. S. (2015). Genetic assignment of large seizures of elephant ivory reveals Africa's major poaching hotspots. *Science, 349*(6243), 84–87. https://doi.org/10.1126/science.aaa2457

WWF. (2020). *Blockchain to strengthen traceability of sustainable seafood supply chains*. World Wide Fund for Nature. https://www.wwf.org

Chapter 15
Wildlife Forensics and Genetic Tools Tracking Trade Routes

Kumudani Bala Gautam, Diksha Kumari, and Akhilesh Kukreti

Contents

Abstract The wildlife forensics is an important branch of forensic science which deals with one of the most challenging types of crime i.e., wildlife trafficking. It presents a significant challenge to the experts because, in contrast to human based crimes, the evidence in these cases is located in greatly altered and deteriorated forms. Also, such criminal activities take place either in transit or consumer level where usually tertiary or quaternary crime scenes are involved. This chapter talks about the significance of the very first step of wildlife crime scene investigation-identification of biological evidence and investigates how genetic technologies can be of great help in wildlife forensics for species identification and tracking geographic location as well as origin. Various genetic technologies for individualisation including mitochondrial DNA, microsatellites, single nucleotide polymorphisms and other DNA sequencing methods have been discussed in association with criminal prosecution. A few key case studies related with elephants, chimpanzees, marine

K. B. Gautam (✉)
Department of Biosciences, Graphic Era (Deemed to be University), Dehradun, Uttarakhand, India

D. Kumari
Defence Institute of Physiology & Allied Sciences (DIPAS), New Delhi, India

A. Kukreti
Government PG College Karanprayag, Chamoli, Uttarakhand, India

A. Poonia et al. (eds.), *Wildlife Trade in India*,
https://doi.org/10.1007/978-981-95-3694-8_15

animals and other keystone species, have been presented to help forensic experts to identify poaching hotspot areas and track the trafficking channels using databases. The chapter also discusses how morphological identification can be combined with technologies like Forensically Informative Nucleotide Sequencing (FINS) and DNA barcoding for identifying processed wildlife evidences. Finally, the authors conclude by highlighting the need for an interdisciplinary approach that integrates genetics, policy interventions, conservation planning, and socioeconomic engagement in order to strengthen wildlife protection policy.

Keywords Wildlife crime · DNA · Keystone species · Genetics

15.1 Introduction

The recognition of evidences is a basic necessity in solving or prosecuting crimes. An investigator's preliminary survey should assess potential evidence at the scene of crime and prioritize each item's relevance. Among different physical evidences biological evidences are taken as prized items. The identification of biological evidences relies on the comparison of *class characteristics* which define a sample and enable it to be categorized alongside similar materials such as body fluids, hair, nails, bones etc. The biological evidences are also categorised by its unique *individual characteristics* so that the matching of evidences and reference sample is highly likely to share a common origin with certainty. One such example is possessing individual characteristics is DNA polymorphism. Biological evidences are always at highest priority since they can create linkage such as victim-to-perpetrator, victim-to-scene, perpetrator-to-scene based on the theory of Locard exchange principle. This principle posits that a cross transfer/exchange of evidences happens during any physical interaction between a culprit and an object or another individual. Hence the biological evidence is usually evaluated for assigning individual with certainty often known as Individualization. For accurate individualization, potential evidences must be judiciously preserved and documented in its original state to ensure just investigation. The biological evidence might get altered and destroyed if scene of crime has not been secured adequately which potentially mislead investigators and complicate efforts to identify the perpetrator accurately. Individualisation in crimes inflicted on human (Human forensics) are rather easy to address then the crime inflicted on biodiversity (Wildlife forensics). To understand the troubled stance in wildlife crime, we first need to know about the significance of crime scenes. There are primarily three types of crime scene namely primary, secondary and tertiary, in few cases quaternary crime scene may also exist. A place or location where initial actual criminal activity took place is known as primary crime scene, while a secondary scene is a place other than primary however related to initial crime and might contain evidences. Whereas a tertiary crime refers to a location which is indirectly connected to the primary scene where the probability of related evidence becomes extremely low. Hence, by moving forward from primary to quaternary, the presence of evidence would become low to negligible, which causes hindrance in the solving of crime through individualisation.

The wildlife crime cases have a tendency to be seized at later stages of the scene. The investigators confiscate wildlife body parts either at transit routes or at consumer's end. Hence, it is often an unattainable task for the enforcement agencies to find an intact primary crime scene where the initial illegal act has been occurred while dealing with wildlife poaching or hunting. Usually, wildlife trafficking activities could be monitored at the tertiary crime scene where the endangered flora or fauna are transformed into articles or derivatives and continuously transported between various locations. However, the seizure may happen sooner or later in tertiary or quaternary scene which involves places such as shops and e-commerce sites. Hence it is rather difficult in wildlife forensics to conduct individualization from the tertiary and quaternary crime scene where the evidence is negligible and number of traded species is high. The challenges are twofold in wildlife forensics: first, an expert has to identify the species that has been traded. Once the identity is established, the location of poaching of that particular species is also an important criterion to pinpoint the vulnerable areas. By determining the vulnerable areas, enforcement agencies can implement monitoring strategies to intercept poaching activities at their source. Therefore, studying the distribution range and molecular makeup of traded wildlife becomes crucial because these species can only be traded from within their natural habitat.

15.2 Genetics: A Tool for Individualisation

To conduct individualisation with higher accuracy a thorough molecular study of an animal is requisite from a given landscape in its habitat. The molecular study enables an expert to quantify genetic diversity and population structuring with unprecedented resolution, both within and across geographical regions, thereby yielding insights that are crucial for reliable species delineation and also for tracking the origin of article/derivative in the illegal wildlife trade networks. The new technological advancement has enriched the field of genetics for the better resolution at the molecular level so that the slightest and subtle variation could be detected for the effective and accurate individualization. In the wildlife crime cases single nucleotide polymorphism (SNP), microsatellite analysis, restriction site-associated DNA sequencing (RAD-Seq) and whole-genome sequencing provide some robust superior alternative techniques rather than traditional morphological assessment. Historically, a traditional taxonomic approach was taken to identify the target specimens that often leads to misidentification between the phenotypically similar and plastic forms which usually hampers the efforts of enforcement officers and misdirects conservation resources. Such misidentification is often seen in a convergent evolution which produce a nearly identical external trait in unrelated species while environmental variation can produce morphological shifts within a single species, hence, both will obscure true biodiversity patterns (Smith & Smith, 1992; Pfenninger et al., 2006). The advancement in DNA sequencing ranging from classical Sanger to high throughput next generation platform is noteworthy, this has revolutionized both, taxonomic resolution and phylogeographic inference. The statistical rigor of phylogenetic and demographic analysis has greatly enhanced by the innovative technologies. These robust computational pipelines have now allowed coalescent based species delimitation and Bayesian clustering methods which provide a framework for testing alternative taxonomic hypotheses, estimating divergence times and demographic histories (Armstrong et al., 2021). Therefore, molecular markers are essential in wildlife forensics and conservation management of endangered and vulnerable species. These genetic tools are requisite for the estimation of population sizes, assessing genetic diversity and understanding gene flow dynamics particularly for species residing in fragmented or isolated habitat (Avise, 1994). Genetic markers often offer information about the genotypic makeup of populations which supports the identification of distinct management units and enables more effective monitoring of species subjected to anthropogenic pressures, including illegal wildlife trade. These markers also aid in identification of evolutionary significant units (ESUs) and facilitate the formation of conservation strategies tailored to the genetic need of targeted species by enhancing the understanding of species diversity and population structure. The analysis of molecular markers will gather information of nucleotide sequence variations resulting from the natural evolutionary processes and environmental adaptation accumulated over the generations. These nucleotide variations are often expressed as mutations in DNA sequences, which serve as a key indicator of genetic differentiation among populations and species (Patwardhan et al., 2014). All these variations are instrumental in estimating genetic divergence which could assist in determining the level of adaptability of a species in different ecological niches. Neutral markers are not directly influenced by natural selection though they offer valuable insights into the underlying patterns of adaptive genetic variation thus providing a baseline for assessing the evolutionary potential of species (Frankham et al., 2002). Apart from wildlife forensics, molecular markers have broad applications across the multiple biological disciplines such as conservation of species or reconstructing evolutionary relationships or deep insights into divergence of lineages over time. In the field of evolutionary biology, molecular markers help to unravel the key processes such as mutation, selection, gene flow and genetic drift which contribute to the understanding of species adaptation and evolution (Avise, 1994; Frankham et al., 2002). Such inclusiveness is vital for predicting the resilience and adaptability of species to ongoing environmental changes and ensure species' long-term survival through informed conservation efforts. An expert should select an appropriate molecular marker according to their specific objective of a study and methodologies employed. Microsatellite, mitochondrial DNA, single nucleotide polymorphism (SNPs) are usually preferred for different genetic studies depending upon the resolution and inheritance patterns. Nevertheless, prior knowledge regarding the polymorphism levels and inheritance patterns of such markers is critical for their effective application in population genetic studies (Røed, 1998). Therefore, appropriate marker selection enhances the reliability of genetic assessment in natural populations particularity while evaluating gene flow, population structure, genetic health and illegally traded wildlife. The mitochondrial and microsatellite are the two widely utilised genetic markers in the field of wildlife forensics.

15.2.1 Mitochondrial Markers

The mitochondrial (mt) genome is approximately 16.5–17.5 kilobases (kb) long, compact and highly conserved circular DNA molecule present inside the mitochondria of cell. This circular genome contains 37 essential genetic elements namely 13 protein-coding genes (PCGs), 22 transfer RNA (tRNA), 2 ribosonal RNA (rRNA) and D-loop. PCGs are involved in oxidative phosphorylation; tRNA genes are responsible for translating mitochondrial proteins and rRNA genes form the structural components of the mitochondrial ribosomes. A highly variable control region or D-loop is a non-coding region which plays a vital role in the regulation of replication and transcription. The D-loop region unveils a higher mutation rate than the rest of the mitochondrial genes. Hence, this region of mitochondria does act as a valuable marker for the population level genetic research. Mitochondrial DNA (mtDNA) has been widely accepted in phylogenetic and phylogeographic studies due to its distinctive genetic characteristics. mtDNA is mostly adopted in studies for its significant feature of being uniparental inheritance. It is a haploid body and an individual can inherit mtDNA exclusively from her/his mother i.e., maternally inherited. The lack of paternal contribution in mtDNA eliminates recombination and provides a perfect traceable genetic lineage through maternal ancestry. This significant property of mtDNA is advantageous for reconstructing evolutionary histories and examining maternal lineages over both contemporary and ancient timescales (Hofreiter & Stewart, 2009).

The utility of mtDNA extends beyond conservation biology and evolutionary biology into applied fields such as forensic science. The most advantageous properties of mtDNA in forensic application is its high copy number in a cell. Unlike nuclear DNA, hundreds to thousands copies of mtDNA is present per cell, hence, significantly increasing the probability of recovering usable genetic material from even highly degraded or limited biological samples. This unique quality of mtDNA is valuable for forensic identifications especially in cases involving old, decayed or trace samples such as hair shafts, bones and teeth where nuclear DNA may be insufficient or completely absent. Furthermore, being maternally inherited provides a consistent and predictable lineage that can aid in confirming identities or tracing geographic and maternal origins (Ma et al., 2018). Overall, the mitochondrial genome remains foundational for molecular research and applications in both evolutionary biology and forensic science. Its structural characteristics, inheritance pattern and mutational dynamics offer robust tools for tracing evolutionary lineages, understanding population history and solving crime cases where conventional nuclear DNA analysis falls short.

15.2.2 Microsatellites Markers

Microsatellites are repetitive DNA sequences commonly well-known as short tandem repeats (STRs). These STRs are composed of short motifs typically di, tri or tetra nucleotide units ranging from 1–6 bp (base pairs) in length. These repetitive units recur multiple times and are generally flanked by unique non-repetitive DNA sequences, which makes such regions easily targetable with specific primers for the successful amplicons (Tóth et al., 2000). The STRs are simple in structure and abundant throughout the genome thus considered as valuable genetic markers for a variety of applications in population genetics, conservation biology and forensic science. Microsatellites are considered highly polymorphic or multi-allelic within a population exhibiting a large number of alleles at each locus due to variations in the number of repeat units. This extensive allelic diversity arises predominantly from the slippage during DNA replication which leads to insertions or deletions of the repeat units. The mutation rate of microsatellites is relatively high, approximately 10^{-4} per locus per generation, making them particularly useful for studying recent evolutionary and demographic events such as population bottleneck, population expansion and migration patterns. Typically, at each microsatellite locus, there is presence of two alleles (bi-allelic), one inherited from each parent. The inheritance of microsatellites follows a co-dominant Mendelian pattern allowing for clear distinction between homozygous and heterozygous individuals. When two identical alleles are at a single locus, such individuals are known as Homozygous whereas Heterozygous individuals carry two different alleles at a single locus. The co-dominant nature facilitates the accurate calculation of allele frequencies, heterozygosity levels and other population genetic parameters which are essential for assessing genetic diversity, inbreeding coefficients and population structure. The high degree of polymorphism in the microsatellite markers makes them exceptionally potent for parentage analysis and kinship studies. Microsatellite markers can unambiguously determine familial relationships and trace lineages within and across populations since an individual typically exhibit unique combination of alleles at multiple STR loci. The discriminative power of microsatellite markers increases with the number of loci analysed allowing researchers and forensic scientists to establish identity with high statistical confidence. In the field of wildlife forensics, STR markers have revolutionized individual identification. Their high variability, co-dominant inheritance and ability to generate distinct profiles/scores for each individual makes them ideal for distinguishing between suspected species or identifying biological remains. Even small quantity or degraded DNA sample can often yield reliable STR profiles due to the short length of the repeat units with the utilization of multiplex PCR techniques. STRs are now standard markers in human forensic database worldwide also known as CODIS (Combined DNA Index System) which is instrumental in criminal investigations, paternity testing and identification of disaster victims (Willems et al., 2014; Gymrek, 2017). The high mutation rate, reproducibility and discriminatory power of STRs makes them crucial tool in applied genetics and wildlife forensics. Thus microsatellites serve as robust and versatile markers applicable across various disciplines ranging from elucidating fine-scale population genetic patterns in wildlife and conservation genetics to solving the complex wildlife forensic cases.

15.2.3 Selection of Appropriate Marker

In the branch of wildlife forensics, Cytochrome *b* (Cyt *b*), Cytochrome c oxidase subunit *I* (CO*I*), 16s rRNA and 12s rRNA genes of mitochondrial region are widely utilised as molecular markers for species identification. These genes are favoured due to their high copy number in cell, maternal inheritance and moderate mutation rate which permits differentiation among closely related species. The CO*I* gene was utilized in DNA barcoding initiative as BOLD- The Barcode of Life Data Systems, whose objective is to create a standardized and accessible platform for identifying species (Ratnasingham & Hebert, 2007). The ribosomal RNA (12S and 16S) and Cyt *b* are conserved genes across taxa yet adequately variable to distinguish species this specific character makes them highly suitable for forensic applications where traditional morphological features are absent or degraded. With the advent of molecular techniques, Forensically Informative Nucleotide Sequencing (FINS) has become foundational and powerful approach to identify species in the cases of illicit wildlife trade. FINS involves the sequencing of conserved genes of mitochondrial or nuclear DNA from the unknown confiscated samples and comparing it to the sequences of reference samples from the known species in the molecular databases. The efficacy of FINS is higher in the circumstances where morphological identification is difficult when confiscated wildlife does appear as processed animal parts, derivatives and products (Bartlett & Davidson, 1992; Guha & Kashyap, 2006). FINS can accurately identify minute or degraded samples such as scales, bones, feathers or powdered forms providing a conclusive species level identification. To enhance the forensic utility of mitochondrial and microsatellite markers, establishment of geographical reference databases of different species from different landscape is prioritised. Along with mitochondrial markers, Short Tandem Repeats (STRs) is increasingly accepted as a critical component in wildlife forensic investigations. STR markers are highly variable and provide better resolution thus utilise not only for the species identification but also for individual identification and recognition of its origin. The establishment of these reference genetic databases facilitate forensic scientists and conservationists to trace the geographic origin of confiscated samples of wildlife, thus, identifying the poaching hotspots and illegal trade routes. The explicit genetic information is instrumental for law enforcement agencies as these dedicated databases will inform targeted protection strategies meanwhile aid in conservation planning, prioritizing surveillance efforts and supports the prosecution of wildlife crime. Therefore, integrating CO*I*, Cyt *b*, 12S, 16S with STR markers provide a comprehensive toolkit for species identification, population tracking and crime scene investigation. This multidisciplinary approach strengthened efforts of multiple stakeholders to combat illegal wildlife trade and supports the sustainable management and conservation of endangered and protected species globally.

15.3 Case Studies

To conduct individualisation with higher accuracy, a thorough molecular study of an animal is requisite from known geographic areas. Wasser et al., 2004 estimated geographic specific allele frequencies of forest Elephant (*Loxodonta africana cyclotis*) and savannah Elephant (*Loxodonta africana africana*) over the entire African elephant range with 16 microsatellite loci. The data generated was used to infer the geographic origin of DNA samples obtained from tusks of unknown origin. This genetic baseline database was also utilised to identify the origin of largest ivory trade across multiple international borders along numerous trade routes. Wasser et al., 2007 documented the identity of multiple ivory pieces, simultaneously, confiscated before the trade ban of 1989. Before the baseline molecular database, it was difficult to identify the poaching hotspots and probable trade routes between Zambia, Malawi, the Democratic Republic of Congo and the Selous Game reserve of Tanzania, which became rather effective after the deployment of mitochondrial marker which could differentiate between forest and savannah elephants and microsatellite which substantiated the individuals within those subspecies. The ability to trace the origin of ivory through genetics enables authorities to focus on specific poaching hotspots, potentially reducing the elephant mortality rates in Africa (Wasser et al., 2008).

Ghobrial et al., 2010 employed a geo-referenced genetic database of chimpanzees from Cameroon. This study estimated 46 chimpanzees at Limbe Wildlife centre using mtDNA, STRs and Bayesian approximation methods for estimating geographic origin and genotypes which revealed varied origin such as Cameroon and areas nearby Cameroon-Nigeria border. This study aimed to identify potential hunting hotspots and transportation routes for live animals which could be critical information for conservation efforts and law enforcement agencies. Findings also indicated that poaching of chimpanzee is widespread, hence there is a necessity for systematic conservation planning as seen in the drastically declined population of western chimpanzees (Heinicke et al., 2019).

Along with the terrestrial animals, molecular variation and phylogenetic approach has been effectively utilized for marine animals as well. DNA profiling methods have been employed to identify species and evaluate the extent of illicit /unregulated exploitation of protected whale stocks in the retail markets of Japan and South Korea (Baker et al., 2000). DNA sequencing was used to amplify the mtDNA from whale products so that identification of various species including minke, fin and humpback whales could be done accurately (Baker et al., 1996). The microsatellite profiling technique facilitated in the identification of unique individuals from the market samples thus revealing substantial number of products from the protected J-stock Minke whales (Dalebout et al., 2002). These studies were evidence to the fact that a variety of whale and dolphin species were found in the commercial markets including protected species (Baker et al., 1996; Grohmann et al., 1999). The outcome suggested that a substantial proportion of Minke whale products are sold from the depleted J-stock which indicates the loop of illegal, unreported and unregulated (IUU) activities (Lukoschek et al., 2009).

Apart from keystone species, the molecular techniques have been used on lower taxa for management of highly trafficked species. Trade for reptile, tortoises, and freshwater turtles for pet, meat, calipee and traditional medicines in South Asia, East Asia and Southeast Asia poses a substantial threat to wild population globally (Mendiratta et al., 2017; Gautam et al., 2023). One such species is Indian Star Tortoise (*Geochelone elegans*) which faces significant threats due to illegal wildlife trade and native habitat loss. Star tortoise is majorly sought in pet trade leading to alarming harvest from the native distribution ranges within the Indian subcontinent where over 55,000 individuals have been confiscated (D'Cruze et al., 2015). The issue with the management of illegal trade and conservation plans of Star tortoise was the lacuna of baseline morphological and genetic data of the species from the native wild population. Sahoo et al., 2024 presented a comprehensive reference database comprising mitochondrial and microsatellite data that can be used for evidence based release and rehabilitation of seized individuals eventually guiding an effective conservation and management strategies.

15.4 Challenges

Many taxa have been collected for meat, food, medicinal or aphrodisiac purposes in contrast to large mammals, keystone species and species used in the pet trade. Therefore, merely depending on the morphology will not be sufficient to identify the species derivative available in the market because trafficked wildlife items may not always be distinguished by their morphological appearance. Conservation Biologist/Forensic expert can identify the traded species precisely as a protective action to comprehend and reduce market activities (Gaubert et al., 2015). When morphological identification lacks reliability while dealing with wildlife body part derivatives, species identification through genetics seeks to compare an unidentified wildlife derivate with a confirmed reference sample whose province or geographical location is well established (Ghobrial et al., 2010). Nevertheless, the absence of verified geo-referenced samples of specific species of interest is a significant constraint for creating reference molecular databases in the field of DNA forensics (Budowle et al., 2005). Bartlett and Davidson (1992) introduced the concept of FINS (Forensically Informative Nucleotide Sequencing), which has been developed to accurately identify the origin of biological specimens when morphological characters are absent due to processing of species into articles or derivative. FINS is crucial for forensic experts, government agencies working for conservation, food department and health regulation. The procedure of FINS consists of four key components: first is isolation of DNA from various biological samples; second is to amplify a specific conserved region gene using PCR; third is determining the nucleotide sequence of the amplicons of a particular gene of interest; and lastly the phylogenetic analysis to identify the most closely related species using molecular reference database. This approach is rapid, reliable and reproducible, hence addresses the constraints of identifying traded species accurately when conventional methods fail. The FINS allows the species identification from a wide range of biological samples, such as wild animal body parts, altered morphological features, processed food as a canned, partially cooked, pickled, salted or smoked products consequently addressing the challenges in the wildlife protection, food safety and conservation management planning. After the introduction of FINS, the method has since been extensively applied in the numerous wildlife forensic investigations (Verma & Singh, 2003; Baker, 2008; Sahajpal & Goyal, 2009; Rajpoot et al., 2017; Yadav et al., 2021). Wildlife forensics now encompasses a broad and potent array of technological, methodological and analytical tools and resources with objective of identifying trafficked species whether higher or lower taxa and potentially determining and analysing trace derivatives with accuracy (Comstock et al., 2003; Gupta et al., 2006). However, there is a pressing need to address the broader socio-economic factors driving poaching and illegal trade, thus effective solutions must integrate analysis with the community engagement and sustainable development initiatives.

15.5 Conclusion

The precise identification of trafficked species is critical for forensic experts and conservationists to comprehend and address global wildlife market. The utilisation of molecular technologies is not only to assist in tracking the trade trends, trade routes or traded species but also to guide conservation plans and policy measures. Accurate identification of illegally traded species is essential for evaluating the extent and consequences of wildlife trade which poses a considerable threat to biodiversity and pushing numerous species to the brink of extinction (Nijman et al., 2011; Hughes et al., 2021). Recently DNA analysis methods have been utilized to detect elasmobranch products in the market uncovering the presence of critically endangered species (Alvarenga et al., 2024). Integrating various data sources such as CITES records, auction information, states record of trade, custom records could improve the comprehension of trade patterns as exemplified by the helmeted hornbill case (Hatten et al., 2024). It has been also observed that the amphibian trade is understudied leaving larger gaps in the conservation efforts, thus requires sufficient monitoring (Hughes et al., 2021). The recognition of traded species enables conservationist to focus on measures and initiatives aimed at reducing the market demand which is vital for safeguarding at the edge species (Donald et al., 2024; Hatten et al., 2024). One effective example is trade prevalence score developed for birds by Donald et al., 2024 which demonstrate how the data integration can effectively shape the conservation strategies. Nonetheless, accurate identification through genetics is crucial. Obstacles such as limited resources and biases in data collection might impede efficient monitoring and conservation initiatives. Hence, emphasizing the need for collaboration among stakeholders and continual enhancements in advances of genetics such as Genotyping by Sequencing, Double Digest Restriction-site Associated DNA and Next Generation Sequencing (GBS, ddRAD, NGS) while analysing trafficked specimens.

References

Alvarenga, M., Bunholi, I. V., de Brito, G. R., Siqueira, M. V. B. M., Domingues, R. R., Charvet, P., et al. (2024). Fifteen years of elasmobranchs trade unveiled by DNA tools: Lessons for enhanced monitoring and conservation actions. *Biological Conservation, 292*, 110543.

Armstrong, E. E., Khan, A., Taylor, R. W., Gouy, A., Greenbaum, G., Thiéry, A., et al. (2021). Recent evolutionary history of tigers highlights contrasting roles of genetic drift and selection. *Molecular Biology and Evolution, 38*(6), 2366–2379.

Avise, J. C. (1994). *Molecular markers, natural history and evolution*. Chapman and Hall.

Baker, C. S., Cipriano, F., & Palumbi, S. R. (1996). Molecular genetic identification of whale and dolphin products from commercial markets in Korea and Japan. *Molecular Ecology, 5*(5), 671–685.

Baker, C. S., Lento, G. M., Cipriano, F., & Palumbi, S. R. (2000). Predicted decline of protected whales based on molecular genetic monitoring of Japanese and Korean markets. *Proceedings of the Royal Society B: Biological Sciences, 267*, 1191–1193.

Baker, C. S. (2008). A truer measure of the market: the molecular ecology of fisheries and wildlife trade. *MolecularEcology, 17*(18), 3985–3998

Bartlett, S. E., & Davidson, W. S. (1992). FINS (forensically informative nucleotide sequencing): A procedure for identifying the animal origin of biological specimens. *BioTechniques, 12*, 408–411.

Budowle, B., Garofano, P., Hellman, A., Ketchum, M., Kanthaswamy, S., Parson, W., van Haeringen, W., Fain, S., & Broad, T. (2005). Recommendations for animal DNA forensic and identity testing. *International Journal of Legal Medicine, 119*, 295–302.

Comstock, K. E., Ostrander, E. A., & Wasser, S. K. (2003). Amplifying nuclear and mitochondrial DNA from African elephant ivory: A tool for monitoring the ivory trade. *Conservation Biology, 17*(6), 1840–1843.

Dalebout, M. L., Lento, G. M., Cipriano, F., Funahashi, N., & Baker, C. S. (2002). How many protected minke whales are sold in Japan and Korea? A census by microsatellite DNA profiling. *Animal Conservation, 5*(2), 143–152.

D'Cruze, N., Singh, B., Morrison, T., Schmidt-Burbach, J., Macdonald, D. W., & Mookerjee, A. (2015). A star attraction: The illegal trade in Indian Star Tortoises. *Nature Conservation, 13*, 1–19.

Donald, P. F., Fernando, E., Brown, L., Busana, M., Butchart, S. H., Chng, S., et al. (2024). Assessing the global prevalence of wild birds in trade. *Conservation Biology, 38*(5), e14350.

Frankham, R., Briscoe, D. A., & Ballou, J. D. (2002). *Introduction to conservation genetics*. Cambridge University Press.

Gaubert, P., Njiokou, F., Olayemi, A., Pagani, P., Dufour, S., Danquah, E., et al. (2015). Bushmeat genetics: Setting up a reference framework for the DNA typing of African forest bushmeat. *Molecular Ecology Resources, 15*(3), 633–651.

Gautam, K. B., Kumar, A., Das, A., & Gupta, S. K. (2023). Himalayan upliftment and Shiwalik succession act as a cradle for divergence in Bengal monitor lizard Varanus bengalensis (Reptilia: Varanidae) in India. *Cladistics, 39*(5), 382–397.

Ghobrial, L., Lankester, F., Kiyang, J. A., Akih, A. E., De Vries, S., Fotso, R., et al. (2010). Tracing the origins of rescued chimpanzees reveals widespread chimpanzee hunting in Cameroon. *Bmc Ecology, 10*, 1–15.

Grohmann, L., Bokermann, I., Sonntag, R., Unseld, M., Hiesel, R., Malek, O., Giese, A., & Brennicke, A. (1999). Whale meat from protected species is still being sold on Japanese markets. *Naturwissenschaften, 86*, 350–351.

Guha, S., & Kashyap, V. K. (2006). Molecular identification of lizard by RAPD and FINS of mitochondrial 16S rRNA gene. *Legal Medicine (Tokyo), 8*, 5–10.

Gupta, S. K., Thangaraj, K., & Singh, L. (2006). A simple and inexpensive molecular method for sexing and identification of the forensic samples of elephant origin. *Journal of Forensic Sciences, 51*(4), 805–807.

Gymrek, M. (2017). A genomic view of short tandem repeats. In *Molecular and genetic bases of disease* (pp. 9–16).

Hatten, C. E., Hadiprakarsa, Y. Y., Lam, J. Y., Mak, J., Toropov, P., & Dingle, C. (2024). Assessing the legal, illegal, and gray ornamental trade of the critically endangered helmeted hornbill. *Conservation Biology, 38*(5), e14358.

Heinicke, S., Mundry, R., Boesch, C., Hockings, K. J., Kormos, R., Ndiaye, P. I., Tweh, C. G., Williamson, E. A., & Kühl, H. S. (2019). Towards systematic and evidence-based conservation planning for western chimpanzees. *American Journal of Primatology, 81*(9), e23042.

Hofreiter, M., & Stewart, J. (2009). Ecological change, range fluctuations and population dynamics during the Pleistocene. *Current Biology, 19*(14), R584–R594.

Hughes, A. C., Marshall, B. M., & Strine, C. T. (2021). Gaps in global wildlife trade monitoring leave amphibians vulnerable. *eLife, 10*, e70086.

Lukoschek, V., Funahashi, N., Lavery, S., Dalebout, M. L., Cipriano, F., & Baker, C. S. (2009). High proportion of protected minke whales sold on Japanese markets is due to illegal, unreported or unregulated exploitation. *Animal Conservation, 12*(5), 385–395.

Ma, K., Zhao, X., Li, H., Cao, Y., Li, W., Ouyang, J., et al. (2018). Massive parallel sequencing of mitochondrial DNA genomes from mother-child pairs using the ion torrent personal genome machine (PGM). *Forensic Science International: Genetics, 32*, 88–93.

Mendiratta, U., Sheel, V., & Singh, S. (2017). Enforcement seizures reveal large-scale illegal trade in India's tortoises and freshwater turtles. *Biological Conservation, 207*, 100–105.

Nijman, V., Nekaris, K. A. I., Donati, G., Bruford, M., & Fa, J. (2011). Primate conservation: Measuring and mitigating trade in primates. *Endangered Species Research, 13*(2), 159–161.

Patwardhan, A., Ray, S., & Roy, A. (2014). Molecular markers in phylogenetic studies: A review. *Journal of Phylogenetics & Evolutionary Biology, 2*(2), 131.

Pfenninger, M., Cordellier, M., & Streit, B. (2006). Comparing the efficacy of morphologic and DNA-based taxonomy in the freshwater gastropod genus radix (Basommatophora, Pulmonata). *Bmc Evolutionary Biology, 6*, 1–14.

Rajpoot, A., Kumar, V. P., Bahuguna, A., & Kumar, D. (2017). Forensically informative nucleotide sequencing (FINS) for the first time authentication of Indian Varanus species: Implication in wildlife forensics and conservation. *Mitochondrial DNA Part A DNA Mapping, Sequencing, and Analysis, 28*(6), 892–900.

Ratnasingham, S., & Hebert, P. D. N. (2007). BOLD: The barcode of life data system (www.barcodinglife.org). *Molecular Ecology Notes, 7*, 355–364. https://doi.org/10.1111/j.1471-8286.2007.01678.x

Røed, K. H. (1998). Microsatellite variation in Scandinavian Cervidae using primers derived from Bovidae. *Hereditas, 129*(1), 19–25.

Sahajpal, V., & Goyal, S. P. (2009). Identification of a forensic case using microscopy and forensically informative nucleotide sequencing (FINS): A case study of small Indian civet (Viverricula indica). *Science & Justice, 50*, 94–97.

Sahoo, S., Kumar, A., Rai, J., & Gupta, S. K. (2024). Unlocking the genetic enigma: A paradigm shift towards evidence-based conservation of Indian star tortoise (Geochelone elegans). *Biological Conservation, 290*, 110421.

Smith, J. S. C., & Smith, O. S. (1992). Fingerprinting crop varieties. *Advances in Agronomy, 47*, 85–140.

Tóth, G., Gáspári, Z., & Jurka, J. (2000). Microsatellites in different eukaryotic genomes: Survey and analysis. *Genome Research, 10*(7), 967–981.

Verma, S. K., & Singh, L. (2003). Novel universal primers establish identity of an enormous number of animal species for forensic application. *Molecular Ecology Notes, 3*, 28–31.

Wasser, S. K., Shedlock, A. M., Comstock, K., Ostrander, E. A., Mutayoba, B., & Stephens, M. (2004). Assigning African elephant DNA to geographic region of origin: Applications to the ivory trade. *Proceedings of the National Academy of Sciences, 101*(41), 14847–14852.

Wasser, S. K., Mailand, C., Booth, R., Mutayoba, B., Kisamo, E., Clark, B., & Stephens, M. (2007). Using DNA to track the origin of the largest ivory seizure since the 1989 trade ban. *Proceedings of the National Academy of Sciences, 104*(10), 4228–4233.

Wasser, S. K., Clark, W. J., Drori, O., Kisamo, E. S., Mailand, C., Mutayoba, B., & Stephens, M. (2008). Combating the illegal trade in African elephant ivory with DNA forensics. *Conservation Biology, 22*(4), 1065–1071.

Willems, T., Gymrek, M., Highnam, G., The 1000 Genomes Project Consortium, Mittelman, D., & Erlich, Y. (2014). The landscape of human STR variation. *Genome Research, 24*(11), 1894–1904. https://doi.org/10.1101/gr.177774.114

Yadav, P., Kumar, A., Sahoo, S., Yadav, N., Hussain, S. A., & Gupta, S. K. (2021). Identification of Gangetic turtles based on species-specific variations on mitochondrial cyt b and nuclear Cmos genes. *Forensic Science International: Animals and Environments, 1*, 100035. https://doi.org/10.1016/j.fsiae.2021.100035

Part VI
Conservation Success Stories and Pathways Forward

Chapter 16
Success Stories of Species Recovery and Conservation in India

Angel Maria Lorance, Aarush S. Nair, Harsh Gulati, and Sarita Rana

Contents

A. M. Lorance · A. S. Nair · H. Gulati (✉)
Department of Zoology, School of Bioengineering and Biosciences, Lovely Professional University, Phagwara, Punjab, India
e-mail: harsh12@kuk.ac.in

S. Rana
Department of Zoology, Kurukshetra University, Kurukshetra, Haryana, India

A. Poonia et al. (eds.), *Wildlife Trade in India*,
https://doi.org/10.1007/978-981-95-3694-8_16

Abstract India has made significant strides in species recovery and conservation, despite ongoing threats such as habitat destruction, climate change, pollution, and wildlife crime. The country's conservation history dates back to the time of Emperor Ashoka, who declared the first protected areas as 'forests without fear' (Abhaya Aranya). However, modern conservation efforts began with the Wildlife (Protection) Act of 1972, which laid the foundation for structured legal protection and conservation initiatives. One of the earliest and most successful projects, *Project Tiger*, launched in 1973, helped stabilize and increase tiger populations through habitat protection, stringent anti-poaching measures, and advanced monitoring techniques.

The chapter explores various species recovery programs across India, highlighting success stories such as the revival of vulture populations, the conservation of the Great Indian Bustard, and efforts to protect the endangered Nilgiri Tahr. The drastic decline in vulture numbers, primarily due to the use of diclofenac in veterinary medicine, was reversed through strict bans and captive breeding programs. Similarly, the critically endangered Great Indian Bustard, which once had a widespread distribution, has benefited from conservation breeding programs, habitat restoration, and anti-poaching measures. The Nilgiri Tahr, a species endemic to the Western Ghats, faces challenges from habitat fragmentation and climate change, yet targeted conservation programs have shown promise in stabilizing its population.

A key aspect of species conservation in India is addressing human-wildlife conflict. With nearly 65% of the population residing in rural areas close to protected landscapes, competition for resources often results in negative interactions between people and wildlife. Conservation strategies now integrate community participation, sustainable livelihood alternatives, and environmental education to mitigate such conflicts. Additionally, illegal wildlife trade and habitat fragmentation due to urbanization and deforestation continue to threaten species survival, necessitating stronger enforcement of conservation laws.

India's legal framework, including the Wildlife Protection Act (1972), the Forest Conservation Act (1980), and the Biological Diversity Act (2002), provides the foundation for conservation policies. Government agencies like the Ministry of Environment, Forest and Climate Change (MoEFCC), the Wildlife Institute of India (WII), and the National Tiger Conservation Authority (NTCA) play crucial roles in implementing conservation measures. In the field of habitat restoration, conducting research, and implementing anti-poaching measures NGOs like WWF India and the Wildlife Trust of India (WTI) play a pivotal role.

Keywords Species recovery · Wildlife conservation in India · Project Tiger · Vulture conservation · Great Indian bustard protection · Nilgiri Tahr conservation · Human-wildlife conflict · Habitat restoration · Anti-poaching measures · Legal frameworks for conservation

16.1 Introduction

Human actions like destroying habitats of animals, changing the climate, pollution, and excessive use of resources threaten the plant and animal world of the world (Johnson et al., 2017). We need to know what works in conserving nature. Successful conservation methods can be used elsewhere to reduce loss of biodiversity. The first Protected Areas were created in India by Buddhist king Ashoka in the form of Abhaya Aranya or 'forests without fear' around 250 BCE (Rangarajan, 2005). Several wild animals and plants, especially tigers are revered in religions that evolved within the Indian sub-continent. The modern conservation era began much after India's independence in the early 1970s, with the enactment of the Wildlife (Protection) Act (1972). At the behest of J. C. Daniel and renowned ornithologist Salim Ali from the Bombay Natural History Society, then the Prime Minister of India, Indira Gandhi played a crucial role in tiger conservation, taking a personal interest in the initiative. With support from the World Wildlife Fund, she launched Project Tiger in 1973 as a centrally sponsored government program, initially covering nine Protected Areas across India.

To measure the progress of a species toward recovery, it is essential to define what constitutes a fully recovered state. However, in some cases, complete recovery may not be feasible. Conservation efforts primarily aim to preserve Earth's biodiversity by preventing extinctions, maintaining viable populations, and facilitating the revival of species experiencing population declines. While much of conservation work has been directed toward halting species loss and reducing decline rates, true success goes beyond just preventing extinctions. Assessment tools like the International Union for Conservation of Nature (IUCN) Red List focus on evaluating species' extinction risks. Although preventing a species from becoming more endangered is a significant achievement (Hoffmann et al., 2010), conservation initiatives often strive for broader goals (Soulé et al., 2003; Sanderson, 2006; Redford et al., 2011). Reintroduction serves as an important conservation strategy, aiming to restore viable populations of species within their native habitats (IUCN/SSC, 2013). However, the success rate for reintroducing large carnivores remains low, with long-term survival presenting significant challenges (Breitenmoser et al., 2001; Hayward et al., 2007; Wemmer & Sunquist, 1988). The ultimate objective of reintroduction is to establish a self-sustaining and viable population in its natural environment (Rodrigues et al., 2006). Therefore, the survival rate is one of the key factors in assessing the reintroduction success, and it depends on birth rate, death rate, immigration, and emigration. It may vary between sexes due to the different lifecycle strategies. Mostly, females are philopatric, while males are more prone to disperse from their natal place (Smith, 1993).

16.2 Major Challenges in Addressing Wildlife Crime in India

16.2.1 Human-Wildlife Coexistence and Conflict

As of the 2021 demographic assessment, approximately 65% of India's population—around 900 million people—resides in rural areas that border protected regions such as national parks, wildlife sanctuaries, and biosphere reserves. These protected zones cover nearly 5% of India's total land area, which is significantly lower than the global average of 16%. This close proximity often leads to intense competition for land and resources between local communities and wildlife. The situation is further complicated by factors such as limited awareness about wildlife conservation, a lack of alternative livelihood options, weak governance, inadequate conservation policies, high unemployment rates, and prevailing socio-economic disparities.

16.2.2 The Evolution of Criminal Behaviour and Global Politics

Wildlife crimes have different forms, such as illegal trade in live animals and plants and commercialization of their parts, such as ornaments or other derivatives. Such crimes are often driven by poachers, middlemen, and terminal consumers, and as such, there is a need for tight regulation and anticipatory control. Political incentives could also drive these crimes, especially international trade in foreign species that are not identified in the existing legal and scientific climates (Hulme, 2015).

16.2.3 Habitat Loss and Fragmentation

Habitat loss and fragmentation are among the largest threats to Indian wildlife. Urbanization, enlargement of agricultural land uses, other infrastructure activities, and forest loss have led to widespread natural habitat degradation. Habitat fragmentation separates populations into isolated, fragmented sites that are more susceptible to extinction due to genetic inbreeding and lack of adaptability to environmental changes (Karanth & Nichols, 1998).

16.2.4 Degradation and Pollution of Natural Habitats

Pollution, caused by industrial waste products, agricultural runoff, or plastic garbage, has a drastic adverse impact on India's wildlife habitats. Pollution of water bodies by pesticides and heavy metals weakens the quality of rivers and wetlands and hence impacts aquatic life as well as terrestrial animals that rely on these water bodies. The impact caused by plastic pollution on sea creatures is especially destructive, as these creatures either consume or become trapped in plastic trash. In addition to this, air pollution can also impact forest ecosystems, changing the diversity and health of diverse species.

16.3 Impacts of Climate Change on Fauna

Climate change is increasingly being recognized as one of the primary causes of biodiversity decline in India. Temperature and precipitation pattern changes are upsetting ecosystems and disrupting life cycles of numerous varied species. For example, alterations in monsoon patterns affect the availability of water and good resources, leading to shifts in species distribution and behavior (Chaturvedi et al., 2011). Coral reefs, being highly sensitive to temperature fluctuations, are increasingly experiencing bleaching events, posing a significant threat to marine biodiversity.

16.4 Legal Framework for Wildlife Conservation in India

India has developed a robust legal framework to safeguard its rich wildlife heritage. At the core of this framework is the Wildlife Protection Act of 1972, which ensures the conservation of wild animals, birds, and plants while establishing protected areas such as national parks and wildlife sanctuaries. Additionally, the Forest Conservation Act of 1980 plays a crucial role in regulating deforestation and managing the diversion of forest land for non-forest purposes, thereby protecting vital habitats. The Biological Diversity Act of 2002 further strengthens these efforts by promoting the conservation of biological resources and ensuring fair and equitable benefit-sharing from their utilization (Ministry of Environment, Forest and Climate Change, 2023).

16.5 Role of Government Agencies and Non-Governmental Organizations

The Ministry of Environment, Forest and Climate Change (MoEFCC) and the Wildlife Institute of India (WII) are primarily responsible for enforcing wildlife conservation laws and policies. These institutions oversee conservation programs, conduct research, and monitor wildlife populations. The National Tiger

Conservation Authority (NTCA), specifically tasked with tiger conservation, has been instrumental in implementing Project Tiger, which has played a key role in stabilizing and increasing tiger populations (NTCA, 2020).

Non-governmental organizations (NGOs) also make substantial contributions to wildlife conservation. Organizations such as the World-Wide Fund for Nature (WWF) India, Wildlife Trust of India (WTI), and Bombay Natural History Society (BNHS) collaborate with government agencies, local communities, and international organizations to tackle various conservation challenges. Their work includes conducting research, restoring habitats, and leading anti-poaching initiatives, all of which are essential to India's conservation efforts (WWF India, 2020).

16.6 Vultures Reborn: India's Fight to Save Nature's Clean-Up Crew (*Gyps indicus*)

The Indian vulture, or long-billed vulture (Gyps indicus), is a food-hunting bird belonging to the Accipitridae family. As an Old-World vulture, it has a diminutive, largely bald head with few feathers, a long bill, and expansive, dark wings. It normally resides on small cliffs and hills in central and southern India.

India is home to nine distinct species of vultures. These include the Indian Griffon Vulture (*Gyps fulvus*), Himalayan Griffon (*Gyps himalayensis*), Bearded Vulture or Lammergeier (*Gypaetus barbatus*), Egyptian Vulture (*Neophron percnopterus*), Red-Headed Vulture (*Sarcogyps calvus*), Oriental White-Backed Vulture (*Gyps bengalensis*), Slender-Billed Vulture (*Gyps tenuirostris*), and Long-Billed Vulture (*Gyps indicus*).

Winding back ten years, the number of wild stocks of three of its species have drastically plummeted. During 2005, the overall population of the genus *Gyps* in India, it was stated to have suffered a decline to a rate of 97%. In year 2002, the three vulture species were classified "Critically Endangered" which is the maximum degree of endangerment, by IUCN—the World Conservation Union because of some reports of recent and widespread reduction in population levels.

Vultures play a crucial role as primary scavengers of carrion in Africa and India. Their removal from the ecosystem could disturb the natural balance among other scavengers and potentially lead to a rise in decomposing animal remains.

The sharp decline in vulture numbers in India was first detected at the Keoladeo Ghana National Park in Rajasthan between the second half of the 1980s and the 1990s where the numbers fell from 816 to 25 (Prakash et al., 2003) thus, surveys in Northern India were conducted. Declines in India has estimated over 97% in a 12-year period.

The Indian vulture population has experienced a severe decline, leading to its classification as Critically Endangered on the IUCN Red List since 2002. Estimates suggest that the number of mature individuals remaining in the wild ranges between 5000 and 15,000. The primary cause of this decline has been linked to the widespread use of the veterinary drug diclofenac, which was administered to cattle for pain relief. When vultures consumed the carcasses of treated animals, they suffered from kidney failure, leading to a drastic drop in their numbers. Vultures that fed on the carcass of dead cattle are suspected to have been poisoned by diclofenac. In 2006, diclofenac was banned in India, Pakistan, and Nepal. There are two other vulture species with which the bird shares its habitat, namely the slender-billed vulture (*Gyps tenuirostris*) and the white-rumped vulture (*Gyps bengalensis*). It was first conceived that either some unidentified viral pandemic sickness was the cause of the drastic population fall for vultures in India or that there was simply a scarcity of food (dead cattle/livestock)—perhaps they have been removed for commercial use. The later part at least was not true. The BNHS, IVRI, and other Indian institutes conducted investigations into the epidemic illness theory in collaboration with scholars studying the topic outside. In 85% of the deceased vultures, it was discovered that, "visceral gout"—a buildup of uric acid in the internal organ surfaces and tissues—was noted (Oaks et al., 2004; Shultz et al., 2004; Swan et al., 2006) concluded that kidney failure, caused by metabolic, infectious, or toxic diseases, was the primary cause of death in vultures.

Data analysis indicates that even a small proportion—approximately 0.8%—of cattle carcasses contaminated with diclofenac can lead to a significant decline in vulture populations. This population decline has had notable socioeconomic and ecological consequences. Vultures play a crucial role in maintaining environmental balance by efficiently disposing of animal carcasses, thereby reducing pollution, limiting the spread of diseases, and controlling populations of opportunistic scavengers.

In the absence of vultures, numerous animal carcasses were left to decompose, exposing water sources to contamination and creating breeding grounds for disease-causing pathogens and pests, particularly rodents. The decline in vulture numbers also led to a sharp increase in feral dog populations, which in turn heightened the risk of rabies transmission to humans. It is estimated that the loss of vultures contributed to an increase of at least 5 million feral dogs, resulting in over 38 million additional dog bites and more than 47,000 rabies-related deaths. The economic impact of these consequences has been estimated at $34 billion. Furthermore, between 2000 and 2005, when vulture populations were at their lowest, human mortality rates reportedly increased by an average of over 4%.

16.6.1 Conservation Actions Undertaken

In March 2006, the Indian government imposed a ban on the use of diclofenac in veterinary medicine to protect vulture populations. As an alternative, meloxicam another nonsteroidal anti-inflammatory drug (NSAID) that metabolizes quickly and is non-toxic to vultures was identified as a suitable replacement. To facilitate this transition, pharmaceutical companies were encouraged to increase the production of meloxicam, ensuring its affordability and availability at a cost comparable to diclofenac. In 2015, the Government of India introduced a regulation to limit the vial size of veterinary diclofenac to 3 ml, effectively reducing the risk of its misuse in livestock treatment. Further research in 2021 identified tolfenamic acid as another NSAID that is safe for vultures, providing an additional alternative for veterinary use.

Since the medicine was banned in 2006, the vulture population had started recovering and remained approximately the same number until 2011. Yet more recent losses in vulture numbers have been recorded in other Indian states. Diclofenac may have been available for veterinary use, although banned and decreased in usage by nearly 50%. Other drugs, aceclofenac and ketoprofen, were also in use and proved deadly. In 2008, the Indian government enacted a law criminalizing the sale of diclofenac as a misdemeanour that is be punishable by law and lead to imprisonment. Further research later found that the population was slowly recovering, and the decline was notably halted in India due to a strict ban on the drug.

A conference aimed at formulating a recovery strategy for vultures in South Asia took place in Parwanoo, Haryana, in February 2004. The event focused on three vulture species: the white-backed vulture, the slender-billed vulture, and the long-billed vulture. The primary objectives of the action plan included eliminating the factors responsible for vulture decline by 2010 and ensuring the establishment of self-sustaining wild populations by 2030. Key areas of focus encompassed the regulation of veterinary drugs, research and monitoring efforts, public awareness and education initiatives, and tracking population trends.

Additionally, a national workshop was held in New Delhi in April 2004 to develop a comprehensive conservation strategy for these vulture species.

16.6.2 Conservation

Efforts to restore the Indian vulture population prompted the establishment of captive breeding programs. These programs, however, have a big challenge: vultures are long-lived birds and have a slow reproductive rate, particularly when in captivity. The end goal is to reintroduce these bred vultures to the wild when their habitat is cleared of diclofenac, the toxic drug that triggered their decline. In 2014, the "Saving Asia's Vultures from Extinction" project was initiated, which seeks to reintroduce these wonderful birds into their natural environment by 2016.

By that year, the Pinjore's Jatayu Conservation Breeding Centre became the first in Asia to attempt releasing captive-bred vultures into the wild. We've also seen small groups of these birds successfully breeding in southern India, particularly in Karnataka and Tamil Nadu. Additionally, specialized breeding centers can be found in West Bengal, Assam, and Madhya Pradesh, along with a few smaller centers linked to zoos.

In order to further conserve these vultures, the Indian Ministry of Environment, Forests, and Climate Change launched the Vulture Action Plan for 2020–25. This initiative aims to reinforce protection measures and establish a framework to ensure that harmful veterinary drugs, beyond just diclofenac, are prohibited.

In 2010, a small colony of Indian vultures, along with three other vulture species, was identified in the Moyar River valley within the Sathyamangalam Wildlife Sanctuary. The discovery included around 20 nests and 40 adult birds. Since the species had not been reported in the area since the 1970s, this finding plays a crucial role in conservation efforts to sustain their population.

Conclusion

In the Indian Subcontinent, *Gyps indicus* and *G. tenuirostris* populations decreased by more than 97% over a period of 10–15 years (Prakash et al., 2007) in the 1990s, which is equivalent to 99% in three generations (BirdLife International, 2020). Between 2003–2011, the total population seemed to be stable (Prakash et al., 2012). However, a decline of 29.5% of *Gyps indicus* in Gujarat, India, during 2005–2010 was reported. Between 2011 and 2015, vulture populations experienced another decline, with an average annual reduction of 11% from 2000 to 2015 (Prakash et al., 2019), resulting in an overall decline of 98.5% over three generations. Reports of Indian Vulture sightings dropped by over 90% between 2000 and 2018, highlighting a severe population collapse (SoIB, 2020). Between 2007 and 2016, surveys conducted in two neighboring Indian states revealed contrasting trends in breeding populations. While one population remained relatively stable, the other showed a continuous decline (McClure et al., 2021). However, following the ban on veterinary diclofenac and the implementation of conservation and protection measures, vulture populations have shown signs of stabilization and gradual recovery over the years (Fig. 16.1).

16.7 A Second Chance for the Great Indian Bustard (*Ardeotis nigriceps*)

Habitat and Distribution

Bustards are a critically endangered genus of birds that depend on grassland habitats. Earlier, they were distributed in dry, semi-arid, and damp grasslands throughout the country. Traditionally, the great Indian bustard was spread throughout Western India, covering 11 states and parts of Pakistan. Its strongholds were originally the Thar desert in the northwest and the Deccan plateau on the peninsula. Today, most of its population lives in Rajasthan and Gujarat. Small populations exist in Maharashtra, Karnataka, and Andhra Pradesh. Bustards are suited to open and flat grounds that provide minimal site obstruction and interference. Therefore, they thrive well in grasslands. In the non-breeding season, they prefer large agro-grass scrubland. During the breeding season, which occurs in summer and monsoon months, these birds gather in well-preserved grasslands featuring a mix of lightly grazed tall grasses, typically under 50 cm in height. They tend to avoid areas with dense vegetation, such as thickets and tall grasses. Among the 22 bustard species found

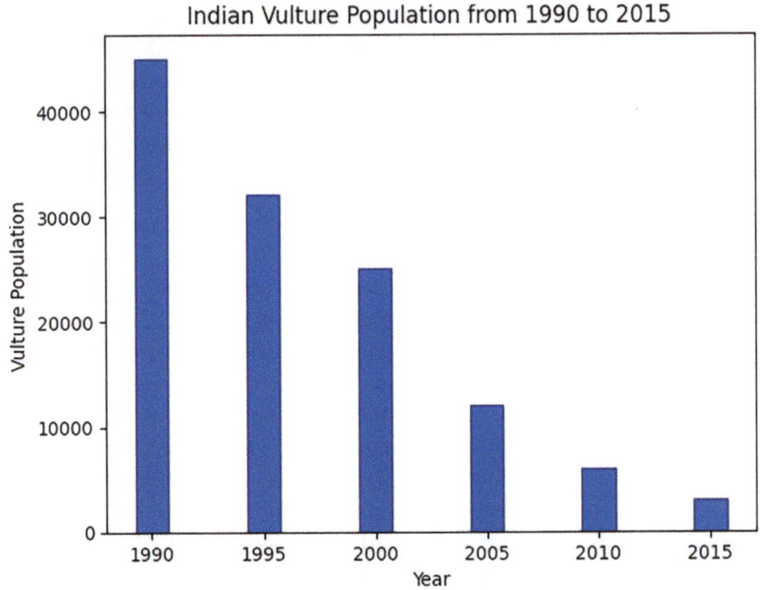

Fig. 16.1 Sharp Decline in Indian Vulture Population (1990–2015): The graph illustrates a drastic decrease in vulture numbers over 25 years, highlighting the urgent need for conservation efforts

worldwide, they rank among the most endangered. The Great Indian Bustard has disappeared from over 90% of its historical range and is now estimated to have a population of fewer than 1000 individuals. The primary cause of their decline is habitat loss due to grassland destruction. India is home to four species of bustards: the Great Indian Bustard, Lesser Florican, Bengal Florican, and Houbara Bustard. The global population of Lesser Floricans is believed to be under 2500, while Bengal Floricans are estimated to number between 400 and 500.

Cause

Hunting poses the greatest threat to this species, followed by instances of poaching outside Protected Areas. Additional dangers include collisions with high-tension power lines, fast-moving vehicles, and attacks by free-roaming dogs in human settlements. Habitat loss and alteration due to large-scale agricultural expansion, mechanized farming, infrastructure projects such as irrigation systems, highways, electric poles, mining, and industrialization further exacerbate the risks. As an indicator species of grassland ecosystems, its survival reflects the overall health and sustainability of this habitat (E. P, n.d.).

Control Measures

Research and Survey: Knowledge on the ecology, behaviour, and threats to the GIB has been very essential in dictating the course of conservational intervention. Such monitoring programs helps to analyse population trends, reproductive success, and migratory patterns as indicators of the effectiveness of the conservation activity (Jangra & Verma, 2024).

Captive Breeding and Reintroduction: Captive breeding initiatives have been created to produce Great Indian Bustards in captivity and then release them into the wild to help natural populations (Dutta et al., 2011).

Threat Mitigation: Several efforts have been adopted to lessen risks to the GIB, such as identifying electricity lines to avoid accidents and implementing anti-poaching and habitat destruction legislation (Jangra & Verma, 2024). Habitat Restoration: Initiatives to repair damaged habitats and construct corridors between fragmented habitats have been implemented to offer optimal conditions for the GIB.

The Ministry of Environment has implemented various measures to safeguard the Great Indian Bustard across India, including Rajasthan. Key initiatives include the following (Ministry of Environment, Forest and Climate Change, 2023):

(i) **Legal Protection**: The Great Indian Bustard is classified under Schedule I of the Wildlife (Protection) Act, 1972, ensuring the highest level of legal protection against hunting.
(ii) Habitat Conservation: National parks and wildlife sanctuaries have been declared as essential habitats of the species to enhance conservation.
(iii) Species Recovery Program: The Great Indian Bustard has been covered under the 'Species Recovery Programme' of the Centrally Sponsored Scheme for Wildlife Habitat Development. This program offers technical and financial support to state governments and union territories to improve the conservation efforts of the species and habitat.
(iv) Conservation Breeding Programmes: Conservation breeding programmes have been initiated in partnership with the forest departments of Rajasthan, Gujarat, and Maharashtra and the technical assistance of Wildlife Institute of India (WII), Dehradun. The programme focuses on breeding a captive population for future release into the wild as well as in-situ conservation.
(v) Conservation Breeding Centers: In collaboration with Rajasthan and Gujarat forest departments, WII, and worldwide experts, sites for the conservation breeding centers of the Great Indian Bustard and Lesser Florican have been located. A satellite conservation breeding facility has been established in Sam, Jaisalmer, Rajasthan.
(vi) Additionally, the Ministry provides financial support to states and union territories under the Centrally Sponsored Scheme to facilitate these conservation initiatives: Development of species Habitats to conserve species, especially the Great Indian Bustard. Request for in situ conservation of the Great Indian Bustard received from the Government of Rajasthan. The Ministry has advised the Rajasthan government that the project must be authorized by the State Board for Wildlife before it can be considered for financing from the State Plan and/or State CAMPA monies (Ministry of Environment, Forest and Climate Change, 2023).

In the summer of 2019, nine eggs were retrieved from a secluded area of the Desert National Park in Rajasthan's Jaisalmer region. They were incubated artificially, and the chicks were nurtured by hand at a workstation near Sam, within the expansive landscape (Sharma, 2020).

On World Environment Day 2013, the Rajasthan state initiated "Project Great Indian Bustard" to locate and enclose breeding grounds of bustards in the existing protected areas, besides developing safe breeding enclosures in non-protected areas.

In 2020, nine chicks were incubated and it became a world record.

A chick in 2024 after a female was artificially inseminated with sperms retrieved from a male that was 200 kilometers away.

The Great Indian Bustard is classified under Schedule I of the Indian Wildlife Protection Act, granting it the highest level of protection. It has been designated as the state bird of Rajasthan and is also included in the National Wildlife Action Plan (2002–2016). The National Forestry Commission, in Recommendation No. 172, suggested launching 'Project Bustard' to focus on conserving this critically endangered species (Ghouse & Indira, 2015). Additionally, the IUCN has urged India to initiate a dedicated conservation project for the species, with captive breeding being a potential strategy to boost its population. More knowledge of GIB behavior is required for ex-situ as well as in-situ conservation. Conserving the Great Indian Bustard needs pluralistic teamwork. This conservation initiative is more than just about a single species; instead, it aims at conserving entire ecosystems. Transcending difficult issues requires collaborative action for long-term conservation success (Ghouse & Indira, 2015).

Conclusion

In a nutshell, the conservation and recovery of the Great Indian Bustard are of utmost importance in sustaining this iconic species, important to the Indian subcontinent's environment. This includes habitat conservation, anti-poaching efforts, and captive breeding programs that could help sustain this species. There is a multiple issue that impacts the bustard population, requiring continuous attention and proper financing for concerted activities among government agencies, local communities, and conservationists. With united efforts, there is hope for a bright future, but urgent and deliberate action is the call of the day to keep the Great Indian Bustard from extinction (Fig. 16.2).

16.8 Leatherback Legacy: A Story of Survival and Conservation (*Dermochelys coriacea*)

The Leatherback turtle (Dermochelys coriacea) is one of the most fascinating organisms in tropical and temperate marine environments, occupying a range from the Pacific Ocean to the North Atlantic, and across the Indian Ocean (Shanker & Kale, 2014). Being the largest living species of marine turtle in the world, it is characterized by its ability to execute some of the most extensive migration patterns

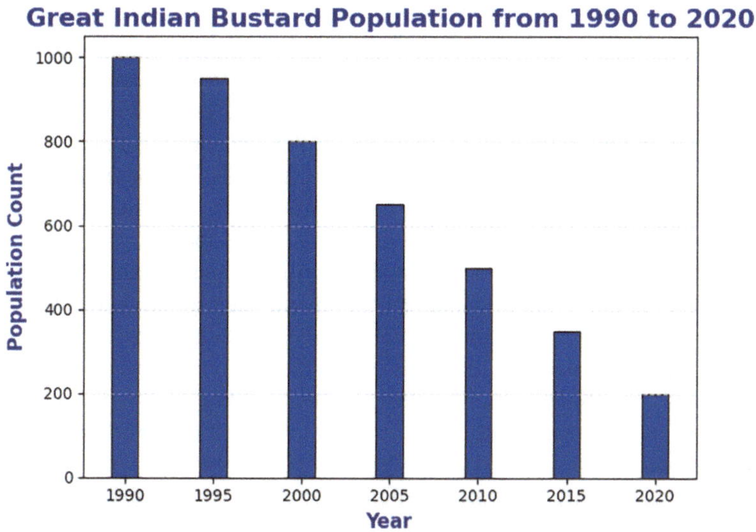

Fig. 16.2 Decline of the Great Indian Bustard Population (1990–2020): The graph depicts the alarming reduction in the population of this critically endangered species over three decades, emphasizing the need for immediate conservation measures

recorded among the turtles. The turtle has been recognized as Vulnerable under the IUCN red list. In addition, the species is accorded considerable legal protection under Schedule I (Part II) of the Indian Wildlife Protection Act of 1972. In India, leatherback turtles mainly nest in the Andaman and Nicobar archipelago (Namboothri et al., 2012). During the last three decades, pioneering research by ANET/MCBT, IISc (CES), and the Forest Department has revealed Little Andaman and the southernmost Great Nicobar Islands as possible nesting sites.

A vital knowledge gap in terms of the ecology and behavior of leatherback nesting involves the impacts of modified habitat dynamics on this species. With the destruction of most nesting beaches in the Andaman Islands and the flooding of those in the Nicobar Islands by the 2004 Indian Ocean Tsunami, a significant reduction in leatherback nesting activity was witnessed (Namboothri et al., 2012). This highlights the imperative for the commencement of long-term ecological work with reference to this species, in addition to a thorough evaluation of its habitat.

Quantification of beach recovery after tsunami occurrences has been carried out, emphasizing the need for quantification to determine the overall magnitude of such recovery. After the tsunami, the Andaman Islands received much more attention than Great Nicobar, which has continued to suffer from neglect due to the difficult accessibility of its nesting sites. Satellite tagging and telemetry studies in Little Andaman have shown progress, with extensive surveys by ANET/CES revealing significant Leatherback nesting activity. Expanding this method further could provide valuable insights. Recent research has also documented nesting on Galathea Beach in Great Nicobar, an area previously unrecorded for such activity, highlighting the importance of ongoing monitoring to track habitat usage changes across the archipelago.

The 2015–2016 SWOT report indicates a lack of sufficient data on the Leatherback subpopulation in this region due to research gaps. The archipelago serves as both a critical nesting ground for India's Leatherback turtles and a potentially significant site in the Northeast Indian Ocean. Addressing these research gaps through comprehensive monitoring and well-structured conservation strategies is essential to improve understanding and protection of the species.

Leatherback turtles are officially listed as 'critically endangered,' with an estimated 34,000 nesting females remaining. The population of Leatherback turtles in the eastern Pacific has reduced from more than 90,000 adults in 1980 to less than 2000 adult females by the year 2000. It is crucial that a similar fate does not befall the Atlantic population (Shanker & Kale, 2014). Indian shores are essential habitats for four nesting/feeding populations of sea turtles: leatherback (*Dermochelys coriacea*), green (*Chelonia mydas*), hawksbill *(Eretmochelys imbricata*), and olive ridley *(Lepidochelys olivacea*). The most prominent nesting places and populations in India are the mass nesting beaches of olive ridley turtles in Odisha, feeding and breeding locations for green and hawksbill turtles in the Andaman and Nicobar Islands, and the Lakshadweep Islands, The most notable nesting sites and populations in India are the mass nesting beaches of olive ridley turtles in Odisha, the feeding and nesting grounds for green and hawksbill turtles in the Andaman and Nicobar Islands and the Lakshadweep islands, and the leatherback turtle nesting population in Little Andaman Island and the Nicobar Islands, all of which are of high conservation importance (Shanker & Kale, 2014). Although all four sea turtle species are protected under Schedule I of the Indian Wildlife (Protection) Act of 1972, their populations in Indian waters are under threat due to unregulated coastal development and accidental entanglement in fishing operations. The olive ridley turtle (*Lepidochelys olivacea*) population in Odisha is particularly at risk, with estimates suggesting that over 100,000 turtles have drowned as bycatch in the last five years.Sea turtles serve as flagship species for various ecosystems, including coral reefs, seagrass beds, open ocean waters, and sandy shorelines. The challenges they face are similar to those impacting other marine and coastal species (Shanker & Kale, 2014).

For many years, sea turtles have confronted several challenges. Manmade factors have marred its existence. First, it had been hunted by the indigenous and hunting population for its meat. In addition, the population has drastically declined with illegal egg trafficking and hunting of the shell and leather (Patankar, 2023). It is also indirectly affected by activities such as infrastructural works through sand mining and coastal deforestation, which bring about habitat destruction and fragmentation to the turtle species. Moreover, bycatch is another issue that affects turtles. This occurs when fishermen accidentally catch them in their fishing gear. However, plastic trash and other debris have started to choke turtles and threaten their existence. In short, these issues have resulted in climate change, leading to the endangered sea turtles (Patankar, 2023).

16.8.1 Legal Protection

16.8.1.1 Wildlife Protection Act (1972)

The Wildlife Protection Act of 1972 focuses on the conservation of wild animals, birds, and plants across India, including sea turtles that nest along the country's western and eastern coasts. To support conservation efforts, India has established four coastal mainland national parks and seventeen protected areas dedicated to marine biodiversity (Patankar, 2023).

16.8.2 National Marine Turtle Action Plan

The Ministry of Environment, Forests and Climate Change (MoEF&CC), through its 'Marine Megafauna Stranding Guidelines' and 'National Marine Turtle Action Plan,' tried to initiate an inter-regional project that connects the Central and state governments, and the civil society to educate and raise awareness about threats to marine animals like sea turtles. Such acts try to take action towards reducing risks, rebuilding damaged habitats, improving public participation, and continuing research and knowledge of marine species (Patankar, 2023).

16.8.3 Conservation Projects

Sea Turtle Project A joint effort of the Government of India and the United Nations Development Program, this project was initiated in 1999. The Wildlife Institute of India is responsible for the organization of these programs. The main objectives include increasing awareness, carrying out evaluations and reports, conducting GIS analysis of nesting sites, conducting TED operations, and conducting workshops for field staff, all for the conservation of Olive Ridley turtles that nest exclusively along the eastern coast of Odisha.

16.8.4 Rehabilitation Program

The Central Marine Fisheries Research Institute initiated a program along the Chennai coast aimed at protecting olive ridley turtle hatchlings, which face high mortality rates due to human activities and predation. So far, the CMFRI field laboratory has successfully facilitated the hatching of 60,410 olive ridley turtles, contributing significantly to the conservation of this endangered species (Patankar, 2023).

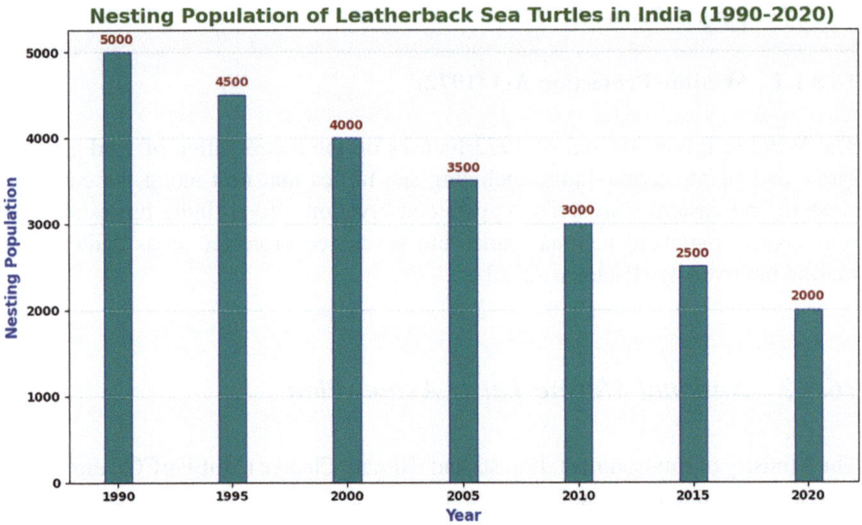

Fig. 16.3 Nesting Population of Leatherback Sea Turtles in India (1990–2020): The graph illustrates the steady decrease in nesting numbers over three decades, highlighting the urgent need for conservation efforts to protect this vulnerable marine species

The Gahirmatha Marine Sanctuary, established in 1997 within the Bhitarkanika Wildlife Sanctuary, is recognized as India's largest nesting site for turtles and Odisha's only dedicated turtle sanctuary. Over the years, it has played a crucial role in the rehabilitation of olive ridley turtles. Meanwhile, Sahyadri Nisarga Mitra has gained recognition among international tourists as a premier birdwatching destination (Patankar, 2023).

One of the most impactful conservation projects on India's western coast, the Marine Turtle Conservation Scheme, was launched to safeguard sea turtle hatchlings along the Konkan coastline. Due to its success, the organization now hosts the renowned 'Turtle Festival' in coastal locations such as Velas, raising public awareness about the importance of protecting young turtles (Patankar, 2023) (Fig. 16.3).

16.9 From Shadows to Stripes: Bengal Tigers Roaring Again (*Panthera tigris*)

The tiger (*Panthera tigris*) is an indicator of what is happening ecologically and the degree to which conservation is working or not. Local extinction of the tiger happened in Panna Tiger Reserve in Central India due to poaching and for many other reasons. The key drivers for the declining populations of tigers include conservation measures, commercial trade of their body parts, hunting, scarcity of prey population, and destruction of their habitat (Nowell & Jackson, 1996; Seidensticker et al., 1999). During the past two centuries, the populations of large carnivores have significantly reduced, shrunk in their geographic range, and become fragmented within their habitats (Harihar et al., 2020). There are a range of human-related threats, from habitat loss to prey depletion and even direct persecution by humans, which threaten the continued survival of these populations (Harihar et al., 2020). As top predators, these species exist inherently at low numbers within food webs. Additionally, their specific life history characteristics make them especially vulnerable to these threats, which often translates into a slow reaction to conservation efforts (Cardillo et al., 2004). In the nineteenth century, India served as the largest habitat for tigers (*Panthera tigris*) (Gee, 1966), yet their numbers continued to dwindle; to combat this decline and halt the alarming reduction in population, the Indian government initiated 'Project Tiger' in 1973 (Panwar, 1982).

Despite conservation efforts, unrestrained poaching and inadequate scientifically guided management approaches resulted in regional extinctions for tigers within Sariska Tiger Reserve and PTR (Chapron et al., 2008; Wildlife Institute of India, 2009). By the time Project Tiger was launched in India, the country's tiger population had drastically declined to approximately 1800 from an estimated 40,000 in the precolonial era (Panwar, 1982). The pugmark census method initially indicated a rise in numbers, with tiger populations reaching around 3500 by the 1990s. This apparent success led to a reduction in conservation efforts, as legal hunting was banned and the domestic market for tiger trophies largely disappeared. However, between 1994 and 1995, poaching significantly impacted the Ranthambore Tiger Reserve, reducing the number of tigers there to fewer than twelve. The crisis deepened with the local extinction of tigers in Sariska Tiger Reserve in 2005, prompting widespread media attention. In response, then-Prime Minister Dr. Manmohan Singh established a Tiger Task Force to tackle the conservation challenges facing India's tigers. To overcome the difficulty of the elusiveness of the species, camera traps work as a suitable method, complemented by radiotelemetry, to monitor animals remotely, such as distinctly marked felids: tigers (Karanth & Nichols, 1998), leopards (Noack et al., 2019), jaguars (Silver et al., 2004) ocelots (Satter et al., 2019), and lynxes (Zimmermann et al., 2013). Advanced technology has significantly enhanced tiger conservation efforts. The E-eye system, equipped with thermal and visual cameras, has been deployed in key tiger reserves to provide real-time monitoring, helping authorities deter poaching activities. Additionally, some reserves have incorporated drones to strengthen surveillance. ExtractCompare, a specialized pattern extraction and comparison software, has been instrumental in maintaining a database of over 100,500 camera trap images, documenting around 8000 individual tigers across India, Nepal, and Bangladesh since 2008. This software assists in comparing camera trap images with confiscated tiger skins to identify poaching hotspots and illegal trade routes. Furthermore, the Wildlife Institute of India (WII) has implemented MSTrIPES, a GPS-based mobile application designed to support patrolling operations in all tiger reserves, ensuring better protection and monitoring of tiger populations. This application ensured that patrol guards covered the required spatial areas and received suitable incentives for their efforts. The MSTrIPES system optimized patrols by identifying weak spots through the analysis of data collected by the patrols combined with ecological studies. Large carnivores require relatively expansive territories to meet their life needs (Majumder et al., 2012; Tucker et al., 2014). Habitat fragmentation, the expansion of agricultural fields, and human activities are recognized as limiting factors for habitat size; furthermore, smaller protected areas are inadequate to sustain a viable long-term population of wide-ranging carnivores in India (Ranganathan et al., 2008; Wikramanayake et al., 1998). The effective conservation translocation through tiger reintroduction in PTR-the only known source population of tigers in the Vindhyan Range-is attributed to effective conservation translocation. A positive population growth rate decreased inter-birth intervals, and high cub survival rates have resulted in rapid population recovery. Effective conservation and management strategies have contributed to the stabilization of the tiger population in Panna Tiger Reserve (PTR), ensuring a strong likelihood of long-term viability. As part of the long-term 'Tiger Reintroduction and Recovery Programme' in PTR, Madhya Pradesh, a total of 25 tigers (7 males and 18 females) were fitted with radio collars between March 2009 and June 2020. The Madhya Pradesh Forest Department authorized the capture and collaring of these tigers (MPFD Letter No./Exp./2009/1205 dated 31/8/09), following the provisions of the Wildlife Protection Act, 1972, Section 11 (1A).

The tigers were tracked and immobilized using the 'Hellabrunn mixture' (125 mg xylazine +100 mg ketamine/mL) (Hafner et al., 1989), which was administered through a Tele-inject projector (Model 4 V.31). The chemical sedation process was conducted under veterinary supervision. Among the collared tigers, fifteen were fitted with high-frequency transmitters from Telonics® Inc., while ten were equipped with VHF/GPS/UHF collars from African Wildlife Tracking® Inc. and Vetronic Aerospace®. These tigers were closely monitored using UHF and satellite tracking, while VHF-collared individuals were tracked continuously by staff and researchers. Monitoring efforts included routine patrols with elephants, camera traps, and radiotelemetry to assess inter-birth intervals (IBI), litter sizes, and cub survival rates. Initially, all adult and sub-adult tigers were radio-collared following their reintroduction. However, as the population grew, not all individuals were tagged, necessitating the development of a camera trap survey method to track non-collared tigers. To analyze the population growth rate, the software Vortex version 10 (Lacy & Pollak, 2014) was used with 100 iterations.

This software is appropriate for modeling species of low fecundity as well as with long lifespans, and it widely used within published reintroduction models. The growth rate (r) of more than 0 is a positive indicator for population growth, while less than 0 is a sign of decline. Similarly, an annual multiplicative growth rate (λ) greater than 1.0 indicates positive population growth ($N\,t + 1 > N\,t$), while a rate less than 1.0 indicates a decline in the population ($N\,t + 1 < N\,t$). The first set of reintroductions comprised three tigers in 2009, including two females and one male. Later, the culling of two semi-wild tigresses was done in 2011 followed by catching a wild female in 2014, and a male in 2015. Out of seven reintroduced animals, the population grew significantly, expanding more than eightfold, to 59 individuals in 2021. The population growth rate was calculated at 0.258 (~26%), while an annual multiplicative growth rate was established at 1.29. A total of 120 cubs were registered from 45 litters produced by 18 females between 2010 and 2021. Thirteen

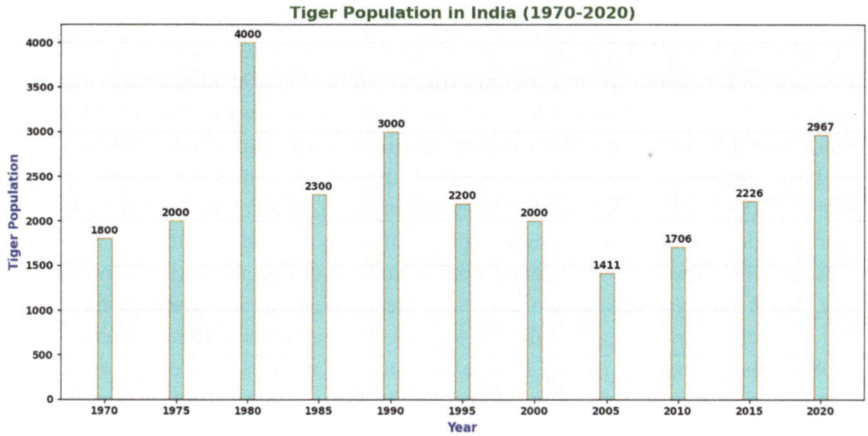

Fig. 16.4 Fluctuating Tiger Population in India (1970–2020): The graph highlights the population trends of tigers over five decades, showing an initial rise, a sharp decline, and a significant recovery in recent years, reflecting the impact of conservation initiatives like Project Tiger

females from the Panna Tiger Reserve's progeny have given birth to 26 litters, resulting in 64 cubs since 2013. The most frequently recorded litter sizes were 2 or 3 cubs, each accounting for 31.1% of occurrences in the reserve. The average litter size was 2.66 ± 0.1 (SE) ($n = 45$, ranging from 1 to 4). The average IBI in the reserve was determined to be 19.16 ± 0.5 (SE) months (with a range from 14 to 24 months). During the course of the study, twelve litters were abandoned by the breeding females, which then eventually delivered supplementary litters between 5 to 8 months after abandonment (Fig. 16.4).

16.10 Restoring the Roar: The Asiatic Lion's Conservation Journey (*Panthera leo persica*)

Asian and world carnivores are on the verge of extinction due to chronic habitat loss and poaching. Conservation history of Asiatic lion of Gir forest, India, has been a success story, and its history can be learned for the conservation and recovery of other endangered predators. Asiatic lions inhabit the last forest cover of the Indian state of Gujarat. Gir and Girnar hill systems of the state harbours the largest dry deciduous forest, thorny forests, and savannas of Gujarat and are the principal habitats of a rich variety of plant and animal species, some of which are part of the list of endangered (Singh & Kamboj, 1996). The reduction in lion populations in their wider range is largely due to poor management of protected areas, human-lion conflict, prey loss, habitat loss, and poaching (Bauer et al., 2022). Forests cover about 9.6% of the land in Gujarat, and a large sum of conservation funds of about 216 million USD is available for numerous projects, including those of the Gir

Protected Area. The income derived from tourism by the Gujarat State Lion Conservation Society (GSLCS) goes entirely towards the conservation of lions. Moreover, support from numerous programs, like the India Eco-Development Project and the Biodiversity Conservation & Rural Livelihood Improvement Project (BCRLIP), guarantees the utilization of varied strategies in conservation while providing funds for basic protection, resource management, and restoration of the habitat (Mishra et al., 2010). The Asiatic lion (Panthera leo persica), reduced to the Gir Forest and nearby regions in the south-western edge of Gujarat's Saurashtra zone, has recovered substantially. Previously threatened with extinction during the early twentieth century, when barely a few dozen survived, the species has increased to more than 400 in recent years. In spite of their designation as endangered (Breitenmoser et al., 2008), conservation has played its role in sustaining their existence and population. In monitoring movements of the lions, Vertex Plus GPS collars (Vectronic Aerospace, GmbH, Berlin, Germany), weighing less than 3% of body weight of a lion to prevent behavior disruption, were employed. For immobilization, both ketamine hydrochloride (2.2 mg per kg body weight) and xylazine hydrochloride (1.1 mg per kg body weight) were injected (Sontakke et al., 2009). The anesthetic was administered through a gas-operated Telinject TM G.U.T 50 dart (Telinject, Inc., Dudenhofen, Germany). For recovery after collaring, an intravenous injection of Yohimbine hydrochloride (0.1–0.15 mg per kg body weight) was administered as an antidote. All radio-collaring procedures were conducted under the supervision of trained veterinary officers and experienced wildlife health professionals. The movements of the collared lions were closely monitored and managed by the Gir Hi-Tech Monitoring Unit.

Geographical location of each radio-collared animal was transmitted by the GPS Plus X software (Vectronics Aerospace GmbH, Berlin, Germany) at the Gir Hi-Tech Monitoring Unit, a science-based technology monitoring system commenced in 2019 at Sasan-Gir, Gujarat. GPS collars featured the following: mortality sensors, drop-off release programmable mechanism, and user-programmable GPS schedule to receive hourly location fixes. Home ranges and core areas of home ranges were estimated using ArcMap (Version 10.8.1, ESRI, Redlands, USA). Home ranges and core areas were estimated at 95% and 50% respectively using Fixed Kernel Density Estimation. They were estimated using the Adehabitat package of the R programming environment (Calenge, 2006; Kernohan et al., 2001). Lions are known to inhabit a variety of environments within their range. Asiatic lions, currently classified as a vulnerable species, are found in diverse landscapes featuring different terrains, vegetation types, and land uses. This is largely due to a nearly fourfold increase in their population since 1968. As of June 2020, approximately 48% of the 674 lions recorded have moved beyond protected areas, spreading across nine districts and 13 forest administrative divisions. Lions residing outside the Gir Protected Area (Gir PA) form satellite populations in distinct habitats, including varied terrains, riverine regions, and coastal ecosystems. Efforts to conserve and restore the species are actively carried out through various government initiatives and non-governmental organizations.

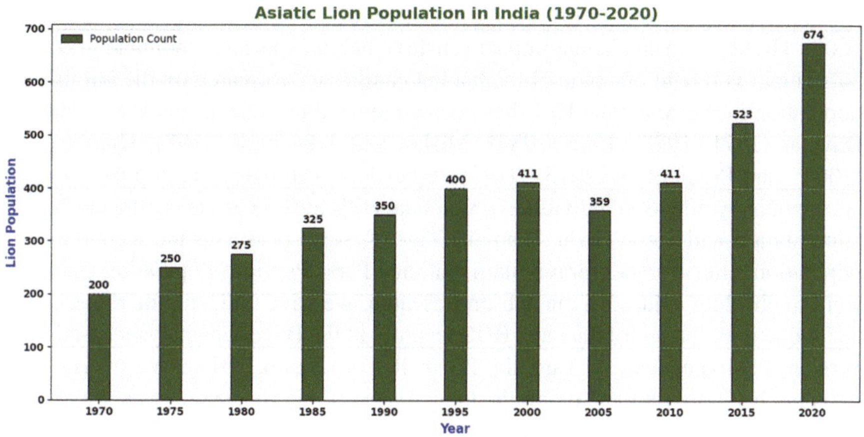

Fig. 16.5 Rising Asiatic Lion Population in India (1970–2020): The graph showcases the steady growth of the Asiatic lion population, with significant conservation successes, particularly after 2010, highlighting the effectiveness of habitat protection and conservation efforts in Gir Forest

Asiatic Lion Reintroduction Project: The Asiatic Lion Reintroduction Project is a government scheme whose aims are the multiplication of the Asiatic lion in the wild and the prevention of extinction. The central government initiated an "Asiatic Lion Conservation Project" with a financial commitment of INR 97.85 crore. The implementation period is three years during which support for enhancing conservation measures is intended under the financial backdrop of FY 2018–19, FY 2019–20, and FY 2020–21.

Wildlife Crime Cell State-level Task Force and Wildlife Crime Cell of Greater Gir area (GGR) have been formed, strengthening protective efforts (Fig. 16.5).

16.11 Scaling Heights: Saving the Endangered Nilgiri Tahr (*Nilgiritragus hylocrius*)

The Nilgiri tahr (*Nilgiritragus hylocrius*) is a mountain ungulate endemic to the Western Ghats of India. It is classified as an Endangered species under the IUCN Red List criterion C2a(i) (Alempath & Rice, 2008) and is also listed in Schedule I of the Indian Wildlife (Protection) Act of 1972, which provides it with the highest level of legal protection (Parliament of India, 1992). Estimates suggest that the global population of Nilgiri tahr ranges between 2000 and 3000 individuals (Daniels et al., 2008; Predit et al., 2015). But no dependable population estimate is available and our facts on species range and conservation status are sporadic. Disjunctive state distribution and Tamil Nadu and Kerala are Nilgiri tahr habitat. Species' habitat area is below 5% of the whole Western Ghats area (Grubb, 2005). Habitat destruction, invasion of alien species, domestic livestock grazing on narrow habitat, poaching, and habitat fragmentation are the significant threats to Nilgiri tahr (Daniels et al., 2008). Highly climate change impact sensitive habitat specialist montane grassland Nilgiri tahr is reliant on some of the habitat conditions. Research on the Nilgiri tahr population in the Anamalai Hills has been conducted by several scholars, including Davidar (1971, 1978), Rice (1984), Mishra and Johnsingh (1998), Kumar et al. (2002), and Predit et al. (2015). Their studies have estimated the population to be approximately 500–600 individuals. Anamalai Hills are one of the significant Nilgiri tahr habitats, and Tamil Nadu Anamalai Tiger Reserve possesses the second largest population, after Kerala's Eravikulam National Park. Previous population estimates were in absolute values. In our present research, we have utilized a mark-recapture technique with the use of package BBRecapture in the Bayesian analysis based on R version 3.3.4 (Fegatelli & Tardella, 2013; R Core Team, 2018). We overlaid the potential habitat with the data on the habitat need of the species such as grasslands, shola forest, and the cliffs, and the landscape as slope classes from a Digital Elevation Model (DEM). Land covered by grass with the slope value of more than 45 degrees fall under steep slopes or cliffs, the same habitats being shown as the habitat of the tahr. We were thus in a position to choose the priority locations where the survey could be conducted. The model was able to identify the mountainous area as well as areas of the development of steep slopes and cliffs in the Western Ghats. In those areas where large-scale disturbance such as hunting and other man-made disturbance, Nilgiri tahr can be seen to stay close to cliffs and not far from cliffs. The cliffs have protective advantages on them from both nature and man's intrusion and thus are very much essential in the identification of the habitat use of all ungulates in mountain habitats. With the field survey, the assessing teams were very well conversant with the issues and problems Nilgiri tahr was undergoing in the concerned area. The issues which are arisen are by preliminary observation as well as hearsay reports, and thus there is a need for further appraisal by qualitative survey that is holistic in nature.

Isolation and Small Populations

Most of them are populations of fewer than 20 animals and are thus highly vulnerable to inbreeding as well as other stochastic risks to extinction. Strategic management intervention can be made to enable some of the low-density populations to increase, as a function of availability of dispersal habitat. Critical assessment of habitat connectivity must be undertaken and provision to enable creation of safe linkages to enable reconnection of fragmented populations wherever feasible.

16.11.1 Impact of Human Disturbances on Small Habitat Patches

Both small and large populations are affected by human activities such as cattle grazing, fuelwood collection, unauthorized tourist visits, and human movement through their habitat. These disturbances, whether individually or collectively, place stress on the Nilgiri tahr and restrict their access to essential daytime foraging areas. To minimize these negative effects, strict regulation and enforcement of habitat usage are necessary.

Disease and Competition
Stock presence is threatening the tahr directly, the presence of which enables disease transmission from domestic to wild herbivores. Stock is also competing with tahr for available resources, with implications that must be addressed of existing knowledge gaps through targeted research.

Poaching
Local community information reveal poaching to be seriously threatening the tahr conservation program. Low density is contributing to the problem, and problems are severe.

Habitat Loss, Degradation, and Fragmentation
The spread of invasive species is reducing the land available for native habitats. For example, regions like Kudukkadi Betta in the southern Nilgiris and Nandala Malai in the Chinnar Wildlife Sanctuary face fragmentation due to these invasions. Additionally, human-induced habitat modifications and increasing anthropogenic pressures are further contributing to habitat loss and fragmentation.

16.12 Addressing and Assessing Key Threats

1. **Isolation of small populations:**

The challenge to the subject demands exploration of tahr population habitat dynamics and genetic diversity, e.g., tahr-open habitat extension by tahr habitat protection and proactive habitat management.

2. **Vulnerability of small patches of habitat to anthropogenic disturbance:**

This can be solved by imposition of community-based conservation regimes, augmented surveillance mechanisms for deterrence of illegal activity, and imposition of routine monitoring regimens for such populations.

3. **The role of disease and interspecific competition**

A careful consideration of the role of livestock grazing, presumptive competitive interactions of tahrs and livestock, and disease transmission among such populations needs to be addressed in effective formulation of conservation measures.

4. **Combating poaching:**

In discouragement of poaching activity, an augmentation of provision of training facilities and equipment to Forest Department staff, as well as encouragement of cooperative patrolling exercises with local communities, needs to be taken care of.

5. **Habitat loss, degradation, and fragmentation issues:**

Land use alteration that will most likely further split tahr populations needs strict regulation. Conservation of tahr needs to respond to this by both long-term intervention and policy reaction. Short-term intervention on habitat conservation measures is also needed.

The findings validate the double-observer survey technique to be statistically reliable and sufficient for Nilgiri tahr population estimation.Our estimate of 510 Nilgiri tahrs, with a 95% confidence interval of 300–858, aligns with previous population assessments. Mishra and Johnsingh (1998) estimated 339 individuals in the Anamalai Tiger Reserve, with a total population ranging from 540 to 645, excluding the Parambikulam Wildlife Sanctuary in Kerala. Their study did not account for the Udumalpet and Amaravati Ranges. Davidar (1978) provided an estimate of approximately 598 individuals within the Reserve. These past assessments fall within our confidence range. However, a direct comparison with Predit et al. (2015) was not possible, as their study was limited to six sites within the Reserve and recorded 107 visible tahrs. Apart from this, a study by Kumar et al. (2002) in the same sites of our survey had counted 186 tahrs and had identified eight subpopulations; the study, however, was characterized by year-round monitoring and did not focus on the prime objective of tahr population count estimation (Fig. 16.6).

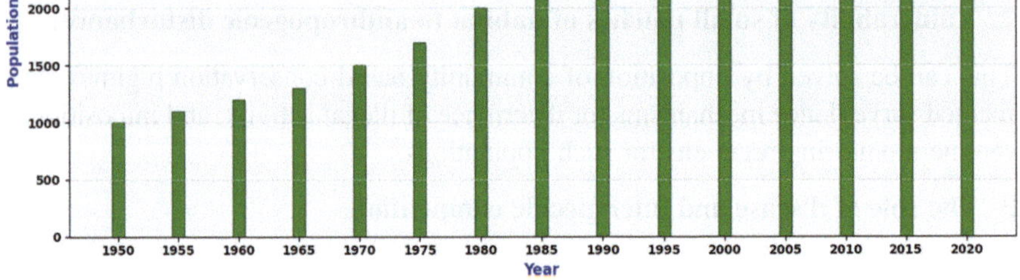

Fig. 16.6 Steady Growth of Nilgiri Tahr Population in India (1950–2020): The graph illustrates the gradual increase in Nilgiri Tahr numbers over seven decades, reflecting successful conservation efforts and habitat protection in the Western Ghats

16.13 Conservation Status of Key Wildlife Species in India

India boasts remarkable biodiversity, hosting a vast array of plant and animal species, including many that are unique to the region and classified as endangered. However, rapid urbanization, deforestation, climate change, and poaching have put many species at risk. Conservation efforts, including wildlife sanctuaries, national parks, and government policies, play a crucial role in protecting these species.

This document outlines important wildlife species found in India, their conservation status, and the primary threats to their survival. The table below highlights key details about some of these species (Table 16.1).

This table provides a quick reference to the current conservation status of key species, highlighting the major threats affecting their survival.

16.14 Detailed Analysis of Key Species

Snow Leopard: The Ghost of the Mountains
The snow leopard (*Panthera uncia*) is a solitary predator adapted to the harsh conditions of the Himalayas. Despite its elusive nature, it faces increasing threats due to poaching, habitat destruction, and retaliatory killings by herders. Conservation initiatives such as the Snow Leopard Project focus on mitigating human-wildlife conflict and strengthening anti-poaching measures.

Indian Pangolin: The Most Trafficked Mammal
The *Manis crassicaudata*, or Indian pangolin, is a nocturnal mammal highly sought after for its scales, which are used in traditional medicine. Illegal wildlife trade has led to its endangered status. Conservation efforts focus on stricter law enforcement, community awareness programs, and its inclusion in global trade bans under CITES.

Gharial: A Critically Endangered Reptile
Once abundant in Indian rivers, the gharial (*Gavialis gangeticus*) is now critically endangered due to habitat destruction, dam construction, and pollution. Conservation programs like captive breeding and river habitat restoration have been initiated to support their recovery.

Himalayan Monal: The Jewel of the Himalayas
The Himalayan monal (*Lophophorus impejanus*), the state bird of Uttarakhand, is currently classified as 'Least Concern.' However, habitat loss and hunting for its vibrant plumage pose risks. Conservation programs focus on preserving its natural habitat in the Himalayas.

Indian Star Tortoise: Threatened by Illegal Pet Trade
The Indian star tortoise (*Geochelone elegans*) is facing a decline due to excessive poaching for the pet trade. While it is classified as 'Vulnerable,' conservationists are advocating for stronger legal protections and habitat conservation efforts.

Table 16.1 Reviving the Wild: Conservation Success Stories of Endangered Species in India

S. no	Species name	Scientific name	Conservation status	Threats	Recovery measures	Government acts/policies	Population trend	Habitat	Key conservation organizations
1	Snow Leopard	Panthera uncia	Vulnerable	Poaching, Habitat Loss	Community-based Conservation, Anti-Poaching Measures	Wildlife Protection Act, 1972	Increasing	Himalayan Region	Snow Leopard Trust
2	Indian Pangolin	Manis crassicaudata	Endangered	Illegal Trade, Habitat Destruction	Awareness Campaigns, Anti-Trafficking	CITES, Wildlife Protection Act, 1972	Stable	Forests and Grasslands	Wildlife Trust of India
3	Gharial	Gavialis gangeticus	Critically Endangered	River Pollution, Sand Mining	Captive Breeding, River Conservation	National Chambal Sanctuary	Increasing	Ganges and Brahmaputra Rivers	Madras Crocodile Bank
4	Himalayan Monal	Lophophorus impejanus	Least Concern	Hunting, Habitat Loss	Eco-tourism, Habitat Protection	Wildlife Protection Act, 1972	Stable	Himalayan Forests	State Forest Departments
5	Indian Star Tortoise	Geochelone elegans	Vulnerable	Illegal Pet Trade	Rescue Operations, Habitat Protection	CITES, Wildlife Protection Act, 1972	Stable	Dry Areas and Grasslands	Wildlife SOS
6	Blackbuck	Antilope cervicapra	Least Concern	Poaching, Habitat Loss	Protected Areas, Breeding Programs	Wildlife Protection Act, 1972	Increasing	Grasslands	WWF-India
7	Red Panda	Ailurus fulgens	Endangered	Deforestation, Hunting	Community Participation, Forest Protection	Biodiversity Act, 2002	Stable	Eastern Himalayas	Red Panda Network

8	Hangul (Kashmir Stag)	Cervus hanglu hanglu	Critically Endangered	Habitat Fragmentation, Poaching	Protected Areas, Breeding Initiatives	Wildlife Protection Act, 1972	Slow Recovery	Kashmir Valley	Jammu & Kashmir Wildlife Department
9	Dugong	Dugong dugon	Vulnerable	Coastal Development, Fishing Nets	Marine Protected Areas, Research	Marine Biodiversity Act	Stable	Coastal Waters	Wildlife Institute of India
10	Golden Langur	Trachypithecus geei	Endangered	Deforestation, Habitat Fragmentation	Forest Protection, Awareness Programs	Wildlife Protection Act, 1972	Stable	Northeastern India	Aaranyak

Blackbuck: A Conservation Success Story
The blackbuck (*Antilope cervicapra*) was once severely hunted but is now classified as 'Least Concern' due to successful conservation initiatives. Efforts such as protected areas and anti-poaching laws have helped its population recover.

Red Panda: A Victim of Deforestation
The red panda (*Ailurus fulgens*), native to the Eastern Himalayas, is endangered due to deforestation and poaching. Conservation strategies include habitat restoration, anti-poaching measures, and eco-tourism initiatives to protect this unique species.

Hangul (Kashmir Stag): On the Verge of Extinction
The Hangul (Cervus hanglu hanglu) is perhaps the most threatened deer species, occurring exclusively in Kashmir. Poaching, habitat fragmentation, and low genetic diversity are high threats. Conservationists point to the necessity of dedicated reserves as well as habitat connectivity initiatives.

Dugong: India's Sole Marine Herbivore
The dugong (Dugong dugon) or sea cow is threatened by coastal degradation, fishing bycatch, and pollution. Conservation efforts target marine protected areas and education campaigns to safeguard this species.

16.15 Conservation Strategies and Future Outlook

India has implemented several conservation strategies to protect its wildlife:

- **Legal Protections**: The Wildlife Protection Act, 1972, provides a legal framework to protect endangered species.
- **Protected Areas:** India has designated several national parks, wildlife sanctuaries, and biosphere reserves to safeguard its diverse flora and fauna. These protected areas serve as critical habitats for numerous species and play a vital role in conservation efforts.
- **Community Involvement**: Programs like eco-tourism and local community participation help in conservation.
- **Captive Breeding & Reintroduction**: Successful breeding programs have helped species like the gharial and blackbuck recover.

Despite these efforts, continued focus on conservation policies, research, and awareness is essential to prevent species extinction.

India's diverse wildlife is an integral part of its ecological and cultural heritage. While conservation efforts have seen success, many species remain critically endangered. Strengthening policies, enhancing habitat protection, and increasing public awareness will be crucial in ensuring the survival of these species for future generations.

References

Alempath, M., & Rice, C. (2008). *Nilgiritragus hylocrius*. In *The IUCN Red List of Threatened Species 2008* (e.T9917A13026736). https://doi.org/10.2305/IUCN.UK.2008.RLTS.T9917A13026736.en

Bauer, H., Dickman, A., Chapron, G., Oriol-Cotterill, A., Nicholson, S. K., Sillero-Zubiri, C., et al. (2022). Threat analysis for more effective lion conservation. *Oryx, 56*(1), 108–115. https://doi.org/10.1017/S0030605320000253

BirdLife International. (2020). Gyps indicus (amended version of 2017 assessment). The IUCN Red List of Threatened Species 2020: e.T22729731A180736159. https://doi.org/10.2305/IUCN.UK.2020-3.RLTS.T22729731A180736159.en

Breitenmoser, U., Breitenmoser-Würsten, C., Carbyn, L., & Funk, S. M. (2001). Philosophy and reality of reintroducing large carnivores in Europe. In J. G. Ewen, D. P. Armstrong, K. A. Parker, & P. J. Seddon (Eds.), *Reintroduction biology of Australian and New Zealand fauna.*

Breitenmoser, U., Mallon, D. P., Khan, J. A., & Driscoll, C. (2008). *Panthera leo ssp. persica.* In *The IUCN Red List of Threatened Species 2008* (e.T15952A5327221). https://doi.org/10.2305/IUCN.UK.2008.RLTS.T15952A5327221.en

Calenge, C. (2006). The package "adehabitat" for the R software: A tool for the analysis of space and habitat use by animals. *Ecological Modelling, 197*(3–4), 516–519. https://doi.org/10.1016/j.ecolmodel.2006.03.017

Cardillo, M., et al. (2004). Human population density and extinction risk in the world's carnivores. *PLoS Biology, 2*(7), e197. https://doi.org/10.1371/journal.pbio.0020197

Chapron, G., Miquelle, D., Lambert, A., Goodrich, J., Legendre, S., & Clobert, J. (2008). The impact on tigers of poaching versus prey depletion. *Journal of Applied Ecology, 45*(6), 1667–1674.

Chaturvedi, R. K., Raghubanshi, A. S., & Singh, J. S. (2011). Effect of small-scale variations in environmental factors on the distribution of woody species in tropical deciduous forests of Vindhyan Highlands, India. *Journal of Botany, 2011*, Article ID 297097. https://doi.org/10.1155/2011/297097

Daniels, R. J. R., Easa, P. S., & Alembath, M. (2008). Distribution and status of the Endangered Nilgiri tahr. *Current Science, 94*(2), 189–191.

Davidar, E. R. C. (1971). A note on the status of Nilgiri tahr, *Hemitragus hylocrius* Ogilby, on the Grass Hills in the Anamalais. *Journal of the Bombay Natural History Society, 68*, 347–354.

Davidar, E. R. C. (1978). Distribution and status of the Nilgiri tahr (*Hemitragus hylocrius*) 1975–1978. *Journal Of The Bombay Natural History Society, 75*, 815–844.

Dutta, S., Rahmani, A. R., & Jhala, Y. V. (2011). Running out of time? The great Indian bustard (*Ardeotis nigriceps*)—status, viability, and conservation strategies. *European Journal of Wildlife Research, 57*(5), 615–625. https://doi.org/10.1007/s10344-010-0472-z

E. P. (n.d.). EIACP Programme Centre "Wildlife & Protected Areas Management." Retrieved from https://wiienvis.nic.in//Database/SRP_8555.aspx#7._Great_Indian_Bustard__(click_on_name_to_go_top)

Fegatelli, D. A., & Tardella, L. (2013). Improved inference on capture-recapture models with behavioural effects. *Statistical Methods and Applications, 22*, 45–66. https://doi.org/10.1007/s10260-012-0209-4

Gee, E. P. (1966). Estimates of tiger populations in India. In *Ecology and conservation of the tiger.*

Ghouse, S. M., & Indira, P. (2015). Conservation strategies of Great Indian Bustard *Ardeotis nigriceps* of Rollapadu Wildlife Sanctuary in India: An overview. *International Journal of Pharmaceutics and Drug Analysis, 3*(2), 27–33. Retrieved from https://www.neliti.com/publications/409216/conservation-strategies-of-great-indian-bustard-ardeotis-nigriceps-of-rollapadu

Grubb, P. (2005). Artiodactyla. In D. E. Wilson & D. M. Reeder (Eds.), *Mammal species of the world: A taxonomic and geographic reference* (3rd ed., pp. 637–722). Johns Hopkins University Press.

Hafner, S., Wiesner, H., von Hegel, G., Halm, S., & Erhardt, W. (1989). The antagonism of ketamine/xylazine anesthesia ("Hellabrunn mixture") in wild zoo ruminants. *Berliner und Munchener Tierarztliche Wochenschrift, 102*(6), 181–184.

Harihar, A., Pandav, B., Ghosh-Harihar, M., & Goodrich, J. (2020). Demographic and ecological correlates of a recovering tiger (*Panthera tigris*) population: Lessons learnt from 13 years of monitoring. *Biological Conservation, 252*, 108848. https://doi.org/10.1016/j.biocon.2020.108848

Hayward, M. W., O'Brien, J., & Kerley, G. I. H. (2007). Practical considerations for the reintroduction of large, terrestrial mammalian predators based on reintroductions to South Africa's Eastern Cape Province. *The Open Conservation Biology Journal, 1*(1), 1–11. https://doi.org/10.2174/1874839200701010001

Hoffmann, M., Hilton-Taylor, C., Angulo, A., Böhm, M., Brooks, T. M., Butchart, S. H. M., et al. (2010). The impact of conservation on the status of the world's vertebrates. *Science, 330*(6010), 1503–1509. https://doi.org/10.1126/science.1194442

Hulme, P. E. (2015). Invasion pathways at a crossroad: Policy and research challenges for managing alien species introductions. *Journal of Applied Ecology, 52*(6), 1418–1424. https://doi.org/10.1111/1365-2664.12470

Jangra, L., & Verma, R. K. (2024). Great Indian Bustard: India's most threatened avian species – Key perils and conservation measures. *Flora (CITES), 4*(5), 1958–1962. https://doi.org/10.5281/zenodo.11428042

Johnson, et al. (2017). Biodiversity losses and conservation responses in the Anthropocene. *Science, 356*(6335), 270–275. https://doi.org/10.1126/science.aam9317

Karanth, K. U., & Nichols, J. D. (1998). Estimation of tiger densities in India using photographic captures and recaptures. *Ecology, 79*(8), 2852–2862.

Kernohan, B. J., Gitzen, R. A., & Millspaugh, J. J. (2001). Analysis of animal space use and movements. In J. J. Millspaugh & J. M. Marzluff (Eds.), *Radio tracking and animal populations* (pp. 125–166). Academic Press.

Kumar, M. A., Singh, M., Srivastava, S. K., Udhayan, A., Kumara, H. N., & Sharma, A. K. (2002). Distribution patterns, relative abundance, and management of mammals in Indira Gandhi Wildlife Sanctuary, Tamil Nadu, India. *Journal of the Bombay Natural History Society, 99*(2), 184–210.

Lacy, R. C., & Pollak, J. P. (2014). *Vortex: A Stochastic Simulation of the Extinction Process.* Version 10.0, Chicago Zoological Society.

Majumder, A., Basu, S., Sankar, K., Qureshi, Q., Jhala, Y. V., Nigam, P., & Gopal, R. (2012). Home ranges of the radio-collared Bengal tigers (*Panthera tigris tigris*) in Pench Tiger Reserve, Madhya Pradesh, Central India. *Wildlife Biology in Practice, 8*(1), 36–49. https://doi.org/10.2461/wbp.2012.8.4

McClure, C. J. W., Rolek, B. W., & Virani, M. Z. (2021). Contrasting trends in abundance of Indian vultures (*Gyps indicus*) between two study sites in neighboring Indian states. *Frontiers in Ecology and Evolution, 9*, 629482. https://doi.org/10.3389/fevo.2021.629482

Ministry of Environment, Forest and Climate Change. (2023, March 27). Conservation plan for Great Indian Bustards. *Press Information Bureau.* Retrieved from https://pib.gov.in/PressReleaseIframePage.aspx?PRID=1911128

Mishra, C., & Johnsingh, A. J. T. (1998). Population and conservation status of the Nilgiri tahr *Hemitragus hylocrius* in Anamalai Hills, South India. *Biological Conservation, 86*, 199–206.

Mishra, C., Parchure, M., Goswami, V., & Bhat, A. (2010). Balancing conservation planning and mitigation: Perspectives from a case study of the Asiatic lions. *Frontiers in Ecology and Evolution.*

Namboothri, N., Swaminathan, A., Choudhury, B. C., & Shanker, K. (2012). Post-nesting migratory routes of leatherback turtles from Little Andaman Island. *Indian Ocean Turtle Newsletter, 16*, 21–23.

National Tiger Conservation Authority (NTCA). (2020). Project Tiger: Impact and Population Stabilization. In Status of Tigers in India (Latest reports and population assessments), *Ministry of Environment, Forest and Climate Change, Government of India. Retrieved from https://ntca.gov.in*, https://ntca.gov.in

Noack, J., Daguet, A., Auliya, M., & Wilting, A. (2019). First density estimation of the Indochinese leopard Panthera pardus delacouri in Cambodia. *Raffles Bulletin of Zoology, 67*, 421–432.

Nowell, K., & Jackson, P. (Eds.). (1996). *Wild cats: Status survey and conservation action plan*. IUCN—The World Conservation Union.

Oaks, J. L., Gilbert, M., Virani, M. Z., Watson, R. T., Meteyer, C. U., Rideout, B. A., Shivaprasad, H. L., Ahmed, S., Chaudhry, M. J. I., Arshad, M., Mahmood, S., Ali, A., & Khan, A. A. (2004). Diclofenac residues as the cause of vulture population decline in Pakistan. *Nature, 427*(6975), 630–633. https://doi.org/10.1038/nature02317

Panwar, H. S. (1982). What to do when you've succeeded: Project Tiger ten years later. *Ambio, 11*(6), 330–337.

Patankar, S. (2023, January 22). How are marine turtles being conserved in India? *Think Wildlife Foundation*. Retrieved from https://thinkwildlifefoundation.com/how-are-marine-turtles-being-conserved-in-india/

Prakash, V., Pain, D. J., Cunningham, A. A., Donald, P. F., Prakash, N., Verma, A., Gargi, R., Sivakumar, S., & Rahmani, A. R. (2003). Catastrophic collapse of Indian white-backed *Gyps bengalensis* and long-billed *Gyps indicus* vulture populations. *Biological Conservation, 109*(3), 381–390. https://doi.org/10.1016/S0006-3207(02)00164-7

Prakash, V., Green, R. E., Pain, D. J., Ranade, S. P., Saravanan, S., & Prakash, N. (2007). Recent changes in populations of resident *Gyps* vultures in India. *Journal of The Bombay Natural History Society, 104*(2), 127–133. https://www.biodiversitylibrary.org/part/154686

Prakash, V., Bowden, C., Cuthbert, R., Lindsay, N., Prakash, N., Routh, A., & Parry-Jones, J. (2012). Husbandry guidelines for 'in range' conservation breeding programs of *Gyps bengalensis*, *Gyps indicus*, and *Gyps tenuirostris*. *Royal Society for the Protection of Birds*.

Prakash, V., Galligan, T. H., Chakraborty, S. S., Dave, R., Kulkarni, M. D., Prakash, N., Shringarpure, R. N., Ranade, S. P., & Green, R. E. (2019). Recent changes in populations of Critically Endangered *Gyps* vultures in India. *Bird Conservation International, 29*, 55–70. https://doi.org/10.1017/S0959270923000394

Predit, M. A., Prasath, V., Mohanraj, M., & Desai, A. (2015). Status and distribution of the Nilgiri Tahr *Nilgiritragus hylocrius*, in the Western Ghats, India. https://doi.org/10.13140/RG.2.1.2085.4486

R Core Team. (2018). *R: A language and environment for statistical computing*. R Foundation for Statistical Computing.

Ranganathan, J., Chan, K. M., Karanth, K. U., & Smith, J. L. D. (2008). Where can tigers persist in the future? A landscape-scale, density-based population model for the Indian subcontinent. *Biological Conservation, 141*(1), 67–77.

Rangarajan, M. (2005). India's wildlife history: An introduction. In *Nature and nation: Essays on environmental history*.

Redford, K. H., Amato, G., Baillie, J., Beldomenico, P., Bennett, E. L., Clum, N., et al. (2011). What does it mean to successfully conserve a (vertebrate) species? *BioScience, 61*(1), 39–48. https://doi.org/10.1525/bio.2011.61.1.9

Rice, C. G. (1984). The behaviour and ecology of Nilgiri tahr *Hemitragus hylocrius* Ogilby, 1838 (Doctoral dissertation, Texas A&M University).

Rodrigues, A. S. L., Pilgrim, J. D., Lamoreux, J. F., Hoffmann, M., & Brooks, T. M. (2006). The value of the IUCN Red List for conservation. *Trends in Ecology & Evolution, 21*(2), 71–76. https://doi.org/10.1016/j.tree.2005.10.010

Sanderson, E. W. (2006). How many animals do we want to save? The many ways of setting population target levels for conservation. *BioScience, 56*(11), 911–922. https://doi.org/10.1641/0006-3568(2006)56[911:HMADWW]2.0.CO;2

Satter, C. B., Augustine, B. C., Harmsen, B. J., Foster, R. J., & Kelly, M. J. (2019). Sex-specific population dynamics of ocelots in Belize using open population spatial capture–recapture. *Ecosphere, 10*(7), e02792.

Seidensticker, J., Christie, S., & Jackson, P. (1999). *Riding the tiger: Tiger conservation in human-dominated landscapes*. Cambridge University Press. https://doi.org/10.1017/CBO9780511565333

Shanker, K., & Kale, N. (2014). Monitoring and conservation of sea turtles in India through a network of partners and index sites. *Project report to USFWS*. Madras Crocodile Bank Trust and Dakshin Foundation.

Sharma, A. (2020, January 2). India succeeds in breeding critically endangered Great Indian Bustard. *The Times of India*. Retrieved from https://weather.com/en-IN/india/news/news/2020-01-02-india-succeeds-breeding-critically-endangered-great-indian-bustard

Shultz, S., et al. (2004). Diclofenac poisoning is widespread in declining vulture populations across the Indian subcontinent. *Proceedings of the Royal Society of London. Series B, Biological Sciences, 271*, S458–S460.

Silver, S. C., Ostro, L. E., Marsh, L. K., Maffei, L., Noss, A. J., Kelly, M. J., Wallace, R. B., Gomez, H., & Ayala, G. (2004). The use of camera traps for estimating jaguar *Panthera onca* abundance and density using capture/recapture analysis. *Oryx, 38*(2), 148–154. https://doi.org/10.1017/S0030605304000286

Singh, H. S., & Kamboj, R. D. (1996). *Biodiversity conservation plan for Gir: A management plan for Gir Sanctuary and National Park* (Vol. I–II). Forest Department, Gujarat State.

Smith, J. L. D. (1993). The role of dispersal in structuring the Chitwan tiger population. *Behaviour, 124*(3–4), 165–195. https://doi.org/10.1163/156853993X00560

Sontakke, S. D., Umapathy, G., & Shivaji, S. (2009). Yohimbine antagonises the anaesthetic effects of ketamine-xylazine in captive Indian wild felids (lion, tiger, and leopards). *Veterinary Anaesthesia and Analgesia, 36*(1), 34–41. https://doi.org/10.1111/j.1467-2995.2008.00427.x

Soulé, M. E., Estes, J. A., Berger, J., & Del Rio, C. M. (2003). Ecological effectiveness: Conservation goals for interacting species. *Conservation Biology, 17*(5), 1238–1250. https://doi.org/10.1046/j.1523-1739.2003.01599.x

Swan, G. E., Cuthbert, R., Quevedo, M., Green, R. E., Pain, D. J., Bartels, P., Cunningham, A. A., Duncan, N., Meharg, A. A., Oaks, J. L., Parry-Jones, J., Taggart, M. A., Verdoorn, G., & Wolter, K. (2006). Toxicity of diclofenac to *Gyps* vultures. *Biology Letters, 2*(2), 279–282. https://doi.org/10.1098/rsbl.2005.0425

Tucker, M. A., Ord, T. J., & Rogers, T. L. (2014). Evolutionary predictors of mammalian home range size: Body mass, diet and the environment. *Global Ecology and Biogeography, 23*(10), 1105–1114. https://doi.org/10.1111/geb.12194

Wemmer, C., & Sunquist, M. (1988). Snow leopards: Is reintroduction the best option? In *International Zoo Yearbook* (Vol. 27, pp. 172–173).

Wikramanayake, E. D., Dinerstein, E., Robinson, J. G., Karanth, U., Rabinowitz, A., Olson, D., Mathew, T., Hedao, P., Conner, M., Hemley, G., & Bolze, D. (1998). An ecology-based method for defining priorities for large mammal conservation: The tiger as a case study. *Conservation Biology, 12*(4), 865–878. https://doi.org/10.1111/j.1523-1739.1998.96435.x

Wildlife Institute of India. (2009). *Status of Tigers and its Prey Species in Panna Tiger Reserve, Madhya Pradesh*. Technical Report. (Noting local extinction of tigers in Panna in 2009 and Sariska in 2004).

WWF India. (2020). Annual Report 2020. WWF India, *WWF India*, https://www.wwfindia.org/news_facts/wwf_publications/

Zimmermann, F., Breitenmoser-Wursten, C., Molinari-Jobin, A., & Breitenmoser, U. (2013). Optimizing the size of the area surveyed for monitoring a Eurasian lynx (*Lynx lynx*) population in the Swiss Alps by means of photographic capture–recapture. *Integrative Zoology, 8*(3), 232–243. https://doi.org/10.1111/1749-4877.12013

Chapter 17
Collaborative Approaches to Wildlife Conservation

Neelam Pawar, Awasthi Rajan, Asha Poonia, and Ravikant Verma

Contents

N. Pawar (✉)
Department of Pharmaceutical Sciences, Chaudhary Bansi Lal University, Bhiwani, Haryana, India
e-mail: preeti.commerce@cblu.ac.in

A. Rajan
Scientific Advisor, Mangrove Foundation of India, Tiruvallur, Tamil Nadu, India

A. Poonia
Faculty of Life Sciences, Chaudhary Bansi Lal University, Bhiwani, Haryana, India

R. Verma
Department of Zoology, CCS Haryana Agricultural University, Hisar, Haryana, India

A. Poonia et al. (eds.), *Wildlife Trade in India*,
https://doi.org/10.1007/978-981-95-3694-8_17

Abstract This chapter explores the multifaceted landscape of collaborative approaches to wildlife conservation, focusing on how diverse actors, institutions, and governance frameworks coalesce to protect biodiversity in increasingly complex socio-ecological contexts. It begins with conceptual frameworks that inform collaboration in conservation, examining actor architecture and institutional configurations that underpin joint efforts. Special attention is given to the evolving role of Public-Private Partnerships (PPPs) and the significant contributions of Non-Governmental Organizations (NGOs), particularly in India.

The chapter delves into various modalities of collaborative governance, including Community-Based Natural Resource Management (CBNRM), Joint Forest Management, and co-management regimes, all of which emphasize shared responsibilities and local participation. It highlights the importance of transboundary and landscape-level collaborations, which transcend administrative borders and foster regional ecological connectivity.

Emerging tools and practices such as knowledge co-production, collaborative research models, digital platforms, citizen science, and surveillance technologies are analyzed for their role in enhancing coordination and data sharing. Furthermore, the chapter reviews legal frameworks and enabling mechanisms that support or hinder collaboration, including issues of legal pluralism, jurisdictional overlaps, and short-term project logic that can undermine long-term governance.

Key constraints such as political capture, elite domination, intersectional exclusions (gender, caste), and epistemic injustices are critically examined to reveal how power imbalances can distort conservation outcomes. The chapter stresses the importance of evaluating not just ecological outcomes but also social and governance dimensions, using integrated frameworks that account for trade-offs, unintended consequences, and ethical dilemmas.

Finally, the chapter outlines innovations and transformative pathways such as One Health, nature-based solutions, and biocultural restoration, advocating for reflexive and adaptive practices in conservation governance. By weaving together theory, practice, and critique, this chapter provides a comprehensive foundation for understanding and advancing collaborative wildlife conservation in diverse contexts.

Keywords Collaborative Wildlife Conservation · Public-Private Partnerships (PPPs) · Community-Based Natural Resource Management (CBNRM) · Governance Effectiveness · Legal and Institutional Frameworks · Biocultural Restoration

17.1 Introduction

Since no organisation can itself handle the complex issues brought on by biodiversity loss, wildlife conservation has increasingly come to depend on cooperation between public and private stakeholders. In order to mobilise resources, put effective conservation strategies into practice, and ensure long-term sustainability, the collaboration between government agencies and the private sector has proven particularly important (Austin et al., 2009). This kind of cooperation improves the ability to keep an eye on wildlife populations, safeguard habitats, and implement laws, particularly in places that are out of the direct jurisdiction of government officials (Haines et al., 2021; Bodin & Crona, 2009).

In order to conserve wildlife, a wide range of stakeholders are essential. Governments make contributions by establishing laws, policies, and enforcement systems. Non-governmental organisations (NGOs) provide advocacy, knowledge, and grassroots involvement. Through land-use policies and corporate social responsibility programs, the private sector—which includes companies in the energy, tourism, and agriculture sectors influences conservation outcomes (Cooper, 2021). Because they provide traditional ecological knowledge and serve as stewards of biodiversity in their areas, local communities who are frequently directly reliant on natural resources are essential to the success of conservation initiatives (Sonant, 2007; Christie et al., 2006).

Addressing complex ecological issues requires an integrated strategy that links these actors. Cooperation promotes shared accountability, lessens conflict, and harmonises conservation objectives with socioeconomic advancement. Accelerated habitat degradation, climate change, and species extinction highlight the need for such integrative strategies. Local voices have frequently been ignored by traditional top-down conservation models, which have also failed to adjust to dynamic socio-ecological systems (Lacher & Wilkerson, 2013).

Unprecedented rates of habitat fragmentation, biodiversity loss, and climate change highlight the ecological urgency of cooperation in wildlife conservation. These global issues are too extensive and complicated for any one government or organisation to handle successfully. Collaborative efforts facilitate more comprehensive and contextually sensitive responses by bringing together a range of resources, perspectives, and areas of expertise. Since many species and ecosystems are at the brink of irreversible tipping points, the need for coordinated action has never been greater (Austin et al., 2009).

Working together, stakeholder groups can improve the sustainable management of many ecological resources that belong to a common pool. Conflicts of interest among stakeholders, however, frequently prevent the benefits of collaborative management from being fully realised. Understanding the trade-offs managers make between various management outcomes and the socioeconomic and location-specific factors that influence these preferences are essential for developing effective strategies for improving collaborative management. Methods that rely solely on quantitative or qualitative techniques frequently fall short in identifying some of the fundamental issues preventing cooperation (Birol et al., 2006; Bodin & Crona, 2009; Boxall & Adamowicz, 2002; Breffle & Rowe, 2002).

17.2 Conceptual Frameworks for Collaboration in Wildlife Conservation

Over the past few decades, the idea of collaboration in wildlife conservation has changed dramatically, reflecting broader changes in ecological knowledge, social theory, and environmental governance. The field has progressively adopted more inclusive and participatory paradigms after being initially framed within state-centric conservation models, in which governments unilaterally created and implemented policies. Many people in the development primarily have the view that problems are fundamentally complex and that they transcend ecological, political, social, and economic boundaries. We can balance the conservation of our goal with cooperative ways if we discuss different actors who are experts. Collaborations in conservation can be defined as the organized interactions between two or more stakeholders who are working toward a common environmental goal. There are topological collaborations that might have combined governments and patriotic models, such as Common Management. Call management can be described as a shared obligation between the government agencies that provide the legislative frameworks and the local community or resource users. The joint governors conservation are private sector stake orders that are border collations of actors who are truly involved in decision making and implementation. Better city models are mostly vacant. The local community will be the focus of the consultative mechanism and the shipping of the conservation outcome through the voices of the community. Humans and ecological components which are in fact interdependent and coalvolving are closely conceived in the concept of social ecological system frameworks, which are also at collaborative topologies. This illustrates how ecological theory emotional conservations necessitate not only an awareness of ecological processes but also a dynamic comprehension of how to impact them. Therefore, these partnerships, local knowledge, enforcing the dapability of managing the trade, and active governors, who are the chief, supply the principles learning it correctly you make decisions and encourage flexibility. According to the main perspective of the Vigilance theories, resonance stories enhance the ability of the psychological and social systems to identify and distribute all functions. Collaboration is necessary for conservation, but it is not a simple or universally beneficial process; rather, it is dynamic and influenced by history past conflicts in power and trust relationships. If we discuss the state sectors and we register, the community members may bother external agencies because of their previous migration, so harmony is also crucial when it comes to the language of collaboration in mall discussions, even when disagreements compromise. Tension may be draws attention to the crucial aspects of critical involvement with politics and cooperation, such as who is knowledgeable, whose agenda is determined, and what benefits are dispersed. Therefore, it is a more complex model to consult the framework of partnerships from partnership to shame the situation. Intricate context and intricacy. When discussing successful partnerships, it is necessary to have institutional agreements, social and political landscapes, governor structures, and the entire collaborations where conservation is carried out. Key ecological experts such as legitimate structures and equity legumes are quite sensitive (Marchini et al., 2025; Sen, 2021; Olatunbosun, 2025; Marchini et al., 2021; Manuku, 2021; Manyani et al., 2024; Keough & Blahna, 2006; Aguirre, 2017).

17.3 Actor Architecture and Institutional Configurations in Collaborative Wildlife Conservation

Complex networks of actors and institutional arrangements are the foundation of collaborative conservation efforts. These arrangements dictate who takes part, how decisions are made, and which results are given priority. Designing fair and successful conservation strategies requires an understanding of the roles, relationships, and power dynamics among various stakeholders (Aguirre, 2017). The architecture of conservation governance is shaped in large part by state institutions. Governments establish the parameters of conservation efforts through legislative frameworks, regulatory bodies, and policy directives. Decentralisation, or the transfer of power and accountability from central agencies to regional and local bodies, has become increasingly prominent in recent years (Bunnefeld et al., 2017). Opportunities for more participatory and context-specific governance models have arisen as a result. Nonetheless, institutional strength, openness, and the political will to support local actors are critical to decentralization's success (Macdonald et al., 2017). In India, the Ministry of Environment, Forest, and Climate Change (MoEFCC) is in charge of managing national parks and wildlife sanctuaries, creating environmental policies, and enforcing laws pertaining to wildlife protection. Laws, rules, and government policies are examples of formal institutions.

Through official and informal channels, community and Indigenous collectives have progressively asserted their agency in conservation. In addition to having extensive ecological knowledge, many Indigenous groups have preserved traditional governance structures that closely correspond with conservation goal (Castree, 2002). Justice can be advanced and conservation results improved by acknowledging and incorporating these systems into formal governance. To avoid tokenism and guarantee that all voices, particularly those of marginalised subgroups, are heard, it is necessary to carefully navigate representational and internal community dynamics issues. Through the preservation of grazing areas, involvement in community-based conservancies, and dissemination of traditional ecological knowledge, the Maasai communities in Kenya help to conserve wildlife (Jepson & Canney, 2003). Local communities, governments, and foreign donors are frequently connected through conservation NGOs and civil society organisations (CSOs). Their influence is not always neutral, even though they can be helpful in advocacy, technical support, and capacity building. By establishing agendas, managing financial resources, and influencing narratives in ways that may obscure regional priorities, certain NGOs play the role of power brokers. Greater accountability and reflexivity are required in the sector as a result of this dual role, particularly when collaborating across cultural and scalar boundaries. In addition to working with local communities to carry out sustainable conservation projects, WWF (World Wide Fund for Nature) works internationally on environmental advocacy, education, and wildlife conservation. Together with governments, non-governmental organisations, and other stakeholders, the Global Environment Facility (GEF) finances extensive biodiversity conservation initiatives in developing nations (Johnsingh & Joshua, 1994). By producing knowledge, creating instruments for observation and assessment, and influencing evidence-based policymaking, the academic, scientific, and epistemic communities support collaborative conservation. Their involvement can improve the efficacy and legitimacy of conservation initiatives, especially when local stakeholders co-produce the research. The traditional and scientific knowledge systems, however, continue to clash. Mutual respect, epistemic humility, and inclusive research methodologies that validate different ways of knowing are necessary to bridge this divide (Zuka & Zuka, 2024).

Through market-based tools like ecotourism, payments for ecosystem services, and conservation finance, the private sector is also becoming more and more involved in conservation. Companies may participate for profit incentives, risk reduction, or corporate social responsibility. These strategies have the potential to spur innovation and new funding, but they also carry risks, such as the commodification of nature, the marginalisation of local voices, and the perpetuation of current power disparities. The long-term sustainability of market-based tools depends on making sure they are socially inclusive and ethically designed. In African nations, eco-lodges and safari tour operators frequently use sustainable tourism methods that support local livelihoods and biodiversity while also raising money for conservation areas (Stoett, 2004).

17.4 Role of Public-Private Partnerships (PPPs) in Wildlife Conservation

A key tool in wildlife conservation, public-private partnerships (PPPs) bridge the gap between the private sector's creativity, investment, and efficiency and the government's capabilities. The role of PPPs in developing sustainable, scalable, and inclusive conservation models has grown in importance as conservation challenges become more complex due to the rapid loss of biodiversity, climate change, and habitat degradation.

Public-private partnerships are successful. It is crucial for conservation efforts to have coordinated collaborations between private organizations, government agencies, local cooperatives, and public institutions like the national park authority, the environmental ministry, and the forest department. Through the use of technology, entrepreneurship, and private sector resources, protected areas and wildlife sanctuaries are managed more efficiently with public scrutiny. Public-private partnerships provide benefits in discussions such as providing the financial resources that allow us to think beyond traditional government support (Lauber & Decker, 2011).

Additionally, they have advanced technology, such as sophisticated animal monitoring systems and GIS mapping infrastructures for ecotourism. Market-based tactics are also used by public-private partnerships to develop their financial initiatives for biodiversity conservation, such as carbon credits and wild-friendly certificates, followed by discussions based on logic.

Additionally, they are improving community participation for well-designed public-private collaborations and developing job educational platforms and live plywood possibilities around conserved landscapes.

Numerous successful projects are underway in India, such as the Madhya Pradesh Forest Department, which has private ecotourism companies and WWEF India that collaborate and are a part of the Satpura landscape tiger partnership. Public sectors provide their protection monitoring for stringent habitat conservations, while private organizations provide job infrastructures (Cooke & Lane, 2019).

Similarly, the Gujarat Forest Department's partnership with private firms for Gir National Park has helped manage tourist inflow and infrastructure without compromising lion conservation (Galluccio, 2015). Globally, Africa's African Parks Network, a non-profit that enters into long-term management contracts with governments, exemplifies large-scale PPPs in managing national parks across countries like Malawi, Rwanda, and Chad, combining donor funding, private investment, and government support. The African Parks Network operates through PPPs with national governments to manage protected areas such as Akagera National Park in Rwanda.

Governments retain ownership of the land while African Parks (a non-profit) manages day-to-day operations, supported by private donors and tourism revenue. This has helped restore wildlife populations and improve park management. The Rhino Impact Bond, developed by the Zoological Society of London (ZSL), involves private investors, conservation NGOs, and the government of South Africa. A performance-based conservation financing model is established whereby investors contribute to rhino conservation initiatives in protected areas and get paid when rhino populations increase. PPPs between local conservancies, private safari companies, and the Namibian government promote wildlife conservation in Namibia by means of ecotourism. The government supports policy, private partners operate lodges and tours, and local communities oversee conservancies. Tourism-related income is put back into community development and conservation.

Fig. 17.1 Framework of public-private partnerships in wildlife conservation

A PPP involving private diving resorts, local governments, and non-governmental organisations like Conservation International aids in the restoration of coral reefs in the Verde Island Passage. Local governments control and keep an eye on marine activities, NGOs offer scientific expertise, and resorts finance and carry out reef restoration (Lewis, 2021; Borah et al., 2009).

There are difficulties in applying PPP models to wildlife conservation, despite their potential. Equity is a key issue, making sure that local communities are not dislocated or left out of decision-making processes. Transparency in governance is another, since ambiguous roles and responsibilities may result in disputes or poor conservation results. Furthermore, some private organisations might put profit ahead of sustainability, which could lead to social marginalisation or ecological degradation (Khan, 1995).

Nonetheless, these obstacles offer chances for creativity. PPP efficacy can be increased by robust legal frameworks, community co-management clauses, open accountability procedures, and recurring impact analyses. Conservation efforts can be made even more democratic by incorporating traditional knowledge, upholding indigenous land rights, and guaranteeing benefit-sharing. Long-term success requires ongoing stakeholder dialogue and capacity-building for both public and private partners (Davis & Halvorson, 2023) (Fig. 17.1).

17.5 NGOs and Their Contribution to Wildlife Conservation in India

When discussing non-governmental organizations that support wildlife conservation and serve as a catalyst for community development, scientific research, policy change, and grassroots initiatives, NGOs are crucial collaborators that serve as a link between government initiatives, local communities, and scientific committees. Of course, India has a rich but endangered biodiversity.

Where are national policies where NGOs collaborate to promote themselves and make changes at their own level? They also carry out conservation projects at various levels, and their most significant contribution is the conservation of endangered species like elephants and tigers, which is actually greatly aided by group The Wildlife Trust of India, Bombay Natural History Society and WWE India. Habitat restoration, species monitoring, rehabilitation of injured wildlife, and anti-punching gasoline are some of the initiatives. NGOs also play a vital role in raising youth knowledge of conservation cultures, environmental education, and their underprivileged ruling committees.

Additionally, they engage in legal battles through the environment and make institutions accountable for environmental infractions for Watchdogs and certain unlawful wild characteristics. NGOs also have a significant impact on environmental education and awareness, fostering a conservation culture among urban dwellers, young people, and underserved rural communities. In addition to supporting legal battles through environmental litigation, they frequently serve as watchdogs, holding institutions responsible for environmental violations and the illegal wildlife trade. NGOs' adaptability and creativity are one of their special advantages. NGOs, as opposed to government organisations, are able to test out new conservation models, like eco-development initiatives or community-managed reserves, which can subsequently be expanded through public policy. By acknowledging their knowledge systems and ensuring their involvement in decision-making, numerous NGOs have strengthened local and Indigenous communities (Boiral & Yuriev, 2024).

Despite limited funding and operational challenges, Indian NGOs have demonstrated resilience and creativity in pursuing conservation goals. Their role is not merely supplementary but foundational to India's wildlife conservation architecture. Going forward, greater collaboration between NGOs, governments, and the private sector will be key to ensuring sustainable, inclusive, and effective conservation efforts across the country (Table 17.1).

17.6 Modalities and Mechanisms of Collaborative Governance in Wildlife Conservation

In the context of wildlife conservation, collaborative governance refers to a wide range of approaches and systems that unite various stakeholders in order to manage natural resources in a way that is inclusive, flexible, and sustainable. Collaborative mechanisms have emerged as effective tools for tackling conservation challenges as a result of increasing ecological pressures and the shortcomings of centralised governance models (Rahmani, 2014). These mechanisms include community-based resource management, co-management arrangements, transboundary

Table 17.1 Overview of key NGOs and their contributions

NGO name	Focus area	Key contributions	Region of operation
Wildlife Trust of India (WTI)	Human-Wildlife Conflict	Rescue and rehabilitation, conservation awareness, policy advocacy	Pan-India
Snow Leopard Conservancy	Snow Leopard Conservation	Habitat protection, research, and community-based conservation	Ladakh, Jammu & Kashmir
Nature Conservation Foundation (NCF)	Forest Ecosystems	Research, community involvement, species conservation	North-East, Western Ghats
Wildlife Protection Society of India (WPSI)	Anti-Poaching	Intelligence on illegal wildlife trade, law enforcement support	North and Central India
Salim Ali Centre for Ornithology & Natural History	Avian Research	Bird migration studies, habitat protection, academic research	Tamil Nadu (Pan-India projects)
Foundation for Ecological Security (FES)	Sustainable Livelihoods	Community governance of natural resources, eco-restoration	Gujarat, Rajasthan, Andhra Pradesh
Aaranyak	Biodiversity and Wildlife Conservation	Research, anti-poaching, environment education	Assam and North-East India

collaborations, public-private partnerships, knowledge co-production, and the growing use of digital platforms and citizen science. Together, these approaches represent a shift towards more democratic, multi-actor systems that prioritize ecological integrity alongside social equity (Rahmani, 2014).

17.6.1 Community-Based Natural Resource Management (CBNRM)

A fundamental model of collaborative governance, CBNRM gives local communities the authority to manage and profit from their natural resources. Communities are viewed in this model as active stewards of forests, wildlife, and water bodies in addition to being beneficiaries. CBNRM improves community livelihoods and ecological outcomes by decentralising decision-making and guaranteeing that local populations directly benefit socioeconomically from conservation (Himawan et al., 2025).

In India, community-managed forests in the Northeast, participatory ecotourism in tribal areas, and the management of sacred groves are all instances of CBNRM. These programs frequently combine modern conservation science with indigenous ecological knowledge to produce context-specific solutions that honour regional customs and values. Additionally, CBNRM promotes a sense of ownership, which is essential for both long-term sustainability and adherence to conservation norm (Decker et al., 2016).

17.6.2 Joint Forest Management and Co-Management Regimes

In order to manage degraded forests together, local communities and forest departments form partnerships under the formalised Joint Forest Management (JFM) mechanism that first appeared in India in the 1990s. Communities take part in forest regeneration and protection under JFM in return for usufruct rights to fuelwood, fodder, and forest products. These agreements, which are formalised through village forest committees and micro-plans, institutionalise grassroots cooperation (Leong et al., 2011).

JFM is still a noteworthy attempt at co-management, despite its mixed results because of problems like uneven power relations and low financial returns. In a broader sense, co-management regimes are cooperative arrangements in which the state or other stakeholders share governance duties with resource users. The success of these models, which seek to strike a balance between ecological objectives and socioeconomic equity, frequently depends on openness, trust, and solid legal support.

17.6.3 Transboundary Conservation and Landscape-Level Collaborations

Because wildlife habitats frequently cross administrative and political borders, cooperation between different jurisdictions is required. In order to manage shared ecosystems and migratory species, transboundary conservation initiatives entail collaboration between various states or nations. The survival of migratory birds, elephants, big cats, and other species that range widely depends on these efforts to preserve landscape connectivity. Restoring the ecological corridor across the buffer zones and protection area is the aim of the Indian Nepal landscape program. Different monitoring techniques, policies in the data sharing framework. A established equilibrium between conservation level techniques, such as the Western Ghat biodata state integrated training process, requires cooperative monitoring processes, which are essentially partnerships.

The significance of non-governmental organizations, government organizations, local communities, and academic institutions is that they often collaborate on landscapes and cross boundaries. They support holistic conservation, which is similar to a core system service that ensures economic times and habitat connectivity (Vernooy et al., 2009).

17.6.4 Knowledge Co-Production and Collaborative Research Models

Incorporating experimental and traditional scientific knowledge into conservation planning and knowledge coproduction is crucial for collaborative governance. These methods utilize the expertise of academic and technical experts to produce more grounded and pertinent solutions while also taking into account the value of local and Indigenous communities.

Every stage of the research process is supported by collaborative research models. This is the parties Patriots where we probably would expand our local actors to demonstrate the findings. I appreciate formalizing the topic, gathering the data, identifying the problems, distributions analysis, and research actually addressing this conservation practical conservation difficulties.

Participatory research on forest management, water governance, and ecosystem services has been pioneered in India by organisations such as the Foundation for Ecological Security (FES) and the Ashoka Trust for Research in Ecology and the Environment (ATREE). These projects show how co-produced knowledge can promote innovation in conservation, foster trust, and heal epistemological gaps (Sharma & Sharma, 2025; Nayak & Berkes, 2008).

17.6.5 Role of Digital Platforms, Citizen Science, and Surveillance Technologies

The majority of patients can be managed more effectively thanks to digital technology, which also fundamentally alters governance. As a result, digital platforms for conservation make it easier for stockholders to communicate, which enhances government accountability.

The digital platforms have made wildlife conservation more accessible to the general public. Additionally, they have been gathering data on their Momentum citizens and collecting crucial information such as species distribution, habitat change, and migration patterns through the biodiversity portal and as eBird, iNaturalist, and the India Biodiversity Portal.

Because policies are informed by crowd source data sets and research studies, public awareness of biodiversity has improved. These days, surveillance technologies like GPS, drones, and camera traps are becoming more and more common. The accuracy of wildlife activities monitoring and antiboaching measures. The geography information center can design the landscape level and habitat assessment because it is truly aided by remote sensing. In addition to improving decision-making, these data are accessible to a wide range of stake orders and offer increased transparency.

These days, there are cloud-based forms in mobile applications that provide real-time reporting of illicit activity when people are living through a conflict or forest fire.

All of them are crucial for wildlife study because they enhance the use of artificial intelligence and provide information through the analysis of satellite data and camera trap photos.

Since these technologies have a lot of potential, they should be handled carefully, especially when working with data ownership. Cities that use privacy-vulnerable communities are the most important consideration when managing this possible technology. Combining conservation systems with role-responsible inclusion, common digital tools, and time-participated government makes them more adaptable. Therefore, rather than aiming for a just outcome for all stakeholders, if we are unable to address a number of variables and the structural determinants for the success of collaborative wildlife conservation. If our legal framework is helpful, then our participation tools institutional mechanism. We can engage a range of actors thanks to the technology industries. In India, we are combining digital platforms with community-focused planning. The Wildlife Protection Act provides the primary legislative framework and habitats for protecting wild animals, plants, and their ecosystems. Local communities in eco-development communities and wildlife conflict mitigation are examples of participatory approaches that are permitted, and there has been a recent shift in the loss of the original exclusion and centralization. Legal clarifications and institutional support requirements create a multi-layered legal framework for these classes, which strengthens collaborative governance for legitimacy, community involvement, and conservation (Gour & Reddy, 2015) (Fig. 17.2).

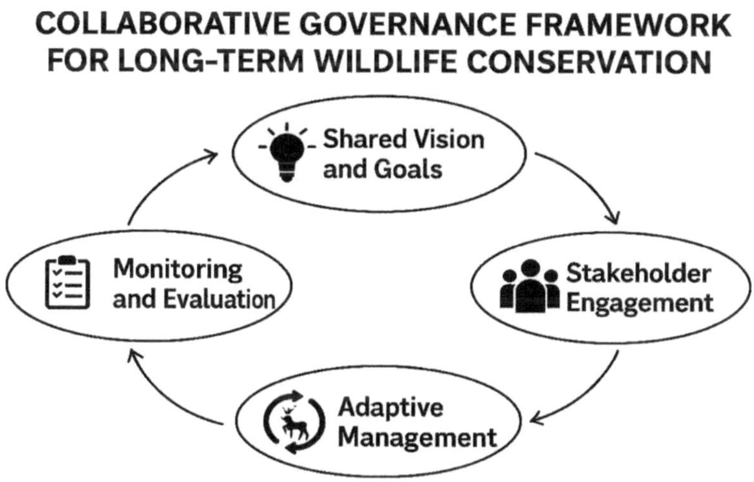

Fig. 17.2 Collaborative governance framework for long-term wildlife conservation

17.7 Enabling Conditions and Determinants of Success in Collaborative Wildlife Conservation

The success of collaborative wildlife conservation depends on a variety of enabling factors and structural determinants rather than just the goals of stakeholders. These include technological advancements, strong institutional mechanisms, effective participatory tools, supportive legal frameworks, and the ability to involve a variety of actors. In India, a mix of digital, community-focused planning, and progressive laws. The main legislative framework for safeguarding wild animals, plants, and their habitats is provided by the Wildlife (Protection) Act. Recent changes have started to allow for more participatory approaches, such as including local communities in eco-development committees and wildlife conflict mitigation, despite its initial more centralised and exclusionary nature. All of these laws provide a multi-layered legal framework that encourages collaborative governance by establishing legal clarity, mandating institutional support for conservation partnerships, and legitimising community participation. The foundation for inclusive and adaptable conservation governance has been established as a result. This chapter goes into great detail about these enabling conditions, focussing on how they promote more equitable, resilient, and sustainable conservation outcomes (Sharma et al., 2024).

17.7.1 Legal Frameworks Supporting Collaborative Conservation

Framework for policy and strong legal backing from our past laws Since India has established a very strong conservative cooperative conservation system, the government's final function or job is to support biodiversity conservation and natural resource governance, which includes acknowledging community rights.

17.7.1.1 Forest Rights Act (2006)

The 2006 Forest Rights Act Another name is Schedule Tribe and Traditional Forest Jewellers, which empower the forest's inhabitants and mark a crucial turning point by acknowledging the historical significance of the forest resources created by these indigenous and tribal groups.

Individuals who participate in communal land rides and gram documents are responsible for managing the preservation of the forest environment through the use of shared forest resources, such as rice, and restoring historical conservation. Create a decentralized system for his government by recognizing their local community and serving as a co-manager of biodiversity (Heise-Pavlov, 2020). In areas like Odisha and Maharashtra, where communities have legal rights with PF, increasing forest cover reduces illicit logging and helps manage the forest. They are also illustrating how legal recognition may boost forest conservation efforts.

17.7.1.2 Biological Diversity Act (2002)

Another important piece of legislation is the Convention on Biological Diversity, which essentially outlines India's commitments to conservation under the Biological Diversity Act of 2002. By decentralizing biodiversity governance, it preserves traditional knowledge and facilitates the sharing of biological resources. The fair use of biological resources to ensure their sustainability and to record the biodiversity in the people's biodiversity register is another function of the biodiversity management committee. Because they give the knock local knowledge folders the institutional basis to participate in the financial and diversity planning, participatory conservations.

17.7.2 Participatory Tools and Practices

Legal recognition alone is not sufficient. Effective implementation of collaborative conservation depends on participatory tools that enable inclusive planning, decision-making, and monitoring.

17.7.2.1 Village-Level Micro-Plans

Locally created conservation and development plans that specify how a community plans to manage its natural resources are known as village-level micro-plans. These plans are frequently created in collaboration with technical specialists, NGOs, and forest departments. Micro-plans take into account socioeconomic requirements (like access to water, livelihoods, and healthcare) as well as ecological priorities (like soil preservation, afforestation, and fire prevention). Micro-planning guarantees that conservation efforts are grounded in site-specific realities, sensitive to context, and in line with local goals. By clearly defining roles and responsibilities for all parties involved, it also improves accountability (Gochhayat, 2011).

17.7.2.2 Participatory Rural Appraisal (PRA)

A collection of instruments and methods known as Participatory Rural Appraisal (PRA) is used to help communities comprehend their own socio-ecological systems. PRA techniques include focus groups, wealth ranking, seasonal calendars, and resource mapping. These resources support communities in prioritising issues, communicating their ecological knowledge, and working with outside facilitators to co-design solutions. By encouraging group learning and trust among neighbours, PRA increases social capital. By guaranteeing the involvement of women, elders, and under-represented groups in the conservation process, it also fosters inclusivity. PRA has been effectively used by NGOs and development agencies for livelihood planning, conflict resolution, and biodiversity assessments (Gadgil, 2003).

17.7.2.3 Biodiversity Management Committees (BMCs)

Biodiversity Management Committees are established at the village, block, and municipal levels in accordance with the BDA. They are responsible for managing local biodiversity, creating People's Biodiversity Registers, and making sure that benefit-sharing guidelines are followed when granting access to biological resources. In order to institutionalise local knowledge and make sure that conservation decisions take community priorities into account, BMCs are essential hubs. Sufficient training, funding, and cooperation with other governing entities, such as panchayats and forest departments, are essential to their success

17.8 Technological Enablers

Cooperative conservation efforts have been greatly enhanced by the incorporation of digital tools and geospatial technologies. Technology increases community engagement, improves monitoring and evaluation systems, and increases planning accuracy all of which act as force multipliers

17.8.1 GIS for Community Mapping

Community-based mapping of forests, wildlife habitats, and water sources has been made possible in large part by Geographic Information Systems (GIS). Communities locate biodiversity hotspots, identify degraded areas, and draw traditional boundaries using GPS-enabled devices. Planning resource use, negotiating with outside agencies, and protecting Community Forest Rights all depend on these maps.

Assessing ecosystem services, visualising landscape changes over time, and incorporating local knowledge into spatial planning are all made easier with the use of GIS data. GIS increases transparency in conservation planning and promotes a common understanding of the landscape by incorporating communities in the mapping process (Choudhary, 2024)

17.8.2 Mobile Apps for Citizen Science and Conflict Monitoring

Apps for tracking wildlife conflicts and citizen science platforms have made conservation more accessible thanks to mobile technology. Users can submit photos, behavioural observations, and species sightings through apps such as eBird India, iNaturalist, and India Biodiversity Portal. These databases are useful for both policymaking and scientific research. In Karnataka, conflict-monitoring applications like WildSeve enable villagers to instantly report livestock kills or crop raids. These tools guarantee prompt compensation, lessen retaliation against wildlife, and allow forest departments to respond quickly. Additionally, local youth and school groups can participate in ecological data collection and awareness campaigns through open-source platforms, utilising technology as a tool for education and conservation (Khan, 2023).

17.8.3 Mechanisms Enabling Collaboration in Wildlife Conservation

When organised through transparent and inclusive mechanisms that connect disparate actors, interests, and knowledge systems, collaboration in wildlife conservation flourishes. These mechanisms serve as institutional and procedural frameworks that help communities, governments, NGOs, scientists, and the private sector communicate, coordinate, and take joint action. Multi-stakeholder platforms that offer participatory spaces for planning and decision-making, like Biodiversity Management Committees or Joint Forest Management Committees, are among the most important mechanisms. These forums ensure that local viewpoints impact conservation priorities and strategies while also fostering trust. The Forest Rights Act (2006) and the Biological Diversity Act (2002) offer the statutory foundation for community involvement, while policy harmonisation ensures coherence across sectors.

Through knowledge-sharing tools such as participatory rural appraisal (PRA), citizen science platforms, and community mapping with GIS, diverse stakeholders can collaboratively generate data and solutions. Conflict resolution strategies that assist in managing competing interests and lowering tensions include facilitated dialogue and grievance redress systems. Finally, benefit-sharing plans, conservation incentives, and public-private partnerships are examples of financial instruments that integrate ecological goals with economic sustainability. Together, these mechanisms not only operationalise collaboration but also ensure that it remains inclusive, adaptable, and grounded in social equity and ecological integrity (Palmer, 1913).

17.9 Constraints and Structural Barriers to Collaborative Wildlife Conservation

The long-standing structures that yours and the conference have created undermine the effectiveness of collaborative life constellation to overcome these limitations. We provide good collaborative conservation models for equitable, sustainable, and inclusive by diversity effectiveness, despite the fact that it is a promising strategy for biodiversity different action. These factors include legal, institutional, migrational, social, and political factors that create contradictions for biodiversity protection (Hydara, 2025).

17.10 Political Capture, Elite Domination, and Extractive Collaboration

When discussing extractive cooperation, different stakeholders with their own goals have access to financing opportunities and land resources for suitable conservation initiatives.

Even though these partnerships can be explored and their experts tend to exploit local resources, they may not adequately benefit the local community despite offering long-term improvements for communist life, equitable benefits, or other means of empowering local communities, such as economic activities created by conservation in terms of tourism and ecotourism carbon of its projects.

Instead of involving the local community and providing specialists without providing a fair benefit, powerful persons can obtain the punchland resources to accomplish their own goals. This can be done through partnerships that typically exploit local resources or long-term improvements to community life.

For instance, in the case of eco-tourism or carbon offset projects, the economic activity generated by conservation areas might not adequately benefit local communities. Instead, they may have to cope with the ecological and social fallout from conservation efforts, such as being displaced or losing access to customary resources. When power is concentrated in a few hands, it undermines the basic principles of cooperation. Communities are either excluded or only included in token ways as a result of this centralised decision-making model (Gurău & Dana, 2018).

17.11 Conflicts Over Representation, Voice, and Legitimacy

One of the main issues with many cooperative conservation projects is lack of representation, which occurs when the diverse needs and interests of all stakeholders—especially marginalised groups—are not adequately reflected. This results in disagreements over legitimacy and voice because decisions are usually made by a small group of actors who may not have the power to adequately represent the variety of local concerns. For example, women, Indigenous peoples, and lower caste groups are often excluded from leadership positions or decision-making bodies, despite their close ties to the land and ecosystems under management. This representation gap exacerbates the fact that more powerful stakeholders silence the voices of those who live in or close to conservation areas. As a result, conservation decisions may fail to consider community needs, leading to resistance, non-compliance, and ultimately the failure of the conservation effort. Furthermore, the legitimacy of conservation policies is also questioned when decision-makers are thought to be external or removed from local realities. Conservation efforts run the risk of being perceived as unlawful and top-down directives if local knowledge is not taken into account or cultural norms are not respected. This would further alienate communities and make successful collaboration more challenging (Radicchi, 2021).

17.12 Legal Pluralism and Jurisdictional Overlaps

Another obstacle to cooperation in wildlife contributions is the existence of several legal systems in a same media. There are conservation laws in place in post-colonial settings like India and other nations, and local government practices, customs, and systems are still in place. The existence of several legal systems in one place actually hinders sincere cooperation in wildlife conservation since these divergent legal systems may lead to judicial and deputy overlaps, which makes coordination difficult. There is a conflict between the Forest Rights Act of 2006 and the Wild Protection Act of 1972, which actually causes obligations to be delayed during the season, causing conservation misunderstandings due to the lost legal contradictions that impact the implementation of cooperative conservation projects. There is conflict within government organizations, and they use the local community and businesses to navigate the complex legal systems in the situation, making it difficult for the local community to find the effective assets for their rights. This is due to natural resources and any legal frameworks that deal with ownerships, management, and various demands for government organizations, businesses, and the local community.

17.13 Short-Term Project Logics Versus Long-Term Governance Needs

Logix, which is actually printed by the Donald Drive, has structure limits for planning, conservation funding, and short-term projects. Projects are chosen based on your timeframe, money, and schedule. They include things like protecting species, resolving social potential, restoring habitat, and government difficulties. As a result, they typically have a short-term outlook, although governance is required for long-term sustainability. Short-term outcomes might be necessary, especially when it comes to resolving systemic issues and ensuring that local institutional capacities are willing to anchor and agree with community engagement over the long term project.

Conservation efforts that focus on short-term goals usually ignore the need for institutional continuity, adaptive management, and community empowerment to guarantee that conservation outcomes are sustained over time. Without long-term planning, conservation projects lose institutional memory, and instead of building on past successes or learning from past mistakes, communities may find themselves starting over with every new project cycle. Progress towards true, community-led conservation governance is hampered by this lack of continuity, which also erodes the basis for long-term cooperation (Münster, 2014).

17.14 Gender, Caste, and Intersectional Exclusions

Certain groups are routinely excluded from leadership roles, resource access, and decision-making processes due to intersectional, gender, and caste exclusions, which are often reinforced by conservation efforts. In India and other countries with strong social hierarchies, women, lower caste groups, and Indigenous communities often face significant barriers to participation in conservation governance.

The roles that women play in managing natural resources, such as gathering forest products, managing water sources, or protecting biodiversity, are often overlooked or undervalued in formal conservation planning, for example, in many traditional communities. Similarly, Dalits and other marginalised groups might not be able to participate in participatory decision-making processes due to historical discrimination and power dynamics. The exclusion of these groups not only compromises social justice but also diminishes the overall effectiveness of conservation efforts. Studies have shown that gender-equitable and inclusive governance models are better at managing natural resources in a sustainable way. Ignoring these social injustices undermines the legitimacy and equity of conservation programs and diminishes the likelihood of true collaboration (Parish Joshi, 2020).

17.15 Epistemic Injustices and Marginalization of Local Knowledge

Epistemic injustice is the term used to describe the systematic undervaluation or rejection of Indigenous, traditional, and local knowledge in favour of Western or scientific knowledge systems. Because outside experts and authorities disregard local communities' extensive environmental knowledge and sustainable resource management practices, this kind of injustice is pervasive in conservation initiatives.

The marginalisation of local knowledge is often linked to the idea that scientific knowledge is better or more objective. However, local knowledge systems that date back centuries are extremely sensitive to the distinct ecological environments in which communities live. Indigenous knowledge, such as the seasonal cycles of animal migration, plant regeneration, or water availability, is often crucial for effective conservation and biodiversity management. Ignoring these knowledge systems costs conservation efforts valuable insights that could increase the effectiveness of conservation strategies.

Moreover, the legitimacy of conservation efforts is weakened by the absence of knowledge co-production between scientists and local communities. Because of epistemic injustice, local people are unable to fully participate in governance and exercise their right to influence conservation policies that have an immediate impact on their lives (Darmawan, n.d.).

17.16 Evaluating Outcomes and Governance Effectiveness in Collaborative Wildlife Conservation

The success of collaborative wildlife conservation projects depends on the ability to systematically evaluate outcomes across ecological, social, and governance dimensions. Effectiveness evaluation is essential for tracking advancement, enhancing tactics, promoting responsibility, and ensuring that conservation models remain inclusive and current. In addition to fundamental metrics, frameworks for efficient evaluation must take into account complexity, different points of view, and continuous learning (Berg-Nordlie, 2017).

17.17 Frameworks for Assessing Ecological, Social, and Governance Outcomes

Collaborative conservation evaluation frameworks must record results that take into account biodiversity goals and the welfare of involved communities. Trends in species populations, habitat restoration, biodiversity indices, and indicators of ecosystem health are examples of ecological metrics. These are essential for determining whether conservation initiatives are meeting their main environmental goals. Better livelihoods, empowerment, fair benefit sharing, and increased community resilience are examples of social outcomes. The flexibility, power sharing, legal recognition, inclusivity, and transparency of an organisation are evaluated by governance metrics. Multifaceted models that enable intricate and interactive assessments are offered by frameworks like the IUCN's Green List of Protected and Conserved Areas, the Social Assessment for Protected Areas (SAPA), and the Adaptive Governance Framework (Claggett, 1999).

17.18 Multi-scalar Approaches to Measuring Effectiveness

From local community projects to national and transboundary initiatives, collaborative conservation takes place at a wide range of scales. Assessments must therefore be multi-scalar, accounting for site-specific variances, local contexts, and global factors like market forces and climate change. Local ecological indicators, traditional governance, and community involvement may be the main focusses of effectiveness at the local level. Connectivity, habitat corridors, and governance coordination become more important at the landscape or regional level. Legal coherence and policy alignment are essential at the national level, even though contributions to biodiversity targets like the Kunming-Montreal Global Biodiversity Framework can be used to gauge progress internationally. Using a multi-scalar lens also makes it easier to comprehend how choices made at one level can have a cascade of positive and negative effects at other levels. Designing integrated, scale-sensitive strategies requires an understanding of these dynamics (Saxena, 2024).

17.19 Trade-Offs, Unintended Consequences, and Ethical Dilemmas

Collaborative conservation frequently involves trade-offs and unforeseen consequences. For example, effective conservation efforts that boost wildlife populations may lead to increased conflict between humans and wildlife. Despite its financial benefits, ecotourism has the potential to harm the environment or turn culture into a commodity. Participatory mechanisms may unintentionally maintain existing power disparities by favouring elites or marginalising women and young people.

Ethical dilemmas also occur, such as choosing between competing conservation values (such as rewilding versus agricultural sustenance) or prioritising species protection over the needs of the community. It is crucial to acknowledge these conflicts. Rather than striving for perfect results, evaluation should focus on how well trade-offs are managed, how inclusive the decision-making process is, and whether marginalised voices are heard and respected (Connery, 1935).

17.20 Reflexivity and Adaptive Learning in Evaluation Practices

When assessing conservation governance, reflexivity the practice of self-awareness, critical reflection, and openness to change is crucial. Reflexive evaluation entails challenging presumptions, going over goals again, and taking into account input from a range of stakeholders. It also means acknowledging the biases and positionality of the evaluator. Conservation systems are made more flexible and responsive through adaptive learning. Adaptive evaluation employs iterative cycles of monitoring, reflection, and adjustment in place of static assessments. This dynamic process is supported by tools like learning-based indicators, outcome mapping, and Theory of Change models. As systems become more capable of handling complexity and uncertainty, adaptive learning enhances resilience in addition to improving results (Minteer & Miller, 2011).

17.21 Innovations and Transformative Pathways in Collaborative Conservation

Incremental change is no longer adequate in the face of global ecological crises. There is a pressing need for inclusive, justice-focused, and ecologically and culturally grounded transformative approaches to conservation. Law, policy, digital tools, and community leadership innovations are opening up new, transformative, and collaborative avenues for governance (Dorr, 1916).

17.22 Integrative Approaches: One Health, Nature-Based Solutions, and Biocultural Restoration

The interdependence of ecosystem, animal, and human health is acknowledged by integrative frameworks such as one Health. One Health establishes a comprehensive platform for cooperation by dismantling the silos between the fields of public health, veterinary sciences, forestry, and conservation. This is especially important for improving ecological resilience and preventing zoonotic diseases.

A practical and scalable approach to addressing climate, biodiversity, and social issues at the same time is provided by nature-based solutions. Examples of projects that benefit local livelihoods and provide ecosystem services include restoring wetlands, preserving forests, and establishing green corridors.

A revolutionary approach to conservation is biocultural restoration, which combines ecological restoration with Indigenous cultural practices. It challenges the historically colonial narrative of conservation as exclusion by re-centering Indigenous knowledge and values. Sacred groves, customary seed banks, and biodiversity-related cultural practices are effective instruments (Cheng & Li, 2024) (Figs. 17.3 and 17.4).

17.23 Conclusion

The evolving landscape of wildlife conservation increasingly necessitates collaborative approaches that transcend traditional, top-down governance models. As demonstrated throughout this chapter, collaboration across diverse actors ranging from government agencies and non-governmental organizations (NGOs) to local communities, private sector stakeholders, and academic institutions has become essential in addressing the multidimensional challenges of biodiversity loss. These collaborative frameworks allow for pooling of resources, knowledge, and capacities while promoting shared responsibilities in conservation efforts. One of the central insights is that actor architecture and institutional configurations play a critical role in shaping conservation outcomes. Public-Private Partnerships (PPPs), for example, bring efficiency and innovation from the private sector into conservation landscapes,

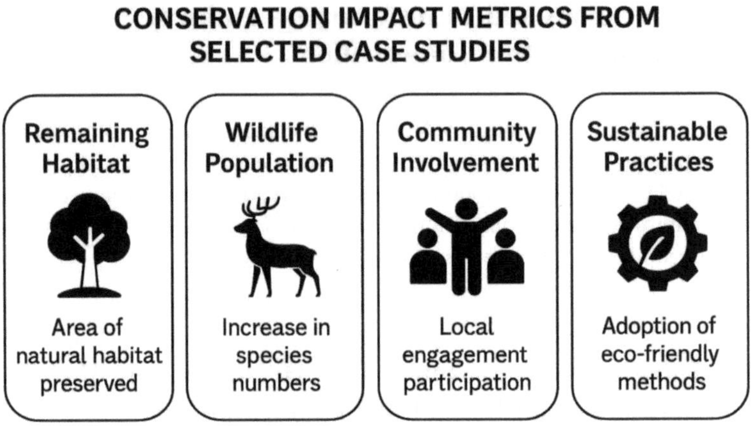

Fig. 17.3 Conservation impact metrics from selected case studies

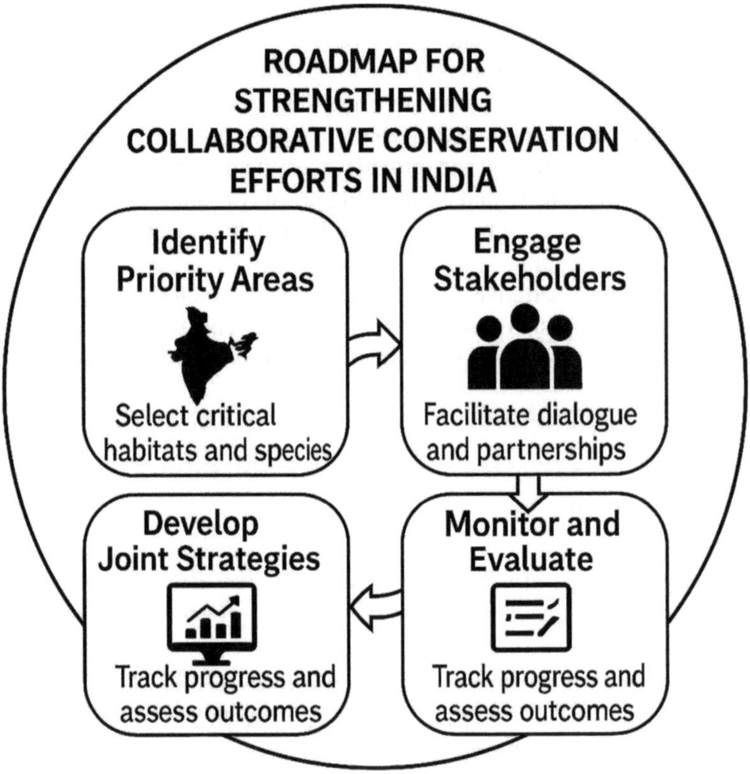

Fig. 17.4 Roadmap for strengthening collaborative conservation efforts in India

while NGOs contribute not only funding and technical expertise but also serve as critical intermediaries between local communities and state mechanisms. Research models for the provision and relevance conservation effect for the research and knowledge-based analysis are developed in collaboration with local knowledge systems and knowledge copper action.

In addition to providing the short funding cycles, this chapter analyzes the structure of the conservation effectiveness that is impacted. Predominance of lawful playrorism in the project.

Social exclusions based on gender, caste, and class make it harder for local knowledge ministers to participate ethically. Participate in the forest management committee-based natural resources management dwelling, which displays the community towers shifting conservations and their local knowledge photos. This is done regularly to produce sustainable resources and clay. In order to manage the environment for administrative boundaries, bridge national boundaries, and promote ecological connective coordination in local policy, transboundaries and landscape level collaborations have good potential.

Other important elements, such as the servant lens, include the digital platform and some new functional tools that provide the basis for conservation governance. These two are essential if we carefully integrate them for already existing systems because they can stop the acceleration of already existing inequalities and access issues. Technology City is in science mobile applications, which are changing the world of data collection and increasing public participation. They also give the transparency of results. We may quickly question how to utilize ecological indicators to assess the effectiveness of collaborative conservation if we examine the government's role in social consequences.

The conservation government must integrate a valuation framework and remove superfluous, independent sequences when building resonance systems that are flexible enough to adapt to changing ecological conditions. This approach is based on adaptive learning.

This chapter highlighted the growing importance of the storm subject and provided for natural basis solutions, one health biocultural resolves.

Possibility of assessments and integration that can result in social agendas and development, and we at Conservation Wildlife Cooperative have solved many of today's ecological issues in one way or another, successfully navigating the long-standing institutional power structures in the environment, and establishing mutual respect and trust.

References

Aguirre, A. A. (2017). Changing patterns of emerging zoonotic diseases in wildlife, domestic animals, and humans linked to biodiversity loss and globalization. *ILAR Journal., 58*, 315–318.

Austin, Z., Cinderby, S., Smart, J. C. R., Raffaelli, D., & White, P. C. L. (2009). Mapping wildlife: integrating stakeholder knowledge with modelled patterns of deer abundance by using participatory GIS. *Wildlife Research, 36*, 553–564.

Berg-Nordlie, M. (2017). Substitution in Sápmi. meta-governance and conflicts over representation in regional indigenous governance. In *Governance in Russian Regions* (pp. 189–218). https://doi.org/10.1007/978-3-319-61702-2_8
Birol, E., Karousakis, K., & Koundouri, P. (2006). Using a choice experiment to account for preference heterogeneity in wetland attributes: The case of Cheimaditida wetland in Greece. *Ecological Economics, 60*, 145–156.
Bodin, O., & Crona, B. I. (2009). The role of social networks in natural resource governance: what relational patterns make a difference? *Global Environmental Change – Human and Policy Dimensions, 19*, 366–374.
Boiral, O., & Yuriev, A. (2024). Sustainability from the top: Revisiting the roles and responsibilities of the Board of Directors. In *Handbook on corporate governance and corporate social responsibility* (pp. 175–190). https://doi.org/10.4337/9781802208771.00024
Borah, J., et al. (2009). Food habits of dholes (Cuon alpinus) in Satpura Tiger Reserve, Madhya Pradesh, India. *Mammalia, 73*(2). https://doi.org/10.1515/mamm.2009.024
Boxall, P. C., & Adamowicz, W. L. (2002). Understanding heterogeneous preferences in random utility models: a latent class approach. *Environmental and Resource Economics, 23*, 421–446.
Breffle, W. S., & Rowe, R. D. (2002). Comparing choice question formats for evaluating natural resource tradeoffs. *Land Economics, 78*, 298–314.
Bunnefeld, N., Nicholson, E., & Milner-Gulland, E. J. (2017). *Decision-making in conservation and natural resource management: models for interdisciplinary approaches*. Cambridge University Press. https://doi.org/10.1017/9781316135938
Castree, N. (2002). False antitheses? Marxism, nature and actor-networks. *Antipode, 34*(1), 111–146.
Cheng, C., & Li, F. (2024). Ecosystem restoration and management based on nature-based solutions in China: Research progress and representative practices. *Nature-Based Solutions, 6*, 100176. https://doi.org/10.1016/j.nbsj.2024.100176
Choudhary, S. (2024) 'The Wildlife (Protection) Act, 1972: A comprehensive study on wildlife conservation in India', SSRN Electronic Journal [Preprint]. doi:https://doi.org/10.2139/ssrn.4848859.
Christie, M., Hanley, N., Warren, J., Murphy, K., Wright, R., & Hyde, T. (2006). Valuing the diversity of biodiversity. *Ecological Economics, 58*, 304–317.
Claggett, S. L. (1999). Short-term and long-term technology needs/matching status at Idaho National Engineering and Environmental Laboratory. https://doi.org/10.2172/758764.
Connery, R. H. (1935). Connery's "governmental problems in wild life conservation" governmental problems in wild life conservation. *The Auk, 52*(4), 472–473. https://doi.org/10.2307/4077569
Cooke, B., & Lane, R. (2019). Market-based instruments and conservation practices on private land. In *Making ecologies on private land* (pp. 93–110). https://doi.org/10.1007/978-3-030-31218-3_6
Cooper, M. E. (2021). Wildlife conservation law. In *Wildlife Biodiversity Conservation* (pp. 101–122). https://doi.org/10.1007/978-3-030-64682-0_5
Darmawan, R. (n.d.). *Three essays on Indonesian political economy: Elite capture, corruption, and female policy makers*. https://doi.org/10.53846/goediss-4839.
Davis, G. E., & Halvorson, W. L. (2023). Long-term research in national parks. In *Science and Ecosystem Management in the National Parks* (pp. 3–10). https://doi.org/10.2307/j.ctv34dm7s3.5
Decker, D., et al. (2016). Governance principles for wildlife conservation in the 21st Century. *Conservation Letters, 9*(4), 290–295. https://doi.org/10.1111/conl.12211
Dorr, G. B. (1916). Wild life and nature conservation in the Eastern States. https://doi.org/10.5962/bhl.title.70084.
Gadgil, M. (2003). India's biological diversity act 2002: An act for the new millenium. *Journal of Biosciences, 28*(2), 145–147. https://doi.org/10.1007/bf02706212

Galluccio, G. (2015). The role of Public-Private Partnerships (PPPs) in scaling up financial flows in the Post-kyoto regime. In *Public private partnerships for infrastructure and business development* (pp. 169–184). https://doi.org/10.1057/9781137541482_10

Gochhayat, S. A. (2011). Project on FRA act 2006: The scheduled tribes and other traditional forest dwellers (recognition of Forest Rights) Act 2006. *SSRN Electronic Journal.* https://doi.org/10.2139/ssrn.1957149

Gour, D. S., & Reddy, P. A. (2015). Need of transboundary collaborations for Tiger Survival in Indian subcontinent. *Biodiversity and Conservation, 24*(11), 2869–2875. https://doi.org/10.1007/s10531-015-0962-0

Gurău, C., & Dana, L.-P. (2018). Environmentally-driven community entrepreneurship: Mapping the link between natural environment, local community and entrepreneurship. *Technological Forecasting and Social Change, 129*, 221–231. https://doi.org/10.1016/j.techfore.2017.11.023

Haines, A. M., Webb, S. L., & Wallace, J. R. (2021). Conservation forensics: The intersection of wildlife crime, forensics, and conservation. In *Wildlife Biodiversity Conservation* (pp. 125–146). https://doi.org/10.1007/978-3-030-64682-0_6

Heise-Pavlov, S. (2020). Improving the management of Australia's biodiversity through better synergy between private and public protected areas – A response to Australia's biodiversity loss. *Pacific Conservation Biology, 26*(1), 3. https://doi.org/10.1071/pc18068

Himawan, H. A., Pramanti, N., & Shahboz, A. (2025). Understanding pro-environmental behavior: A comparative analysis of rural and Urban Communities for Environmental Education and conservation. *Environment Education and Conservation, 2*(1). https://doi.org/10.61511/educo.v2i1.2025.1712

Hydara, M. (2025). PRA Report: Insights and actions from a participatory rural appraisal in Jah Kunda, The Gambia. https://doi.org/10.2139/ssrn.5126788.

Jepson, P., & Canney, S. (2003). Values-led conservation. *Global Ecology and Biogeography, 12*, 271–274.

Johnsingh, A. J. T., & Joshua, J. (1994). Conserving Rajaji and Corbett National Parks – The elephant as a flagship species. *Oryx, 28*, 135–140.

Keough, H. L., & Blahna, D. J. (2006). Achieving integrative, collaborative ecosystem management. *Conservation Biology, 20*, 1373–1382.

Khan, J. A. (1995). Conservation and management of Gir Lion Sanctuary and National Park, Gujarat, India. *Biological Conservation, 73*(3), 183–188. https://doi.org/10.1016/0006-3207(94)00107-2

Khan, S. (2023). *Protecting biodiversity: Legal frameworks for conservation and Ecosystem Preservation* [Preprint]. https://doi.org/10.31219/osf.io/adkft.

Lacher, I., & Wilkerson, M. L. (2013). Wildlife connectivity approaches and best practices in U.S. state wildlife action plans. *Conservation Biology, 28*(1), 13–21. https://doi.org/10.1111/cobi.12204

Lauber, T. B., & Decker, D. J. (2011). Developing adaptability: The promise and pitfalls of collaborative conservation. *Human Dimensions of Wildlife, 16*(4), 219–221. https://doi.org/10.1080/10871209.2011.585437

Leong, K.M., Emmerson, D.P. and Byron, R. (Rudi) (2011) The new governance era: Implications for collaborative conservation and adaptive management in department of the interior agencies, Human Dimensions of Wildlife, 16(4), pp. 236–243. doi:https://doi.org/10.1080/10871209.2011.585436.

Lewis, M. K. (2021). PPPs and megaprojects. *Rethinking Public Private Partnerships* [Preprint]. https://doi.org/10.4337/9781789906400.00014.

Macdonald, C., Gallagher, A. J., Barnett, A., Brunnschweiler, J., Shiffman, D. S., & Hammerschlag, N. (2017). Conservation potential of apex predator tourism. *Biological Conservation, 215*, 132–141. https://doi.org/10.1016/j.biocon.2017.07.013

Manuku, M. (2021). Resilience of social-ecological systems: At the limits of hegemonic masculinity. In *Social-Ecological Systems (SES)* (pp. 95–109). https://doi.org/10.1007/978-3-030-76247-6_5

Manyani, A., et al. (2024). The evolution of social-ecological systems (SES) research: A co-authorship and co-citation network analysis. *Ecology and Society, 29*(1). https://doi.org/10.5751/es-14694-290133

Marchini, S., et al. (2021). Planning for human-wildlife coexistence: Conceptual framework, workshop process, and a model for transdisciplinary collaboration. *Frontiers in Conservation Science, 2*. https://doi.org/10.3389/fcosc.2021.752953

Marchini, S., et al. (2025). Planning for human-wildlife coexistence: Conceptual framework, workshop process, and a model for transdisciplinary collaboration, Frontiers. Available at: https://www.frontiersin.org/journals/conservation-science/articles/10.3389/fcosc.2021.752953/full

Minteer, B. A., & Miller, T. R. (2011). The new conservation debate: Ethical foundations, strategic trade-offs, and policy opportunities. *Biological Conservation, 144*(3), 945–947. https://doi.org/10.1016/j.biocon.2010.07.027

Münster, U. (2014). Working for the forest: The ambivalent intimacies of Human–Elephant Collaboration in South Indian wildlife conservation. *Ethnos, 81*(3), 425–447. https://doi.org/10.1080/00141844.2014.969292

Nayak, P. K., & Berkes, F. (2008). Politics of co-optation: Community forest management versus joint forest management in Orissa, India. *Environmental Management, 41*(5), 707–718. https://doi.org/10.1007/s00267-008-9088-4

Olatunbosun, A. (2025). Wildlife conservation and legal frameworks for protecting Endangered Species. https://doi.org/10.2139/ssrn.5006660.

Palmer, T. S. (1913). *The Auk, 30*(1), 128–128. https://doi.org/10.2307/4071935

Parish Joshi. (2020). Assessing the role of ecotourism in promoting wildlife conservation. In *Unified visions: Collaborative paths in multidisciplinary research* (Vol. 2). https://doi.org/10.25215/8198189815.08

Radicchi, A. (2021). Citizen science mobile apps for soundscape research and public spaces studies. In *Geographic Citizen Science Design* (pp. 130–148). https://doi.org/10.2307/j.ctv15d8174.14

Rahmani, A. R. (2014). Nature chronicles of India: Essays on wildlife. *Bombay Natural History Society (BNHS), 111*(2), 124. https://doi.org/10.17087/jbnhs/2014/v111i2/71747

Saxena, S. (2024). Gender, caste, and street vending in India: Towards an intersectional geography. *Area, 56*(3). https://doi.org/10.1111/area.12939

Sen, A. (2021). Decentralizing conservation processes through rights-based frameworks. In *A Political Ecology of Forest Conservation in India* (pp. 117–160). https://doi.org/10.4324/9781003007852-5

Sharma, E., & Sharma, M. (2025). A case study of tribal communities in Northeast India. In *Emerging trends and practices in gastronomy and culinary tourism* (pp. 125–156). https://doi.org/10.4018/979-8-3693-7530-3.ch005

Sharma, R., et al. (2024). Private protected areas exhibit greater bias towards unproductive land compared to public protected areas. https://doi.org/10.31223/x55t4r.

Sonant, K. (2007). Promoting collaborative conservation. *The Wildlife Professional, 1*(2), 16. https://doi.org/10.4004/1933-2866(2007)1[16:pcc]2.0.co;2

Stoett, P. (2004). Wildlife conservation: Institutional and normative considerations. In *International Law and Sustainable Development* (pp. 501–518). https://doi.org/10.1163/9789047406709_031

Vernooy, R., et al. (2009). Toward centres of excellence for CBNRM (community-based natural resource management). In *Collaborative learning in practice* (pp. 1–18). https://doi.org/10.1017/upo9788175968639.003

Zuka, S., & Zuka, B. (2024). Market-Community Wildlife Collaborative Management in Malawi: Subjectivities and shifting configurations of protests and celebrations. https://doi.org/10.2139/ssrn.4688652.

Chapter 18
Policy Recommendations for Combating Wildlife Trade

Surender Mehra

Contents

Abstract Illegal wildlife trade remains one of the most persistent threats to India's rich biodiversity. While our country has some of the strongest legal provisions in the world and is a committed party to global agreements such as CITES, the challenges on the ground are far from over. Enforcement agencies are often confronted with highly organised criminal networks, rapidly changing trade routes, and the growing misuse of online platforms for illicit wildlife transactions.

S. Mehra (✉)
NITI Aayog, Government of India, New Delhi, India
e-mail: surenmehra@nic.in

A. Poonia et al. (eds.), *Wildlife Trade in India*,
https://doi.org/10.1007/978-981-95-3694-8_18

The chapter evaluates emerging challenges in online and transboundary wildlife trade, highlighting loopholes in digital marketplaces and porous borders, to outline clear and practical recommendations for strengthening India's fight against wildlife crime. The chapter proposes enhanced collaboration with technology companies, forensic advancements, and improved intelligence-sharing across South Asia based on field realities and policy experience. It analyzes the efficacy of state-level Wildlife Crime Control Units, the role of comprehensive crime databases, and the importance of sensitizing judicial officers on wildlife protection laws and prosecution complexities. It also calls for better alignment between international obligations and national laws, inter-agency coordination, capacity building, strengthening forensic capabilities by use of modern forensic science, and fortifying surveillance at borders and transit points to prosecute cross-border wildlife crimes. Special focus is placed on strengthening border surveillance, tackling social media–driven trade, and building the skills of frontline staff through targeted training programmes. Recognising that enforcement alone is not enough, the chapter also emphasises engaging with communities dependent on wildlife trade, recommending rehabilitation programs, alternative income schemes, and targeted education to reduce poaching incentives.

Collectively, these multilayered interventions are presented as a roadmap to address the socio-economic, legal, and operational complexities of wildlife trafficking, with the goal of stronger, more coordinated, and forward-looking approach to combating wildlife trade in India to preserve India's biodiversity and ensuring sustainable development.

Keywords Illegal wildlife trade · Policy recommendations · CITES implementation · Transboundary wildlife crime · Online wildlife trafficking · Wildlife forensics · Community-based conservation · Alternative livelihoods · Inter-agency coordination · Judicial sensitization

18.1 Background

Around the world, the illicit wildlife trade is expanding, generating over USD $20 billion per year. The data provided by a book titled "State of India's Environment 2017" published by non-profit "Centre for Science and Environment" informed that between 2014 and 2016 poaching and wildlife crimes increased up to 52 percent (Pandey & Sengupta, 2017). By the end of 2016, India had recorded over 30,382 cases of wildlife crime and animal mortality. The range of species targeted has also widened—rising from about 400 species in 2014 to 465 species in 2016. That year was particularly grim for our tiger population, with 50 tigers lost to poaching—the highest figure in the past decade. Peafowl too suffered heavily; between 2015 and 2016, 340 peacocks were killed for illegal trade, a staggering 193% jump from 2014 (Pandey & Sengupta, 2017). Other species under severe pressure include blackbuck, nilgai (blue bull), chinkara, elephant, leopard, rhinoceros, spotted deer (chital), and wild boar. Reptiles and lesser-known species are no safer—turtles and pangolins face relentless threats. In just two years, authorities seized 37,267 turtles, averaging nearly 100 turtles confiscated every single day in 2016 (Pandey & Sengupta, 2017). In January 2017, the STF of Uttar Pradesh Police seized around 6430 endangered soft shell and flap shell turtles from a house in Amethi district (Pandey & Sengupta, 2017). "*State of India's environment: Through the eyes of numbers*" published by non-profit "Centre for Science and Environment" informed Elephant deaths from electrocution have surged, accounting for 72% of all elephant fatalities between 2018 and 2023, driven by both accidental contact with poorly maintained power lines and deliberate illegal wiring in farmlands (Pandey, 2024). Crimes registered under the Environment (Protection) Act went up by 30.6% between 2021 and 2022. At the present pace of disposal, it would take the courts more than 34 years—over three decades—to clear the pending cases under this law (Pandey, 2024).

Pangolins have become the most illegally trafficked animal in the world for their meat and scales. Pangolin parts are in high demand in China for its presumed medicinal value, remains at high threat, with over 1,00,000 pangolins captured illegally every year. The body parts of pangolin are not only restricted up to the ground market but also available in online shopping via the internet which is now fuelling illegal trade of wildlife (Gomez et al., 2023). As per the union environment ministry, 106 online shopping platforms are being used for illegal wildlife trade in India (Dasgupta, 2016). The website list includes most of the popular shopping sites. In India, pangolin species are given the highest level of protection under the Wild Life (Protection) Act, 1972. Anyone caught hunting a pangolin or possessing its scales can be punished with imprisonment for three to seven years, along with a fine of at least ₹10,000 under the provisions of the Act. According to the Trade Records Analysis of Flora and Fauna in Commerce (TRAFFIC) seizure reports, between 2009 and 2013, over 3000 pangolins were hunted in India (TRAFFIC, 2018). Since each pangolin carries about 1.5 kg of scales, the quantity of scales seized during this period points to roughly 3350 animals being killed. On the black market, pangolin scales can fetch more than $3000 per kilogram, and there have even been instances of these scales being used to make coats (The Telegraph, 2016).

18.2 Harmonization of International Conventions and Domestic Laws

India became a signatory to the Convention on International Trade in Endangered Species of Wild Fauna and Flora (CITES) in July 1974 and subsequently deposited the instrument of ratification in 1976 (Poonia et al., 2022). A significant legislative development occurred with the enactment of the Wild Life (Protection) Amendment Act, 1986 (Act No. 28 of 1986), which was passed by the Parliament on May 23, 1986. This amendment introduced several changes to the Wild Life (Protection) Act, 1972, including the insertion of Chapter VA (Poonia et al., 2022).

In alignment with the objectives of CITES and in response to the alarming decline in the elephant population, the Central Government exercised its powers under Section 61(1) of the principal Act and, through a notification dated October 24, 1986, transferred the Asian elephant (*Elephas maximus*) to Schedule-I of the Act, by inserting it as Entry 12B. This categorization accorded the Indian elephant the highest degree of legal protection under domestic law (Government of India, 1986).

Pursuant to this inclusion, any act of hunting Asian elephants attracted stringent penal provisions, including a minimum term of three years of imprisonment, as prescribed for offences involving Schedule-I species. The said legislative action effectively imposed an absolute prohibition on the killing of Indian elephants and resulted in a complete ban on trade and commercial activities involving the species.

Further, in case of *M/S Ivory Traders & Manufacturers Association and others Vs. Union of India and others, AIR 1997 in civil Writ Petition No 1016 of 1992 With CWP Nos. 1272/92, 1631/92, 1749/92, 1303/92, 1964/93* Government of India submitted its intention in Hon'ble Delhi High Court (M/S Ivory Traders & Manufacturers Association & Others v. Union of India & Others, 1997) by stating as follows;

> *Superficially this may be so, but when an article is manufactured from ivory it is impossible to distinguish whether that article is manufactured from mammoth ivory or from elephant ivory. The petitioner is in no position to guarantee that no ivory derived illegally from the Indian elephant would be sold in the garb of the mammoth ivory because there is no method by which one can distinguish the article made from Indian ivory and mammoth African ivory...*

While concluding the important and contentious matter in the case, Hon'ble Delhi High Court held that the words ivory imported into India occurring in section 49B (1) (ia) would "include all descriptions of imported ivory, whether elephant ivory or mammoth ivory" (M/S Ivory Traders & Manufacturers Association & Others v. Union of India & Others, 1997).

In this Landmark judgment Hon'ble Delhi High Court further held that there is no fundamental right to trade in imported ivory, as it is pernicious and dangerous to the elephant. The trade in imported ivory has led to gradual decline of the African elephants. The court held that even if trading in ivory is treated as a fundamental right under Article 19(1)(g) of the Constitution, the ban imposed by the Act serves the public interest and upholds the moral duty set out in Article 48A. It further ruled that prohibiting the trade of imported ivory and products made from it does not violate Article 14 and cannot be considered unreasonable or arbitrary (M/S Ivory Traders & Manufacturers Association & Others v. Union of India & Others, 1997).

Even with a global ban on the ivory trade, both Asian and African elephants continue to be poached in alarming numbers. Every year, tens of thousands are killed for their tusks. In the 1980s alone, it is estimated that around one lakh elephants were slaughtered annually, with some regions losing as much as 80% of their herds (World Wide Fund for Nature [WWF], n.d.). Rhino horn trade remains highly profitable for poachers, as the horns are prized in traditional Chinese medicine. In response to this threat, the department has stepped up anti-poaching measures in national parks and wildlife reserves across the country. A 2000 survey estimated Nepal's protected areas to hold 612 rhinos, but with nearly 50 deaths reported each year, their mortality rate has been rising at a worrying pace (Talukdar et al., 2025). To strengthen efforts against such crimes, India formally became a member of the South Asia Wildlife Enforcement Network (SAWEN) during the Third Asia Ministerial Conference on Tiger Conservation. Through SAWEN, India will work alongside seven other South Asian countries to tackle cross-border wildlife crime, monitor poaching and illegal trade trends, and address related threats to biodiversity within and across the region.

The Wild Life (Protection) Amendment Act, 2022 adds a dedicated framework to implement India's obligations under CITES. The Amendment received parliamentary approval in December 2022 and key provisions (including Section 49M) came into force from 1 April 2023. The amendment introduces a new chapter for regulation of international trade in CITES listed species, creates a national mechanism of management and scientific authorities, brings all CITES listed species into the Act's schedules (Schedule IV), and adds requirements for permits, registration and reporting for live specimens (Government of India, 2022). These key legal additions & mechanisms incorporated through this amendment are elaborated as below;

(i) **New Chapter on Regulation of International Trade:** The Act now includes provisions specifically regulating import/export/possession/transfer of CITES listed species and their parts/products; this establishes national legal backing for CITES permit-based trade controls (Government of India, 2022).
(ii) **Management Authority & Scientific Authority:** The Central Government may designate a *Management Authority* (to issue import/export permits, monitor trade and maintain records) and a *Scientific Authority* (to provide scientific advice and monitor sustainability/detriment findings) as required under CITES. Every person engaged in trade of scheduled specimens must report transactions to the Management Authority (Government of India, 2022).
(iii) **Incorporating CITES species in Schedule IV:** All species listed in the three CITES Appendices were incorporated under Schedule IV of the Wild Life (Protection) Act, 1972 (bringing exotic and internationally listed species within statutory protection and trade controls) (Government of India, 2022).
(iv) **Registration & reporting for live specimens:** Section 49 M (notified April 1, 2023) requires registration of possession/transfer and reporting of births and deaths of living specimens of species that are CITES-listed (as placed in Schedule IV) (Government of India, 2022).

18.3 Policy Innovations for Addressing Online and Transboundary Trade

Several organisations, including the world's leading wildlife forensic laboratory run by the United States Fish and Wildlife Service, have offered to help countries analyse wildlife crime data by providing forensic science support. However, very few nations have made full use of these facilities. INTERPOL, through its Environmental Crime Programme, also provides access to international-level forensic capabilities for investigating such crimes (Talukdar et al., 2025; Government of India, 2022; INTERPOL, n.d.-a, -b, 2019; Wildlife Crime Control Bureau, n.d.-a).

On 5 June 2009, the World Customs Organization (WCO) launched ENVIRONET—a global, real-time communication platform to support the fight against cross-border environmental offences (World Customs Organization, 2009). The WCO also maintains an internet-based seizure database through its Customs Enforcement Network (CEN), which has included wildlife seizures as one of its 13 monitored commodities since 2002 (World Customs Organization, 2009). This database records non-personal information such as the date and place of seizure, species involved, quantity, origin, destination, mode of transport, concealment method, and related CITES documents. Customs agencies across the world contribute seizure data voluntarily. The WCO's Regional Intelligence Liaison Offices regularly analyse this information and publish assessment reports, which are also accessible to all member Customs administrations for further analysis (World Customs Organization, 2009).

Without reliable information on wildlife crimes, it is impossible to design effective prevention strategies, and any efforts made are likely to fail. Inadequate data will not support the successful prosecution of offenders. Therefore, the first step in examining wildlife offences is to gather relevant crime statistics and analyse any available data on the frequency and extent of such offences. In some regions, these statistics are recorded separately for each specific offence, whereas in others, they are grouped under broader categories such as forest offences, environmental crimes, or import–export violations (UNODC, "Crime Investigation," in *Criminal Justice Assessment Toolkit*) (United Nations Office on Drugs and Crime, 2006).

To ensure proper monitoring, systems such as stock inventories, tagging or marking, and detailed documentation are used to track plants, animals, and their derivatives from the source to the final vendor—and sometimes even beyond. At every stage of trade, there must be proof that the specimen or material has been sourced legally. This includes the necessary permits, markings, and receipts that confirm payment of required duties and taxes.

18.3.1 Combating Online Illegal Wildlife Trade

The illegal online trade in wildlife presents a growing challenge for biodiversity conservation and law enforcement in India. There are few cases of online illegal wildlife trade on various platforms, which have been detected by wildlife and other law enforcement agencies. Wildlife traffickers often use *WhatsApp, Telegram, and Chatrooms* to conceal identities and trade information. The Wildlife Trust of India (WTI), an NGO highlighted terms like *"latoo"* (turtle or tortoise) and codes used to evade detection (Chatterjee, 2015). The major cities like Chennai, Mumbai, Kolkata, Bengaluru, Agra, and parts of West Bengal are critical nodes in tortoise and turtle trafficking networks which are often linked to the global pet trade (Kalra et al., 2023a). In one case, Delhi police and Forest officials seized 50 Indian roofed turtles (*Pangshura tecta*), 45 spotted pond turtles, 3 Indian eyed turtles, and 2 Indian soft-shell turtles and multiple other turtle species from suspects who had been trafficking them via *WhatsApp* and *Telegram* (The Hindu, 2024).

The Wildlife Crime Control Bureau (WCCB) has also detected various cases of illegal online wildlife trade commonly involving ivory, pangolin scales, and exotic pets through social media and e-commerce platforms. In addition, around 300 listings on an online portal for sea cucumbers were uncovered and action initiated against the sellers (The New Indian Express, 2022; Wildlife Crime Control Bureau, n.d.-b). In another incident, wildlife enforcement officials uncovered the sale of Schedule-I species protected corals (*Seafan*) and monitor lizard genital organs, marketed for occult rituals through *Instagram*. Following digital surveillance, WCCB and the Haryana Forest Department conducted a raid at the offender's premises in Faridabad and seized the illicit goods, and made an arrest (Wildlife Crime Control Bureau, n.d.-b; The Times of India, 2025). Two individuals were arrested in Ujjain (Madhya Pradesh) by Directorate of Revenue Intelligence (DRI) and forest department after trying to finalize the sale of leopard skins (with heads attached) and a wild boar tusk through *WhatsApp* group (Press Information Bureau, 2025). The Directorate of Revenue Intelligence (DRI) also monitors social media and dark-net activity, intervening in several cases including seizures of exotic animals and peacock feathers.

In another case, over 44 kg of pangolin scales, 43 dead tokay geckos, and two elephant tusks were seized in Meghalaya, leading to eight arrests. The investigation began with monitoring *YouTube* channels where traders posted videos and contact numbers. Undercover messages via *WhatsApp* helped turn the leads into a full-fledged law enforcement operation. Some *Facebook* pages in Kolkata found selling protected animals: civets, green pigeons, turtles, tortoises, and rare birds. The platforms enabled widespread illegal sales of many species listed in Schedules of the Wild Life (Protection) Act, 1972, with sellers receiving animals from across regions like Odisha, the Northeast, and Maharashtra (The Independent, 2021). In Malkangiri district, Odisha, forest officers intercepted a vehicle carrying 1078 live freshwater turtles destined for illegal sale. The intelligence-driven operation led to the arrest of three individuals, demonstrating enforcement agencies' vigilance at trafficking hot spots and border routes (Deccan Chronicle, 2025).

Each example reflects different enforcement strategies, from digital entrapment and monitoring of platforms to field interception and cross-agency coordination. These cases of online trade in wildlife trade, highlight both the evolving nature of online wildlife crime and the multi-layered response mechanisms being deployed to combat it. In response, the Government of India has undertaken a series of legislative, institutional, and technological measures aimed at curbing this emerging threat. These enforcement efforts reflect a multi-pronged approach that integrates environmental law, cybercrime enforcement, inter-agency coordination, and international cooperation.

Online wildlife trafficking is adaptive and traffickers change codewords, platform choices, and logistics. The most effective response is multi-layered and coordinated: prevention and demand reduction, continuous online intelligence, rapid multi-agency operational response, robust digital forensics, platform cooperation, and visible prosecutions. WCCB should be the nodal coordinator with clear SOPs linking forest departments, police cyber cells, DRI/Customs and forensic labs.

18.3.2 Role of Forensics in Combating Transboundary Wildlife Trade

In wildlife crime analysis, identification of species from unknown biological samples is the foremost application of wildlife forensics (Verma & Singh, 2003; Gupta et al., 2005). Other three applications, which are lesser or rarely used are, 'Identification of Geographic Origin' (More relevant in case of implementation of international legislations such as CITES), 'Individual Identification' (when it is required to demonstrate that various articles seized are from same individual animal) (Gupta et al., 2011a), 'Familial Identification' (mainly used for differentiation of captive bred or wild animals) (Ogden et al., 2009). Most of the literature available in relation to the subject of present research is mainly concerning the role of forensics in law enforcement, but how this wildlife forensics reporting has impacted the cases in court of Law. There are few studies conducted for analysing the data collected by data mining techniques (Gorr, 2003; Gorr & Harries, 2003; Kaur & Bhatia, 2016). These studies are based on the forecasting of crime in specific locations using the crime data. For wildlife crime analysis no such detailed studies have been conducted. One toolkit is available for the analysis of wildlife and forest crime data (United Nations Office on Drugs and Crime, 2012). Availability of a nationwide wildlife crime data would be useful in assessment of wildlife crimes and their hotspots across the country.

Though general evidence is collected during the investigation by forest officers or other investigative agencies, if such evidence is supported with scientific interpretation and presentation, this comes as great help for investigation agencies for successful prosecution in the Court of Law. Today Wildlife Forensics is not only restricted to a subject matter of genetic research, but it has emerged as a very important tool for 'Scientific Interpretation and Presentation' of evidence in investigation of wildlife crime ultimately leading to a successful prosecution in court of law. Evidence collection is not only an important but rather essential component in any Crime Investigation and prosecution in Court of Law. These days wildlife forensics dealing with analysis of samples related to wildlife crime has many facets, such as species identification, place where crime took place and time when it was committed etc. All these questions are required to be conclusively replied during investigation for any successful prosecution in court of law. Mostly, it is the species identification, which is used in cases of poaching in order to identify trace evidence in field or from suspect's possession (Gupta et al., 2005, 2011a, b; Vipin et al., 2016).

Scientific analysis and modern forensic investigation play a key role in wildlife law enforcement because establishing the possession of seized material and identifying the same as part of any wild animal mentioned in schedules of the Wildlife (Protection) Act, 1972 are legal requirements for any trial case to be concluded in the favour of prosecution. Therefore, possession of animal articles and proof of the same being identified in scientific investigation as part of any species mentioned in Schedules of the Wild Life (Protection) Act, 1972 leads to conviction. In the judgement in case of *Babu Lal and another vs. State (Delhi administration) 1982CRILJ41, 20(1981)DLT354*, it was held by Hon'ble Delhi High Court that the possession of the accused was proved and thereafter a presumption of unlawful possession was drawn against the accused in terms of section 57 of the Wild Life (Protection) Act, 1972 (*Babu Lal & Another*, 1982). Likewise, in another landmark judgement in the case of *Sansar Vs. State 1994 I AD Delhi 13*, it was held by the Hon'ble Supreme Court that when the possession, custody and control of the accused over the animal articles, Cured and uncured trophies has been established by the prosecution, the presumption is that the accused is guilty unless and until the accused disproves the same (*Sansar v. State*, 1994).

The wildlife forensic techniques have been discussed in some landmark judgments of the Courts of Law while establishing law of the land in the field of Wildlife law enforcement. In a landmark judgment in case of *M/S Ivory Traders & Manufacturers Association and others Vs. Union of India and others, AIR 1997 in civil Writ Petition No 1016 of 1992 With CWP Nos. 1272/92, 1631/92, 1749/92, 1303/92, 1964/93*, the Hon'ble Delhi High Court discussed the forensic investigation techniques at length (M/S Ivory Traders & Manufacturers Association & Others, 1997). Hon'ble court used excerpts from book of Doughlas H. Chadwick which explained the difference between ancient tusks from ones modern as below;

> *Fortunately, scientists at the National Fish and Wildlife Forensics Laboratory in Ashland, Oregon, discovered a method to distinguish ancient tusks from modern ones. Using a scanning electron microscope, they focus on the tooth's characteristic crosshatched patterns, called Schreger lines. These are formed by tiny denial tubules, which turn out to be twice as dense in mammoths and mastodons as in modern elephants. As a result, the Schreger lines meet at angles of less than 90 degrees in the bygone species but more than 110 degrees in existing elephants, a minor but unmistakable difference. Forensic techniques can also distinguish proboscidean ivory from that of hippos, wart hogs, and walruses. Conservationists hope that advances in Chemical "fingerprinting" techniques will soon*

enable specialists to identify which particular elephant population a tusk came from, on the basis of DNA from tissues coating the base of the tooth.

When a buyer intends to buy a curio, he is not interested to know whether it was created from elephant ivory or mammoth ivory. An average buyer also does not have the expertise or the knowledge to distinguish between articles made from mammoth ivory and Indian ivory. To him the translucent whiteness of the ivory matters. He buys it purely on aesthetic considerations or as a statue symbol. To give permission to trade in Articles made from mammoth ivory would result in laundering of Indian ivory. The identification can be made by experts in the field or those who have experience in this line and not by a layman who sets out to buy an ivory article.

The Hon'ble court while deciding the matter concluded that those traders, who claimed to be handling ivory imported from Africa or mammoth ivory, were in fact dealing with ivory sourced from India, which led to the unlawful poaching of Indian elephants. The hon'ble court also held that Indian legislation intended to include mammoth ivory as far as prohibition in elephant ivory is concerned.

In the notable case of *Pyarelal vs. State, 1995 Cr.L.J. 2075 at 2076 (SC); AIR 1995 SC 1159; Supp. (2) SCC 684*, the apex court upheld the evidentiary credibility of a trained and experienced forest officer (Press Information Bureau (PIB), 2023a). The Court observed that the officer's testimony effectively established the accused's possession of prohibited wildlife trophies, and this position was consistently affirmed by all subordinate courts involved in the case. This judicial recognition underlines the importance of timely and scientifically validated wildlife forensic evidence and reinforces the admissibility and probative value of expert testimony from forest officials under the Wild Life (Protection) Act, 1972.

Study of some typical cases from WII's Forensics lab has revealed certain interesting facts about some other closed cases on the basis of scientific investigations. In one instance, a tiger (Panthera tigris) was poached from a zoological park in India in the year 2000, where its skin and claws were taken away by the offenders. Unfortunately, no biological samples were collected from the deceased tigress at that time. However, in 2005, a claw embedded in a pendant was recovered from a local youth. The genetic profile derived from multiple loci of this claw was compared with samples from the supposed biological relatives of the tigress. Computer-assisted genetic analysis confirmed that the claw belonged to the missing individual from the suspected family (Gupta et al., 2005).

Such genetic tools have also greatly supported wildlife management efforts. During routine forest patrols, an elephant tusk was discovered and secured by forest personnel (Gupta et al., 2011b). Soon after, a carcass of a male elephant was located with one tusk missing, raising suspicion of illegal hunting for ivory. To establish a connection, small ivory fragments from the found tusk and the carcass were subjected to DNA profiling using microsatellite markers. The results confirmed that the ivory from both incidents originated from the same elephant (Gupta et al., 2011b).

18.4 Policy Innovations to Combat Transboundary Illegal Wildlife Trade

The 2022 amendment to the Wild Life (Protection) Act, 1972 has significantly strengthened India's framework for implementing CITES obligations by introducing a stronger permit and compliance regime. Trade in CITES-listed species now requires domestic permits, continuous monitoring, and mandatory reporting, enhancing the traceability of live animals and regulated parts. Internationally, the amendment closed earlier legal gaps that had placed India in *Category 2* under the CITES National Legislation Project; following its enactment, the CITES Standing Committee upgraded India to *Category 1* for full compliance. For traders, zoos, pet-owners, and breeders, the changes mean mandatory registration and permits for all CITES-listed species, reporting of births and deaths of live specimens, and stricter documentation for cross-border movement. For enforcement agencies, the amendment provides clearer statutory powers to regulate imports and exports, seize illegally traded specimens, and operate under a formal *Management Authority* and *Scientific Authority* framework to ensure informed and coordinated decision-making in line with CITES requirements (Press Information Bureau (PIB), 2023a, b; UNEP-FAOLEX, 2022).

18.5 Communities Dependent on Wildlife Trade and Livelihood Programs for Rehabilitation

India's illegal wildlife trade has historically drawn heavily on the skills, networks, and traditional knowledge of certain nomadic and semi-nomadic hunting communities. Many of these groups, such as the *Pardhi, Bahelia, Bawaria, Moghiya, Ghihara, Nut*, and *Kanjar* have deep-rooted hunting traditions that once served subsistence or cultural purposes but have, over time, been absorbed into organized poaching and trafficking networks and been co-opted into profit-driven wildlife crime. In many cases, historical stigmatization and economic marginalization contributed to the continuation of such activities. These communities possess exceptional tracking, trapping, and bushcraft skills passed down over generations. They know animal behaviour, migratory patterns, and habitats, enabling them to locate and capture high-value species like tigers, leopards, pangolins, and freshwater turtles with minimal detection. Poaching methods include steel traps, snares, poison baits, and silent hunting weapons, reducing noise and visibility. Many of these communities face chronic poverty, landlessness, and social exclusion, a legacy of being labelled *"criminal tribes"* under the British-era *Criminal Tribes Act, 1871*. These communities have long-held traditional roles in hunting, which have, over time,.

Pardhi tribe engaged in poaching lions (Gir), tigers, and other wildlife using steel-traps, often supplying wider trafficking networks. They are still active in and around tiger reserves like Panna and Sariska. ***Bahelia***, a nomadic hunter-gatherer caste historically involved in big cat poaching, frequently identified in tiger and leopard trafficking cases. *Bawaria*, another traditional hunter community interestingly, from a non-forest state of Haryana in north India, involved in trapping and trading wildlife parts. They have been identified in organized poaching networks targeting tigers and small mammals active in Uttarakhand, Uttar Pradesh, Rajasthan, and Madhya Pradesh (Kalra et al., 2023a, b).

Some other tribes such as *Kanar*, Nut and *Moghiyas* has been found to act as traders and intermediaries, often targeting turtles and small mammals in northern India. Masaed tribe is known for snake charming and involved in trading small mammals. *Baddiyes* are found to be involved in trade of big cat products and acted as middlemen. *Ghihara* community also known as *Ghihara Network* represented by Sansar Chand, a key wildlife crime syndicate leader, managed a trafficking network responsible for thousands of animal deaths of Tigers, leopards, otters etc. He exploited marginalized communities around Sariska and surrounding areas for wildlife crime (Kalra et al., 2023a, b).

While earlier crimes were opportunistic and local, modern involvement often includes targeted poaching for specific species on order, use of motorbikes, mobile phones, and encrypted messaging apps to coordinate with buyers and multi-species trafficking i.e. shifting from big cats to reptiles, birds, and even marine species as enforcement tightens.

Hunter communities in India have transitioned from subsistence hunting to playing pivotal roles in commercial wildlife crime. Their unique skill sets, mobility, and integration into trafficking networks make them highly effective poachers, posing a significant challenge to conservation enforcement. Addressing their role, the issue requires a dual strategy of strong enforcement against ongoing crimes, and long-term rehabilitation through livelihood and education programs to break the cycle of dependence on poaching. In India, various existing and emerging alternative livelihood programs have been implemented or proposed to support communities traditionally dependent on wildlife crime, with the aim of reducing their reliance on illegal activities while promoting conservation and sustainable development. Forest Department of various states and conservation organizations are increasingly emphasizing the need to engage poaching-prone communities in eco-tourism, homestay operations, and nature guiding. These programs not only offer sustainable livelihoods but also foster community led conservation efforts. There are several successful initiatives by both government agencies and Non-Governmental Organisations (NGOs) have already demonstrated the potential for transformation.

Further, few best practices in rehabilitation of hunting communities in India are elaborated as below;

(i) In Madhya Pradesh, the *Pardhi* community, historically involved in hunting, has transitioned to eco-tourism-based livelihoods through initiatives supported by NGOs and government agencies such as NABARD. Through nature guiding and cultural tourism in areas like *Oosera,* the *Pardhis* now use their deep ecological knowledge to promote wildlife conservation. Similar efforts have been implemented by the some Non-Profit Organizations (NPOs) which has introduced alternative income-generating activities such as apiculture and handicrafts as part of their corridor restoration projects in the Panna Tiger Reserve (AESA Network, n.d.; Last Wilderness Foundation, 2022).

(ii) The Periyar Tiger Reserve (PTR), Kerala once faced significant poaching pressure, particularly from organized hunting groups targeting sambar deer, gaur, and elephants for meat, tusks, and other parts. Many poachers came from marginalized local communities, including former shifting cultivators and forest fringe villagers, who depended on forest resources for income. The key Rehabilitation measures by the Forest Department started in the 1990s with the formation of Eco-Development Committees (EDCs). These EDCs included former poachers, woodcutters, and illicit grazers as members and provided structured, legal employment opportunities in forest-based tourism and protection work. The alternative livelihood options were provided through training in eco-tourism services: guided treks, bamboo rafting, boating, interpretation, skill development in handicrafts, spice cultivation, and organic farming in buffer-zone villages. In addition, former hunters trained as nature guides, using their tracking skills to spot wildlife for tourists. Many ex-poachers became part of Patrolling and Protection Squads (e.g., *Forest Watchers*, *Vana Samrakshana Samithis*). They assist forest staff in anti-poaching patrols, fire prevention, and wildlife monitoring. PTR's community-based tourism model (e.g., *Bamboo Rafting*, *Tiger Trail*, *Border Hiking*) channels a share of income directly to EDC members. This created a strong incentive to protect, rather than exploit, wildlife. This not only led to public acknowledgment of ex-poachers who turned protectors and reduced stigma and improved *community–department* trust. This initiative of the Forest Department led to drastic reduction in poaching incidents within Periyar TR and over 300 former poachers and their families rehabilitated into legal livelihood (Shaji, 2020; Periyar Tiger Conservation Foundation, n.d.; The News Minute, 2015).

(iii) Wildlife SOS, for instance, has implemented inclusive conservation models targeting communities such as the *Kalandars,* traditionally known for exploiting sloth bears in street performances. Members of these communities have been rehabilitated through training and employment as bear caregivers, educators, and conservation advocates. Similarly, former snake charmers have been retrained as snake rescuers working with rapid response units, while traditional elephant handlers (mahouts) are now engaged in ethical elephant care. The Wildlife Trust of India (WTI) has also supported the *Kalandars* by helping them establish micro-enterprises such as transport and waste management services. Another notable example is the *Irula* Snake Catchers' where Industrial Cooperative Society, enables members of the *Irula* community to harvest snake venom legally for antivenom production, turning traditional knowledge into a legal and valuable occupation (Wildlife SOS, n.d.; Wildlife Trust of India, n.d.; Wikipedia Contributors, 2025).

(iv) Further, Eco-Development Committees (EDCs) model of community participation from the Periyar TR the provisions of Tiger Conservation Foundations in Tiger Reserves was brought through the Wild Life (Protection) Act, 1972, as amended in 2006. The provision of Act provides for the constitution of Tiger Conservation Foundations (TCFs) under Section 38X, in line with the National Tiger Conservation Authority (NTCA) guidelines. These autonomous, not-for-profit bodies function at the Tiger Reserve level to facilitate conservation, eco-tourism, and community development. The TCF is empowered to receive grants, donations, and eco-tourism revenues, and to channel these funds into habitat improvement, anti-poaching activities, research, and welfare programmes for local communities that support conservation goals.

Collectively, these efforts demonstrate that with the right support and opportunities, communities once involved in wildlife crime can be successfully integrated into conservation and development frameworks, benefiting both people and biodiversity.

18.6 Policy Recommendations

18.6.1 Strengthen Legal Framework

Effective implementation of the *Wild Life (Protection) Act, 1972* to address wildlife trade explicitly to ensure severe penalties in prosecution by investigation and faster judicial processes for wildlife crime cases.

18.6.2 Setting up of State Wildlife Crime Control Unit

There is an urgent need for a special Wildlife Crime Control Unit at state level like the one constituted in the state of Madhya Pradesh. The Special Task Force (STF) for Wildlife Protection is a specialized unit within the Madhya Pradesh Forest Department. The main objective of this unit is wildlife law enforcement by combating Wildlife Crime with focussing on preventing and investigating offenses related to poaching, illegal trade of wildlife and their parts, and other activities that harm wildlife. Its main functions include preventing and investigating poaching and illegal trade, gathering intelligence, conducting raids and making arrests, ensuring inter-state coordination against smuggling, utilizing modern technology including wildlife forensics, and building the capacity of forest officials in wildlife law enforcement and effectively taking up prosecution in trial courts. The MP STF has field units in sensitive areas prone to wildlife crime.

Further, Wildlife Crime Control Bureau, MoEF&CC vide an advisory dated November, 14, 2014 has also suggested the state governments to set up Wildlife Crime Control Units (WCCU) for effectively combating the wildlife crime. The unit may be headed by a senior officer in the Wildlife Wing of the forest department.

18.6.3 Development of Wildlife Crime Database and Framework for Monitoring of Prosecution

A well-structured wildlife crime database and monitoring framework at state level Wildlife Wings of various forest departments to be managed by a dedicated Cell under Wildlife Crime Control Unit (WCCU) is crucial for effective prosecution. This cell should be linked to territorial and wildlife regions units. Such a database should compile detailed information on incidents, seized items, forensic analysis results, and case progress. With such a robust monitoring framework, this system enables authorities to track patterns, identify hotspots, and assess the efficacy of legal interventions. This approach will strengthen wildlife law enforcement efforts with targeted interventions.

18.6.4 Sensitisation among Judicial Officers About Wildlife Protection Laws and Prosecution Issues

The study observed that another reason for less effective prosecution is incomplete understanding of the Wild Life (Protection) Act, 1972 among judicial officers also. This legislation is a specialized statute and requires expert knowledge to convene legal proceedings. Misunderstanding or improper application of its provisions often has been reflected in cases of under penalty in schedule-I cases, grant of bail, acceptance of confession before forest officers, acceptance of forensic investigation report as evidence etc. The synthesis of trial court proceedings shows that in many instances, specific provisions of the Wild Life (Protection) Act, 1972 have not been referred to. On account of this, many cases lead to either acquittal of offenders or under penalty in even cases of conviction. Therefore, there is a need for regular training and sensitization programmes for judicial officers handling wildlife crime cases to understand urgency and legal provisions by the State Judicial Academies or Forest Departments.

18.6.5 Enhance Inter-Agency Coordination

Better coordination involving the Forest Department, Wildlife Crime Control Bureau (WCCB), Department of Revenue Intelligence (DRI), Customs, Police, Cybercrime Units, and Interpol. A shared, real-time Wildlife Crime Information Database for intelligence exchange will further supplement the efforts. Multi-agency collaboration both at national and international levels, and between source, transit and end-use destinations, should be enhanced to tackle the international and organized criminal networks involved in smuggling wildlife articles across borders.

18.6.6 Strengthen Forensic Capabilities

As it has been observed in this study that the timely and accurate forensic analysis of evidence is crucial for successful prosecution, therefore, the enhancement of the infrastructure and expertise including manpower in the wildlife forensic cell at Wildlife Institute of India is key to expeditious reporting of forensic investigation. The forensic capabilities should further be integrated with digital forensics capabilities. Techniques such as image metadata analysis, blockchain traceability, and encrypted communications tracking may adopted to strengthen the evidentiary basis for legal action.

18.6.7 Tackle Online and Social Media-Based Trade

Develop partnerships with social media platforms (*WhatsApp, Telegram, Facebook, Instagram*) to monitor, detect, and block illegal wildlife trade accounts. Training of enforcement personnel in cybercrime cells in wildlife law enforcement and digital forensic investigation. The online commerce platform must be mandated to include mandatory seller verification, transaction audits, and platform accountability for illegal listings.

18.6.8 Community Engagement and Alternative Livelihoods

The Eco-Development Programs (like those in Periyar Tiger Reserve) to rehabilitate poacher communities and incentivize communities to act as wildlife stewards through profit-sharing from eco-tourism and conservation-linked enterprises. These EDCs should be supported through State and Central funding schemes such as Project Tiger (PT), Project Elephant (PE) and Integrated Development of Wildlife Habitats (IDWH) etc.

18.6.9 Strengthen Border and Transit Point Surveillance

Border and transit point surveillance may be enhanced by installing advanced cargo-scanning systems and mandate wildlife trade awareness training for customs and transport staff. Further, sniffer dogs trained for wildlife contraband at airports, seaports, and railway stations may also be deployed by concerned enforcement agencies.

18.6.10 Capacity Building and Specialized Training

The mainstreaming of the subject of combating wildlife crime should be done by institutionalizing annual training for forest officers, police, and judiciary on wildlife laws, forensic identification, and case documentation.

18.6.11 Public Awareness and Citizen Participation

Certain targeted awareness campaigns on the illegality and ecological impacts of wildlife trade among citizens along with confidential reporting systems and rewards for informants providing actionable intelligence.

References

AESA Network. (n.d.). *Ensuring alternative livelihoods to traditional hunting communities by supporting conservation of small forest patches through ecotourism.* Retrieved August 12, 2025, from https://aesanetwork.org/good-practices-45-ensuring-alternative-livelihoods-to-traditional-hunting-communities-by-supporting-conservation-of-small-forest-patches-through-ecotourism/

Babu Lal & Another v. State (Delhi Administration), 1982 Cri LJ 41; 20 (1981) DLT 354 (Delhi High Court).

Chatterjee, B. (2015, June 1). *Web of illicit wildlife trade spreads in India.* Hindustan Times. Retrieved August 12, 2025, from https://www.hindustantimes.com/india/web-of-illicit-wildlife-trade-spreads-in-india/story-qIJ7EbJ1yl1AAAJcTR5unO.html

Dasgupta, S. (2016, July 22). *Amazon, eBay involved in illegal wildlife trade: Indian government.* Eco-Business. Retrieved from https://www.eco-business.com/news/amazon-ebay-involved-in-illegal-wildlife-trade-indian-government/

Deccan Chronicle. (2025, August 6). *Over 1,000 turtles rescued in Malkangiri, 3 arrested.* Deccan Chronicle. Retrieved August 12, 2025, from https://www.deccanchronicle.com/nation/over-1000-turtles-rescued-in-malkangiri-3-arrested-1895884 (deccanchronicle.com).

Gomez, L., Joseph, T., Heinrich, S., et al. (2023). Illegal trade of pangolins in India with international trade links: An analysis of seizures from 1991 to 2022. *European Journal of Wildlife Research, 69*, 85. https://doi.org/10.1007/s10344-023-01708-9

Gorr, W. L. (2003). Short-term forecasting of crime. *International Journal of Forecasting, 19*(4), 579–594.

Gorr, W. L., & Harries, R. (2003). Introduction to crime forecasting. *International Journal of Forecasting, 19*(4), 551–555.

Government of India. (1986). Notification S.O. 837(E), dated 24 October 1986. *The Wild Life (Protection) Act, 1972.* Ministry of Environment and Forests, New Delhi: The Gazette of India, Part II, Section 3(ii).

Government of India. (2022). *The wild life (protection) amendment act, 2022.* Ministry of Law and Justice. The Gazette of India. Retrieved from https://egazette.nic.in

Gupta, S. K., Verma, S. K., & Singh, L. (2005). Molecular insight into a wildlife crime: The case of a peafowl slaughter. *Forensic Science International, 154*(2–3), 214–217.

Gupta, S. K., Bhagavatula, J., Thangaraj, K., & Singh, L. (2011a). Establishing the identity of the massacred tigress in a case of wildlife crime. *Forensic Science International: Genetics, 5*(1), 74–75. https://doi.org/10.1016/j.fsigen.2010.09.026

Gupta, S. K., Thangaraj, K., & Singh, L. (2011b). Identification of the source of ivory idol by DNA analysis. *Journal of Forensic Sciences, 56*(6), 1343–1345. https://doi.org/10.1111/j.1556-4029.2011.01753.x

INTERPOL. (2019). *Global wildlife enforcement: Prospectus* [PDF]. International Criminal Police Organization. Retrieved August 11, 2025, from https://www.interpol.int/content/download/5179/file/WEB_Wildlife%20ProspectusMarch2019.pdf.interpol.int

INTERPOL. (n.d.-a). *Our response to environmental crime*. International Criminal Police Organization. Retrieved August 11, 2025, from https://www.interpol.int/Crimes/Environmental-crime/Our-response-to-environmental-crime.interpol.int

INTERPOL. (n.d.-b). *Wildlife crime*. International Criminal Police Organization. Retrieved August 11, 2025, from https://www.interpol.int/en/Crimes/Environmental-crime/Wildlife-crime.interpol.int

Kalra, S., Davies, A., Martin, R., & Poonia, A. (2023a). Insights from the media into the bird trade in India: An analysis of reported seizures. *Oryx*, (1), –9. https://doi.org/10.1017/S0030605322001594

Kalra, S., Poonia, A., Sharma, R., Kaur, R., Sharma, P., Kumar, A., & Gulia, N. (2023b). Wildlife conservation through local community engagement in India. *Uttar Pradesh Journal of Zoology, 44*(2), 13–24. https://doi.org/10.56557/upjoz/2023/v44i23398

Kaur, P., & Bhatia, R. (2016). Web mining research: A study. *Research Cell: An International Journal of Engineering Sciences, 22*, 253–257.

Last Wilderness Foundation. (2022). *Protection of wildlife corridor via reforestation and alternative livelihood programme, Panna Tiger Reserve.* Retrieved August 12, 2025, from https://thelastwilderness.org/projects-2/protection-of-wildlife-corridor-via-reforestation-and-alternative-livelihood-programme-panna-tr/ (thelastwilderness.org).

M/S Ivory Traders & Manufacturers Association & Others v. Union of India & Others, AIR 1997 Delhi 267 (Delhi High Court, Mar. 20, 1997).

M/S Ivory Traders & Manufacturers Association & Others v. Union of India & Others, AIR 1997 Delhi, Civil Writ Petition No. 1016 of 1992 with CWP Nos. 1272/1992, 1631/1992, 1749/1992, 1303/1992, 1964/1993 (India).

Ogden, R., Dawnay, N., & McEwing, R. (2009). Wildlife DNA forensics—Bridging the gap between conservation genetics and law enforcement. *Endangered Species Research, 9*(3), 179–195. https://doi.org/10.3354/esr00144

Pandey, K. (2024). *State of India's environment: Through the eyes of numbers*. Centre for Science and Environment. https://cdn.cseindia.org/attachments/0.19910600_1717407003_kiran-pandey-soe.pdf

Pandey, K., & Sengupta, R. (2017). State of India's environment 2017: In Figures, ISBN: 978-81-86906-04-0. Published by Centre for Science and Environment 41, Tughlakabad Institutional Area New Delhi 110 062 Phones: 91-11-40616000 Fax: 91-11-29955879 E-mail: cse@cseindia.org website: www.cseindia.org. As assessed from https://share.google/8qhxjoBhxYyELV9Q8, http://library.tiss.edu/sites/all/themes/clean_theme/pdfdoc/sie2017.pdf on 11th August, 2025.

Periyar Tiger Conservation Foundation. (n.d.). *Periyar Tiger Reserve (Wings)*. Retrieved August 12, 2025, from https://www.periyarfoundation.in/wings.html.

Poonia, A., Jangra, M., Sharma, R., Ghanghas, A., & Sindhu, V. (2022). Trends in CITES listed birds trade in South Asian countries in view of evolution of Indian laws during last four decades. *International Journal of Ecological Economics and Statistics, 43*(1) http://www.ceser.in/ceserp/index.php/ijees/article/view/6848

Press Information Bureau. (2025, May 6). *DRI Nagpur unit conducts operation in Ujjain—Seizes two leopard skins with heads and one wild boar horn; two arrested* [Press release]. Press

Information Bureau, Government of India. Retrieved August 12, 2025, from https://pib.gov.in/PressReleasePage.aspx?PRID=2127233.

Press Information Bureau (PIB). (2023a, November 12). *Consequent to amendment of Wildlife Act, placement of the CITES legislation of India in Category 1 of the National Legislation Programme*. Press Information Bureau, Government of India. Retrieved August 12, 2025, from https://pib.gov.in/PressReleasePage.aspx?PRID=1976624

Press Information Bureau (PIB). (2023b, February 6). *The Wild life (Protection) Amendment Act, 2022 has come into force since 1st April 2023*. Press Information Bureau, Government of India. Retrieved August 12, 2025, from https://pib.gov.in/PressReleaseIframePage.aspx?PRID=2011928

Sansar v. State, 1994 I AD (Delhi) 13 (Supreme Court of India).

Shaji, K. A. (2020, March 5). *Periyar Tiger Reserve, a trendsetter in converting poachers to protectors*. Mongabay India. Retrieved August 12, 2025, from https://india.mongabay.com/2020/03/periyar-tiger-reserve-a-trendsetter-in-converting-poachers-to-protectors/

Talukdar, B. K., et al. (2025). The recovery of the greater one-horned rhinoceros in India and Nepal. In M. Melletti, B. Talukdar, & D. Balfour (Eds.), *Rhinos of the world. Fascinating life sciences*. Springer. https://doi.org/10.1007/978-3-031-67169-2_11

The Hindu. (2024, July 11). *Delhi police rescue 100 live turtles; nab smuggler in East Delhi*. The Hindu. Retrieved August 12, 2025, from https://www.thehindu.com/news/cities/Delhi/delhi-police-rescue-100-live-turtles-nab-smuggler-in-east-delhi/article68390394.ece (thehindu.com)

The Independent. (2021, May 6). *Online investigation leads to wildlife seizure of live Chinese pangolins and tokay geckos*. The Independent. Retrieved August 12, 2025, from https://www.independent.co.uk/stop-the-illegal-wildlife-trade/illegal-wildlife-trade-internet-facebook-b1879656.html

The New Indian Express. (2022, September 4). *Wildlife crime goes digital, WCCB increases vigilance*. The New Indian Express. Retrieved August 12, 2025, from https://www.newindianexpress.com/thesundaystandard/2022/Sep/04/wildlife-crime-goes-digital-wccb-increases-vigilance-2494617.html.

The News Minute. (2015). *Once poachers and smugglers, they now protect forests*. The News Minute. Retrieved August 12, 2025, from https://www.thenewsminute.com/kerala/once-poachers-and-smugglers-they-now-protect-forests-33328

The Telegraph. (2016, March 15). *Pangolins: 13 facts about the world's most hunted animal*. The Telegraph – Science. Retrieved from https://www.telegraph.co.uk/science/2016/03/15/pangolins-13-facts-about-the-worlds-most-hunted-animal/

The Times of India. (2025, July 17). *Parts of endangered species seized in raid on astrologer's office in Faridabad*. The Times of India. Retrieved August 12, 2025, from https://timesofindia.indiatimes.com/city/gurgaon/parts-of-endangered-species-seized-in-raid-on-astrologers-office-in-faridabad/articleshow/122591930.cms

TRAFFIC. (2018). *Scale of pangolin trade in India: 2009–2017* [Factsheet]. TRAFFIC India. https://www.traffic.org/site/assets/files/2647/factsheet_scale_of_pangolin_trade_in_india_2009_2017.pdf

UNEP-FAOLEX. (2022). *Wild Life (Protection) Amendment Act, 2022 (No. 18 of 2022)*. Retrieved August 12, 2025, from https://leap.unep.org/en/countries/in/national-legislation/wild-life-protection-amendment-act-2022-no-18-2022.

United Nations Office on Drugs and Crime. (2006). *Crime investigation* [module within *criminal justice assessment toolkit*]. In Criminal justice assessment toolkit. UNODC. Retrieved August 11, 2025, from https://www.unodc.org/documents/justice-and-prison-reform/cjat_eng/3_Crime_Investigation.pdf

United Nations Office on Drugs and Crime. (2012). *Wildlife and forest crime analytic toolkit* (Revised edition). United Nations. Retrieved August 12, 2025, from http://www.unodc.org/documents/Wildlife/Toolkit_e.pdf (unodc.org).

Verma, S. K., & Singh, L. (2003). Novel universal primers establish identity of an enormous number of animal species for forensic application. *Molecular Ecology Notes, 3*(1), 28–31.

Vipin, Sharma, V., Sharma, C. P., Kumar, V. P., & Goyal, S. P. (2016). Pioneer identification of fake tiger claws using morphometric and DNA-based analysis in wildlife forensics in India. *Forensic Science International, 266*, 226–233. https://doi.org/10.1016/j.forsciint.2016.05.024

Wikipedia Contributors. (2025). Irula Snake catchers' industrial cooperative society. In *Wikipedia,* The Free Encyclopedia. Retrieved August 12, 2025, from https://en.wikipedia.org/wiki/Irula_Snake_Catchers%27_Industrial_Cooperative_Society

Wildlife Crime Control Bureau. (n.d.-a). *Wildlife crime investigation manual (annex: Wildlife forensics manual).* Wildlife Crime Control Bureau, Government of India. Retrieved August 11, 2025, from https://wccb.gov.in/WriteReadData/userfiles/file/Wildlife%20Crime%20Investigation%20Manual.pdf

Wildlife Crime Control Bureau. (n.d.-b). *WCCB newsletter.* Ministry of Environment, Forest and Climate Change, Government of India. Retrieved August 12, 2025, from https://wccb.gov.in/Content/Newsletter.aspx

Wildlife SOS. (n.d.). *Equitable conservation: Wildlife SOS' tribal rehabilitation programme for Kalandars.* Retrieved August 12, 2025, from https://wildlifesos.org/equitable-conservation-wildlife-sos-tribal-rehabilitation-programme-for-kalandars/

Wildlife Trust of India. (n.d.). *No Mast Kalandar.* In *Resource Centre.* Retrieved August 12, 2025, from https://www.wti.org.in/resource-centre/resource-centre/WTI

World Customs Organization. (2009, June 5). WCO launches ENVIRONET to connect Customs and enforcement partners in the fight against environmental crime [Press release]. Retrieved August 11, 2025, from https://www.wcoomd.org/en/media/newsroom/2009/june/environet-fight-against-environmental-crime.aspx

World Wide Fund for Nature [WWF]. (n.d.). *Threats to African elephants.* WWF Knowledge Hub. Retrieved on 11.8.2025, from https://wwf.panda.org/discover/knowledge_hub/endangered_species/elephants/african_elephants/afelephants_threats/

MIX
Papier aus verantwortungsvollen Quellen
Paper from responsible sources
FSC® C105338

If you have any concerns about our products,
you can contact us on
ProductSafety@springernature.com

In case Publisher is established outside the EU,
the EU authorized representative is:
Springer Nature Customer Service Center GmbH
Europaplatz 3, 69115 Heidelberg, Germany

Printed by Libri Plureos GmbH
in Hamburg, Germany